APPLE

Production and Value Chain Analysis

Dr. Nazeer Ahmed, is Vice Chancellor, Sher-e-Kashmir University of Agricultural Sciences and Technology of Kashmir (SKUAST-K), J&K. Dr. Ahmed served in various capacities as Scientist, Senior Scientist, Associate Director Research, Professor and Head, Dean Agriculture, Director Resident Instruction cum Dean Post Graduate Studies at SKUAST-K and Director, ICAR-Central Institute of Temperate Horticulture, Srinagar. During his 32 years of service, he has contributed significantly as a teacher, researcher, extension worker and administrator. Dr. Ahmed has published more than 300 research papers in National and International Journals and released 34 varieties and hybrids in Temperate Horticultural crops. Dr. Ahmed has guided many Master's and Ph.D. scholar's and is recipient of several prestigious awards and honours including Rajbhasha Gaurav Award, Dr. R. S. Paroda Award, Dr. M. H. Marigowda National Endowment Award, Dr. Kriti Singh Gold Medal and Vijay Shree Award.

Dr. Shabir A. Wani is presently serving as Head, School of Agricultural Economics & Horti-Business Management and also I/C Kashmir and Ladakh Agriculture Watch Center, SKUAST-K. Srinagar. Dr. Wani has served in various capacities as Scientist, Senior Scientist and Professor in the discipline of Agricultural Economics at SKUAST-K. During his 27 years of service, he has guided many students, authored 6 book chapters in edited books, 8 policy papers, published more than 75 papers in peer reviewed journals, formulated J&K - Comprehensive State Agriculture Plan (c-SAP) and Carrier Advance Schemes (CAS) for SKUAST-K Scientists and Teachers, besides conducted many SKUAS-K initiated livelihood studies in isolated and disadvantageous areas of state. Dr. Wani is twice recipient of DST SERB (GoI), prestigious International Travel Grant award.

Dr. W. M. Wani is presently serving as Professor-Cum-Chief Scientist in the Division of Fruit Science, Sher-e-Kashmir University of Agricultural Sciences and Technology of Kashmir (SKUAST-K), J&K. Dr. Wani has served as Scientist in ICAR – Central Institute of Temperate Horticulture, Srinagar and as Senior Scientist and Professor in the discipline of Fruit Science at SKUAST-K. During his 35 years of service, he has significantly contributed in the field of Horticulture as a Teacher, Researcher, Extension worker, guided many students and published his work in reputed journals. Presently Dr. Wani is actively engaged in popularizing High Density Orcharding in Kashmir and in this regard has also visited many European countries to learn lessons from their experiences.

APPLE

Production and Value Chain Analysis

Editors

Nazeer Ahmed

S. A. Wani

W. M. Wani

2018

Daya Publishing House®

A Division of

Astral International Pvt. Ltd.

New Delhi – 110 002

ISBN **9789387057524 (International Edition)**

Publisher's Note:

Published by   :   **Daya Publishing House®**
A Division of
Astral International Pvt. Ltd.
– ISO 9001:2015 Certified Company –
4736/23, Ansari Road, Darya Ganj
New Delhi-110 002
Ph. 011-43549197, 23278134
E-mail: info@astralint.com
Website: www.astralint.com

Prof. Ramesh Chand

Member, NITI Aayog

नीति आयोग

National Institution for Transforming India

National Institution for Transforming India (NITI Aayog), New Delhi

Government of India

E-mail: rc.niti@gov.in

Office (Ph.): 23096756/23096774

Foreword

When India attained Independence from the colonial rule, it faced serious problem of hunger and under nutrition. A part of the shortage was met through imports. By the mid-1960s food shortage turned quite serious and imports also became very difficult. In the wake of this country embraced the strategy to attain self-sufficiency in food-grains production. The green revolution technology which became available during late 1960s helped the country to achieve breakthrough in productivity and production of staple food. In a few years India started experiencing higher growth in food grain production as compared to the growth in its population. This made the country self-sufficient in staple food production followed by surplus in cereals.

After sufficient availability of rice and wheat the demand for food started diversifying towards fruits and vegetables which also offered higher incomes compared to food grains. At the same time growth rate and income from cereals came under stress due to various reasons. It was then considered appropriate to promote diversification towards fruits and vegetables for raising income of the producers and to match supply with the changes in demand.

The scope of diversification towards horticultural crops varies across region. Horticulture, particularly fruit cultivation is both ecologically and economically superior to other crops in mountain regions. The north-western Himalayan States of India are found to have immense potential to raise income and employment opportunities in the Horticulture sector. The climatic conditions in this region particularly in the state of Jammu and Kashmir offer good scope for cultivation of a variety of temperate fruits like apple, cherry, pear, peach, plum, apricot, almond and walnut. Accordingly, apple industry has emerged strong and established its credibility in improving farmer's income, generating employment and in enhancing exports, besides providing household livelihood security. With the growing consumer awareness about healthy eating, and established perceptions about apples as a healthy and flavorful fruit, the Indian market for apples is expanding. To meet this rapidly growing demand through domestic supply, there is scope in the Himalayan states for both horizontal and vertical expansion of apple cultivation. This require access of farmers to modern production technology and efficient value chains in apple. In this direction the present edited book "Apple: Production and Value Chain Analysis" is a valuable addition to the knowledge on production and marketing of apple. The book covers status, achievements and future R and D strategies on different issues concerning Indian Apple vis-à-vis global advancement. The information contained in this book will meet the technical and economic requirements of researchers, teachers, students, policy planners, development workers and farmers of Indian apple industry. The editors need to be commended for their efforts to provide such useful and sought-after information.

Prof. Ramesh Chand

Preface

During the first few five year plans, priority was assigned to achieve self-sufficiency in food-grains production. However, over the years, agricultural growth is not keeping pace with other economic sectors and is lagging far behind than that of the manufacturing and service sectors. The share of agriculture in GDP has fallen steeply although dependence on agricultural sector for livelihood remains quite high. Government of India has set a policy target of doubling farmers' income by 2022. For increasing farmers' incomes some of the important options delineated included agricultural diversification towards more remunerative commodities, such as horticulture, livestock and fish. In the north-western Himalayan States of India, the obligation of providing income and employment opportunities fall heavily on the Horticulture sector. Owing to various mountain specificities characterizing the region constraints are imposed for raising productivity of field crops and generating income for smallholder. The region offers good scope for cultivation of horticultural crops, covering a variety of temperate fruits like cherry, pear, peach, plum, apricot, almond and walnut in general and apple is considered most important in particular.

In the north-western Himalayan States of India, apple industry has emerged as an important sector for diversification towards horticulture and has established its credibility in multiplying farm income. Apple crop has witnessed most significant increase in the region contributing to the shifts from paddy and other field crops at a very high pace with bright prospects towards this sector. Important reason being handsome returns to the farmers with shortage of water at summer season, apple cultivation compared to field crops requires it at lesser levels. Apple production has developed as an industry and more and more land is apportioned to this sector each year. While the cereal based production system provides household nutritional security. It is ability of apple based value chain to generate sufficient income to provide livelihood security even to smallholder. Apart from the government schemes, the more profound factor for diversification of regions agriculture towards apple sector is driven by comparative advantage principle. Among the temperate fruits, apple is coming up in a big way through horizontal and now vertical expansion. To meet the growing demand, Govt. of India has even relaxed apple import norms and has allowed in bound shipment of fruits through sea port and air ports in Kolkata, Chennai, Mumbai and Cochin.

Timely and reliable information as well as analysis is vital for planning and decision making. In this direction the present edited book "Apple: Production and Value Chain Analysis" is an initiation of a process to create a knowledge house; wherein galaxy of researchers/contributors with diverse areas of specialization have deliberated upon various issues characterizing Indian apple. The book is a synthesis of 30 lead papers contributed by R and D workers/experts/policy planners from different institutions/organizations. It covers status, achievements and future R and D strategies on different issues concerning Indian Apple vis-à-vis global advancement in the value chain. The issues covered range from research needs, innovative technologies, genetic resources, nursery management and crop improvement, disease and pest management, mechanization, pre and post harvest management apart from economics, finance, marketing and trade.

We are highly thankful to Prof. Ramesh Chand, Hon'ble Member, NITI Aayog, Government of India, by sparing his valuable time and contributing forward of this publication. We also take this opportunity to thank all the contributors from different universities, institutes and organizations from India and abroad who have responded to our request to share results of their independent studies. It is hoped that knowledge shared by the contributors in the book will be of interest and benefit the researchers, teachers, students, policy makers, development workers and farmers involved in apple cultivation.

Nazeer Ahmed

S. A. Wani

W. M. Wani

Contents

List of Contributors

Ahmed, N.
Vice Chancellor, Sher-e-Kashmir University of Agricultural Sciences and Technology of Kashmir (SKUAST-K), Srinagar – 1900 25, J&K.

Bhat, Khalid Mushtaq
Associate Professor, Division of Fruit Science, Sher-e-Kashmir University of Agricultural Sciences and Technology of Kashmir (SKUAST-K), Srinagar – 1900 25, J&K.

Bazaz, Sajjad
J&K Bank, Corporate Headquarters, M.A.Road, Srinagar, J&K.

Chen, Xi
Rutgers University, New Brunswick, NJ USA.

Hussain, Barkat
Scientist, Division of Entomology, Sher-e-Kashmir University of Agricultural Sciences and Technology of Kashmir (SKUAST-K), Srinagar – 1900 25, J&K.

Jan, Aarifa
ICAR-Central Institute of Temperate Horticulture, Old Air Field, Rangreth, Srinagar 190007.

Lone, G.M.
Associate Professor, Division of Entomology, Sher-e-Kashmir University of Agricultural Sciences and Technology of Kashmir (SKUAST-K), Srinagar – 1900 25, J&K.

Kirmani, Nayar Afaq
Associate Professor, Division of Soil Sciences, Sher-e-Kashmir University of Agricultural Sciences and Technology of Kashmir (SKUAST-K), Srinagar – 1900 25, J&K.

Kanth, Raihana Habib
Professor and Head, Division of Agronomy, Sher-e-Kashmir University of Agricultural Sciences and Technology of Kashmir (SKUAST-K), Srinagar – 1900 25, J&K.

Khan, Farooq A.
Head, Division of Basic Sciences, Sher-e-Kashmir University of Agricultural Sciences and Technology of Kashmir (SKUAST-K), Srinagar – 1900 25, J&K.

Khan, Junaid N.
Head, Division of Agricultural Engineering, Sher-e-Kashmir University of Agricultural Sciences and Technology of Kashmir (SKUAST-K), Srinagar – 1900 25, J&K.

Masoodi, F. A
Professor and Head, Department of Food Sciences and Technology, University Of Kashmir, Srinagar-J&K.

Mir, Javid I.
Scientist, ICAR-Central Institute of Temperate Horticulture, Old Air Field, Rangreth, Srinagar -190007, J&K.

Mir, M.A.
Professor and Head, Department of Food Technology, Islamic University of Science and Technology, Awantipora-J&K.

Mukhtar, Malik
Scientist, Research Centre for Pesticides Residues and Quality Analysis, Division of Entomology, Sher-e-Kashmir University of Agricultural Sciences and Technology of Kashmir (SKUAST-K), Srinagar – 1900 25, J&K.

Naqash, Farheen
Ph.D. Scholar, School of Agricultural Economics and Horti-Business Management, Sher-e-Kashmir University of Agricultural Sciences and Technology of Kashmir (SKUAST-K), Srinagar – 1900 25, J&K.

Padder, Bilal A.
Scientist, Plant Virology and Molecular Plant Pathology Laboratory, Division of Plant Pathology, Sher-e-Kashmir University of Agricultural Sciences and Technology of Kashmir (SKUAST-K), Srinagar – 1900 25, J&K.

Paray, M.A.
Associate Professor, Research and Training Centre for Pollinators, Pollinizers and Pollination Management, Division of Entomology, Sher-e-Kashmir University of Agricultural Sciences and Technology of Kashmir (SKUAST-K), Srinagar – 1900 25, J&K.

Qazi, Nissar Ahmad
Professor and Head, Division of Plant Pathology, Sher-e-Kashmir University of Agricultural Sciences and Technology of Kashmir (SKUAST-K), Srinagar – 1900 25, J&K.

Sajad, H. Baba
Scientist, School of Agricultural Economics and Horti-Business Management, Sher-e-Kashmir University of Agricultural Sciences and Technology of Kashmir (SKUAST-K), Srinagar – 1900 25, J&K.

Shaheen, F.A.
Scientist, School of Agricultural Economics and Horti-Business Management, Sher-e-Kashmir University of Agricultural Sciences and Technology of Kashmir (SKUAST-K), Srinagar – 1900 25, J&K.

Sharma, M.K.
Associate Professor, Division of Fruit Science, Sher-e-Kashmir University of Agricultural Sciences and Technology of Kashmir (SKUAST-K), Srinagar – 1900 25, J&K.

Sharma, Ravinder
Professor, Department of Social Sciences, College of Forestry, Dr. Y S Parmar University of Horticulture and Forestry, Nauni, Solan, H.P.

Singh, D.B.
Director, ICAR-Central Institute of Temperate Horticulture, Old Air Field, Rangreth, Srinagar – 190 007, J&K.

Summuna, Baby
Division of Plant Pathology, Sher-e-Kashmir University of Agricultural Sciences and Technology of Kashmir (SKUAST-K), Srinagar – 1900 25, J&K.

Wani, Mushtaq A.
Associate Professor, Division of Soil Science, Sher-e-Kashmir University of Agricultural Sciences and Technology of Kashmir (SKUAST-K), Srinagar – 1900 25, J&K.

Wani, S.A.
Professor and Head, School of Agricultural Economics and Horti-Business Management, Sher-e-Kashmir University of Agricultural Sciences and Technology of Kashmir (SKUAST-K), Srinagar – 1900 25, J&K.

Wani, W.M.
Professor and Head, Division of Fruit Science, Sher-e-Kashmir University of Agricultural Sciences and Technology of Kashmir (SKUAST-K), Srinagar – 1900 25, J&K.

Chapter 1

Situation Analysis of Apple in India

Nazeer Ahmed, S. A. Wani and W. M. Wani

Sher-e-Kashmir University of Agricultural Sciences
and Technology of Kashmir, Srinagar
E-mail: vc@skuastkashmir.ac.in

1. Production Scenario

(a) World Scenario of Apple

About 76 million tonnes of apples were grown worldwide in 2012, with China producing almost half of this total (49 per cent). The United States, with more than 5 per cent of world production is the second-leading producer. Other important global players are Turkey (3.8 per cent), Poland (3.8 per cent), and India (2.9 per cent).The largest exporters of apples in 2009 were China, the U.S., Turkey, Poland, Italy, Iran, and India while the biggest importers in the same year were Russia, Germany, the UK and the Netherlands. In the United States, more than 60 per cent of all the apples sold commercially are grown in Washington. Imported apples from New Zealand and other more temperate areas are competing with U.S. production and increasing each year. Most of Australia's apple production is for domestic consumption. Imports from New Zealand have been disallowed under quarantine regulations for fire blight since 1921.Other countries with a significant production are Brazil, Argentina, Ukraine, Germany and South Africa (Figure 1.1).

Trends in Area, Production and Productivity of Apple

The Area, production and Yield of apple at the world level has recorded compound growth rate of 2.4, 3.5 and 1.1 per cent per annum from 1973-74 to 2011-12 respectively. During the same period the area, production and yield of

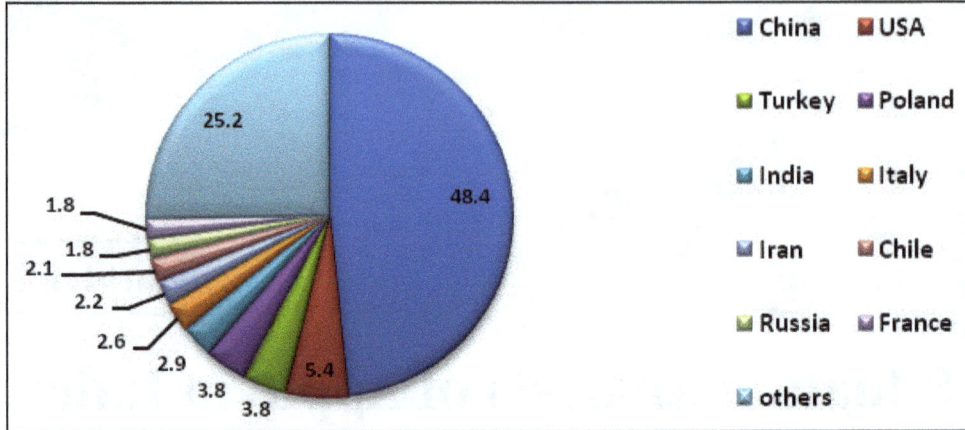

Figure 1.1: Top Ten Apple Producing Countries.

Source: FAO, 2013.

apple in Asia recorded the growth rate of 3.9, 6.3 and 2.4 percent per annum respectively, while as India recorded the growth rate of 2.4, 3.5 and 1.1 per cent respectively (Table 1.1).

Table 1.1: Region-wise Compound Growth Rates in Apple (1973-74 to 2011-12)

Regions	Area	Production	Yield
India	2.4**(0.044)	3.5*(0.090)	1.1(0.088)
Asia	3.9*(0.181)	6.3*(0.093)	2.4*(0.133)
World	1.2(0.088)	2.4*(0.044)	1.2(0.088)

Source: FAO, 2013.

* Significant at 1 per cent level and ** Significant at 5 per cent level of significance.

(b) National Scenario of Apple

Jammu and Kashmir, Himachal Pradesh, Uttarakhand and Arunachal Pradesh are the major apple producing states of India. The two important states namely J&K and Himachal Pradesh accounts for 92 per cent of the total production and about 85 per cent of the total area under apple in India. As far as productivity of apple is concerned J&K has the highest productivity (8.6 tonnes/hectare) followed by Himachal Pradesh (3.9 tonnes/hectare) and Uttarakhand (2.16 tonnes/hectare) (Figure 1.2).

(c) State Scenario of Apple

Apple is the principle fruit crop of Jammu and Kashmir and accounts for 51 per cent of total area of 2.72 lac hectare under all temperate fruits grown in

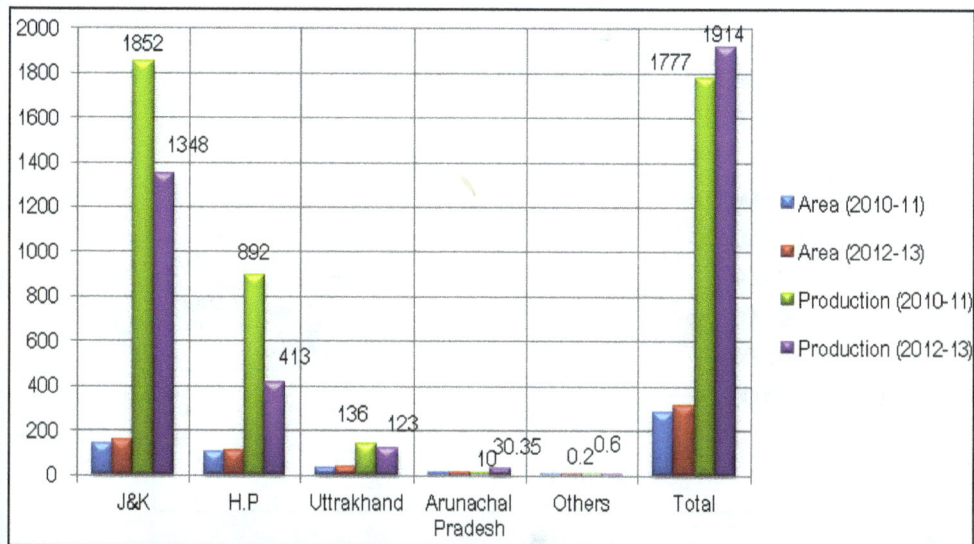

Figure 1.2: State-wise Area and Production of Apple during 2011-12 and 2012-13

Source: **National Horticulture Board (NHB), New Delhi, 2014-15.**

this state. The annual apple production in the state is 12-15 lac M tones. Average yield of commercially important apple cultivars per unit area is the highest in the country ranging between 10-12 tonnes/ha, but it compares poorly to the yields of >40 tonnes/ha in horticultural advanced countries of the world. Climatic and other agro-ecological factors of Kashmir are ideally suited to the cultivation of many apple varieties. Alternate bearing, defective pruning and training, use of seedling rootstock of unknown performance, lack of proper nutrients and water management, deficiency of suitable pollinizers and ineffective control of pests and diseases are the main causes of low productivity (Figures 1.3 and 1.4).

(d) Area, Production and Yield of Apple in Jammu and Kashmir

The dynamics of apple industry in the state has been presented in the Figure 1.5. The analysis revealed that area increased by 2.8 percent from 1980-981 to 2010-11, while as production by 4 percent during the same period. Inspite of many fold increase in area and production, yield has remained almost stagnant at around 9 metric tonnes during the past two decades. However, by the concerted efforts of the farmers, yield has picked up and during the year 2010-11 it was recorded more than 13 metric tonnes per hectare. Area under apple has witnessed a continuous increase since 1980s. During 1980-81 area under this crop was 60286 hectare, which increased to 141717 hectare in 2010-11. Production of this crop has also shown the same pattern as its production has increased by more than three times. As far as productivity is concerned, it has remained around 9 MT/ha from 1980-81 to 2000-01. However, during the last decade it has witnessed a marginal increase.

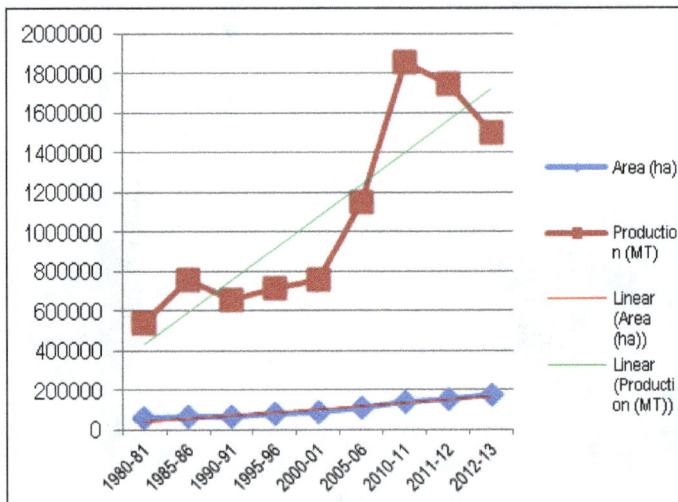

Figure 1.3: Area and Production of Apple (MT/Ha).

Figure 1.4: Productivity of Apple (MT/ha).

Source: **Department of Horticulture, Govt. of J&K.**

Table 1.2 summarizes the exponential growth rate of area, production and yield of apple in Jammu and Kashmir from 1981-82 to 2010-11 The data has been divided into three sub-periods from 1981-82 to 2010-11. *i.e.* 1981-82 to 1990-91,1991-92 to 2000-01, and 2001-02 to 2010-11, besides overall period from 1981-82 to 2010-11.This decomposition of periods were considered appropriate to estimate the structural changes that have taken place in respect of area, production and productivity. The area, production and yield witnessed growth momentum of 2.8 per cent, 4 per cent and 1.2 per cent respectively during the overall period. During

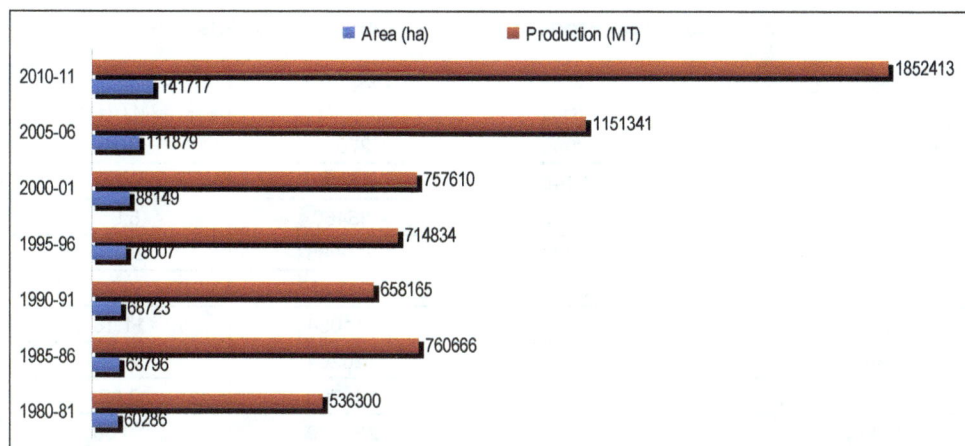

Figure 1.5: Area and Production of Apple in J&K.

Source: **Directorate of Horticulture, Govt. of J&K 2012-13.**

the bygone decade apple industry of state showed an overriding performance by achieving higher trajectory in area expansion besides production and productivity.

Table 1.2: Exponential Growth Rates of Area, Production and Yield of Apple in J&K

	Area	Production	Yield
1981-82 to 1990-91	1.5*	2.7***	1.2***
1991-92 to 2000-01	2.8*	3.0*	0.2***
2001-02 to 2010-11	5.2*	6.5*	1.3***
1981-82 to 2010-11	2.8*	4.0*	1.2*

*Significant at 1 per cent; **Significant at 5 per cent; ***Significant at 10 per cent.

(e) District-wise Scenario

Apple production in the state extends across several districts regardless of the agro climatic suitability and productivity. Given the current levels of productivity, districts such as Srinagar, Ganderbal, Budgam, Anantnag and Kulgam do not seem well-suited for inclusion in the apple value chain on account of low productivity. In the first stage the focus should be on districts having a higher productivity, so that investments have a probability of producing better returns. The average yield across the state was about 13 tons/ha, which was less than 300 per cent the level of the top apple producing countries. The yield gap is large and the potential for increase in yield makes the investments in the sector attractive (Table 1.3).

(f) Historical Background of Apple in Jammu and Kashmir State

The fruit culture in Kashmir valley dates back to the times of King Nara (1000 BC), King Lalitaditya (700 AD) and Harsha (1089 AD). Kashmir has been the home of

Table 1.3: Area, Production, Yield – Across Districts

Sl.No.	District	Area (Ha)	Production (MT)	Productivity (MT/Ha)
1	Srinagar	2990	27886	9.33
2	Ganderbal	5145	46733	9.08
3	Budgam	14585	85460	5.86
4	Baramula	25985	659965	25.40
5	Bandipore	5408	74504	13.78
6	Kupwara	18885	211054	11.18
7	Anantnag	16523	126264	7.64
8	Kulgam	15842	143795	9.08
9	Shopian	21615	235129	10.88
10	Pulwama	10913	112854	10.34

Source: NABASCONS, 2013-14.

large varieties of fruits. Ancient records of Kalahans Rajtarangi and that of Alberunis mention the existence of numerous varieties of fruits of the state. History of fruit growing in Jammu and Kashmir dates back to even 2000 B.C, when apples were recorded indigenous and grown for local consumption only. Apples of Kashmir have found mention in Tuzki Jahangiri, while Lawrence has called Kashmir Valley as fruit country in his famous book, "The Valley of Kashmir". However, horticulture started in an organized form in around 1865 AD when Ermns, head gardener of Public Works Department in France, after preliminary survey, introduced some apple fruit plants and planted at Cheshma Shahi Srinagar in 1875 AD. The collection of about 25000 wild fruit stocks by one Mr. Gollan and their plantation at Baghi Sundri near Sopore, for grafting and distribution in state orchards, marked the beginning of establishment of nursery which provided the corner stone for the development of horticulture specially apple crop in the state.

(g) Employment Potential

The Employment Potential of apple is enormous. Besides strengthening nutritional security, it has potential to provide livelihood security system. Apple crop needs intensive cultivation, yield more, have wide agro-climatic adaptability and need processing. Diversification of apple crop will generate more gainful employment on account of following:

☆ Apple crops need 600-800 man days per hectares as against only 150-200 for field crops.

☆ Apple crops have 8-10 times more productivity than field crops.

☆ Apple crops can be grown on wasteland and dry lands successfully.

☆ Adoption of advanced technologies like high density plantation will require large quantities of quality plant material, leading to the employment of youth in public/private nurseries. Further opportunities for employment, through adoption of advanced grading and packing system and other post harvest handling activities, will also increase.

(h) Nutrition Value of Apple

Apples are full of healthy antioxidants, fiber, vitamins and minerals. One medium sized apple contains 95 calories and 4.4 g of dietary fiber. In addition, an apple is a good source of potassium, phosphorus, calcium, manganese, magnesium, iron and zinc. Apples also contain vitamins A, B1, B2, B6, C, E, K, folate, and niacin. Apples come in different shapes and sizes, so the amount of calories and vitamins in apple varies. Best of all, apples contain no fat, sodium or cholesterol.

(2) Marketing Scenario of Apple

(a) Existing Marketing Infrastructure and Facilities

In order to provide marketing facility for agriculture/horticulture produce at the doorsteps of the farmers/growers, the various programmes have been initiated by way of establishing three terminal fruit and vegetable markets mainly for apple *viz.* Parimpore (Srinagar), Sopore (Baramulla) and Narwal (Jammu) which are functional and the trade is carried on in these markets.

Table 1.4: Fruit and Vegetable Markets in J&K State

Sl.No.	Market Category	Market Location	Total Markets
A	Terminal markets	Narwal*, Parimpora*, Sopore*, Jablipora[1]	4
B	Satellite markets	Shopian*, Pulwama*, Kulgam*, Baramulla*, Handwara*, Udhampur*, Chari Sharief*, Kupwara[1], Kathua[1], Rajouri[1], Bisnah[1], Akhnoor[1], Batingoo[1], Zazna[1], Leh[1], Kargil[1], Pryote Doda[1], Poonch[1], Samba[1], Sunderbani[1], Aglar (shopain)[1]	21
C	Apni Mandi	Pachhar*, Nunmai Kulgam[1], Mandi Poonch[1], Pouria, Reasi[1], Gharian (Udh)[1], Tapyal[1], Raya Bagala[1], Dyala Chak[1], (Chadwal) Bhaderwah Bani[1]	11
	TOTAL		37

Source: Directorate of Horticulture, Govt. of J&K.
* Functional markets, 1: Markets under process to be made functional.

A new terminal fruit and vegetable market is being established at Jablipora Anantnag for which 441 Kanals of land has been acquired. In addition to this the Department of Horticulture is also developing 21 Satellite F and V markets in the state out of which seven are completed and functional, six are likely to be completed and made functional, five are in progress and land acquisition process for three F and V markets is in progress. Furthermore, eleven Apni *Mandis* are also under the process of development out of which three are completed, six are in progress and land acquisition for two Apni *Mandis* is under process (Table 1.4). Most of the existing Satellite and Terminal markets have been set up during late nineties and are being strengthened on piecemeal basis from time to time. These markets lack basic infrastructure in terms of adequate auction platforms, rest houses for farmers/outside buyers, parking areas for trucks, toilets, proper drainage systems, electrification, sanitation, fencing, adequate land for shops, *etc.* For example, in Sopore Mandi the total handling capacity is about 320 trucks per day, whereas outflow touches 500 trucks/day in peak season leading to traffic jams, temporary closure of market for fresh arrival. The traders in all the markets are ready to pay the *Mandi* tax provided the basic infrastructure is created and improved.

(b) Market Practice

The price discovery process in the local markets within the state are not transparent. The markets do not have a price dissemination mechanism and it is difficult to know the prevailing prices on any given day. While there is some understanding of the fees payable to the market intermediaries, there are no norms and enforcement of such norms. High commissions and fees payable to intermediaries tend to get blurred with other fees and charges; often adjusted in the price thereby making the realisation uncertain. While APMC law is passed in the State, it has not been implemented. The law needs amendments in line with the changes suggested by the Central Government and adopted by several states. This has led to the proliferation of unlicensed traders, agents acting in the market with non-transparent auction procedures. The APMCs do not earn revenues through collection of *Mandi* tax and are unable to improve the conditions in the *Mandi.*

The major destination of Kashmir apples is the Azadpur *Mandi*, which is a buyers' market and designed to be so. Manipulation of prices by traders in the *Mandi* is resorted through, stopping apple trucks at the border of entry in to Delhi, use of cold stores to alter supply of apples in the *Mandi;* keeping away small buyers with artificially high price quotes and later reducing prices to low levels to benefit preferred buyers and use of proxies in auctions. The markets wihin the state are comparatively better in price determination and transparency. Growers with pre-harvest contracts (PHC) access the markets easier, but lose out on full benefit of market prices on account of their taking money in advance. Free growers find it difficult to enter markets even when the demand is brisk and the commission agents prefer their 'captive growers' with PHC. Setting up satellite markets has helped growers (especially the free ones) in marketing. Farmers who market apples through cooperatives realise higher prices. Trade margins range from 42 to 73 per cent in the different channels of marketing.Price discovery by grower would be more realistic and effective if he is able to hold back and store his produce for some time. The farmer needs to have conditions (local storage and financial capacity to hold) under which distress selling can be checked. To improve and reform market practice, some measures are required. The first is that of organising farmers into producer collectives or cooperatives. These organisations of farmers will help in aggregation of inputs and outputs and improve their bargaining power

(c) **Marketing Channels**

Apple is produced by large number of small farmers scattered around the valley whereas, the consumers are located throughout the country. Small produce, lack of time ability, knowledge of marketing system and liquidity potential *etc.*, prevent them to undertake direct marketing of apple. The marketing system for apple is highly complex and is composed of different marketing channels for distribution of apple in different markets in each channel varying number of functionaries are involved and numerous specialized business activities called marketing functions are to be performed by them. Following are the commonly encountered and well identified channels in the valley (Table 1.5).

Table 1.5: Marketing Arrangement of Apple

Channel-I	Producer-Wholeseller/Commission agent-Retailer-Consumer
Channel-II	Producer-Pre-harvest contractor- Wholeseller/Commission agent-Retailer-Consumer
Channel-III	Producer-Commission agent- Wholesaler-Retailer-Consumer
Channel-IV	Producer-Pre-harvest contractor- Commission agent- Whole seller-Retailer-Consumer
Channel-V	Producer-Post harvest contractor/potential growers-Commission agent- Wholesalers-Retailer-Consumer

Source: Field survey, 2015.

(d) **Marketing Cost of Apple**

Pre packing cost includes cost over picking, assembling and grading. The apples are picked manually by skilled labour and assembled at a plain place mated with paddy straw and then grade it again by skilled labours. The Figure 1.6 revealed that the total pre packing cost accounts for 8.73 per cent of total marketing cost.

Packing cost includes cost on packing material and labour charges. The cost on packing box alone accounts for 24.8 per cent of total marketing cost. Since Delhi market receives 65 per cent of total apple produce exported to other states from Kashmir and again because of trade monopoly of Delhi market in terms of huge advances to apple orchardists the transportation cost of apple has been estimated with respect to Delhi market. The transportation cost per box of apple was found to be Rs 56.54 per cent of total marketing cost and freight charges from road head to Delhi, accounted for 32.96 per cent of the total marketing cost.

The marketing efficiency of different channels as presented in the Table 1.6 revealed that the channel-III (0.73) turns out to be economically more efficient, followed by channel –I (0.68) and least efficient is channel-V (0.29). It was observed that producer got maximum share of consumers rupee in the channel where, produce was directly marketed to whole seller.The contractor in turn trade their produce to wholesaler at higher prices than producer, because of higher bargaining power. An orchardist could earn maximum share of consumers' price in the channel

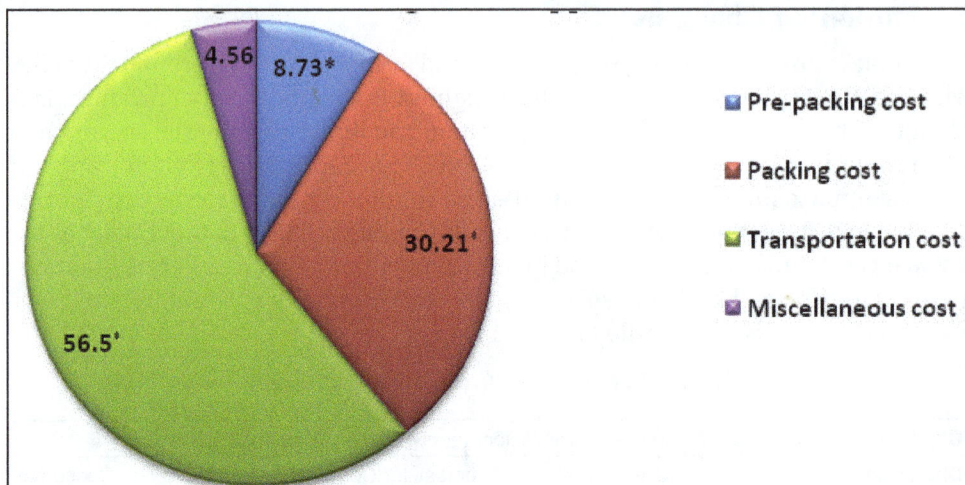

Figure 1.6: Marketing Cost of Apple (Per cent of Total).
Source: **Directorate of Horticulture, Govt. of J&K, 2013-14.**
*W.R.T Delhi market.

Standard box contains 18 Kgs. of apple-pie-packing cost includes picking (5.35 per cent), assembling (1.13 per cent) and grading charges (2.25 per cent).[1]Packing cost includes cost of packing box (24.80 per cent), wrapping paper (1.69 per cent), paddy straw (0.9 per cent), nails (0.56 per cent), packing (1.13 per cent), closing and assembling of boxes (0.56 per cent) and labelling and stencilling cost (0.56 per cent).[2]Transportation cost includes loading and unloading charges (1.63 per cent), Orchard to road head (5.63 per cent), Forwarding charges (5.63 per cent), Freight to Delhi (32.6 per cent), State toll tax (8.51 per cent), Loading at road head (0.79 per cent), Unloading (1.01 per cent), octroi (0.56 per cent) and communication (0.05 per cent).

where he sells his produce directly to the whole seller. However, lack of liquidity potential, ignorance of market demand etc capitalises into distress sale. Liberal cheap credit facility along with other incentives to apple growers would definitely increase their bargaining power (Table 1.6).

Table 1.6: Marketing Margins and Efficiency of different Channels in apple

Functionary	Marketing Channels				
(Values in terms of per cent of consumers price)					
	I	*II*	*III*	*IV*	*V*
Price received by producer	68.17	28.92	43.35	39.46	28.92
Expenses incurred by producer	22.24	-	35.92	11.61	-
Producers net margin	45.93	43.35	37.19	27.85	28.92
Commission agents margin	-	-	7.31	7.31	7.31

Functionary	Marketing Channels				
	(Values in terms of per cent of consumers price)				
	I	II	III	IV	V
Expenses incurred by pre-harvest contractor	-	22.24	-	-	35.92
Sale price of pre-harvest contractor	-	90.07	-	-	73.86
Pre-harvest contractors margin	-	4.09	-	-	9.02
Expenses incurred by post harvest contractor	-	-	-	17.02	-
Sale price of post harvest contractor	-	-	-	74.17	-
Post harvest contractors margin	-	-	-	10.29	-
Expense incurred by whole seller	0.48	0.48	3.06	3.06	3.06
Sale price of whole seller	72.54	72.54	78.92	78.92	78.92
Whole sellers margin	3.88	3.88	2.75	1.80	2.12
Expenses incurred by retailer	12.08	12.08	12.57	12.57	12.57
Retailer margin	15.38	15.38	8.51	8.51	8.51
Sale price of retailer/consumers price	100	100	100	100	100
Marketing efficiency	0.68	0.43	0.73	0.39	0.29

(e) Price Behaviour of Apple

Apple (*Malus pumila*) is commercially the most important temperate fruit and is fourth among the most widely produced fruits in the world after banana,

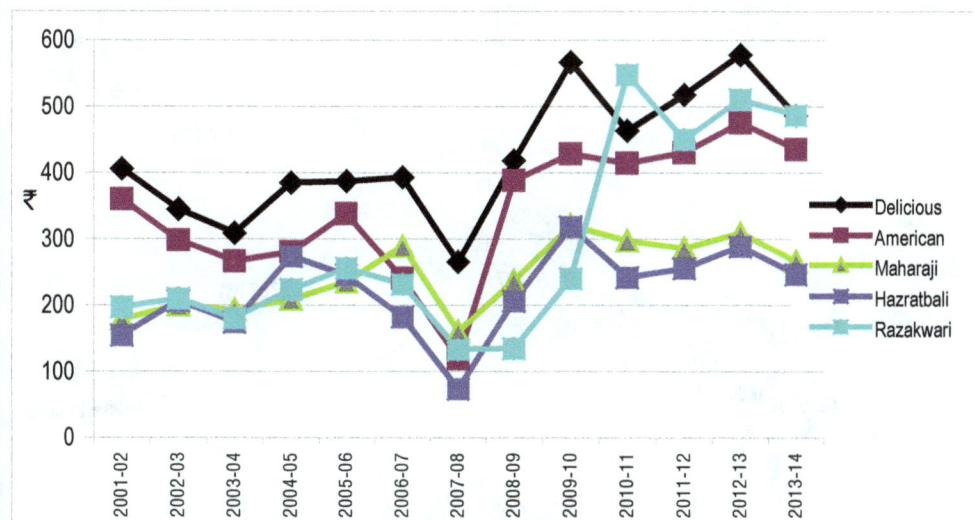

Figure 1.7: Variety-wise Wholesale Rates of Apple (Rs per box).

Source: Directorate of Horticulture, Govt. of J&K 2015-16.

orange and grape. China is the largest apple producing country in the world. It can be observed from the Figure 1.7, market price was highest for delicious variety of apple followed by American.

3. Export and Import Scenario of Apple

India's share in the total world apple production is merely 2.9 percent. Only around 1.6 per cent of the country's production gets exported. The trend in export of apple from India during the period 1999-2000 to 2001-02 is given in the Figure 1.8.

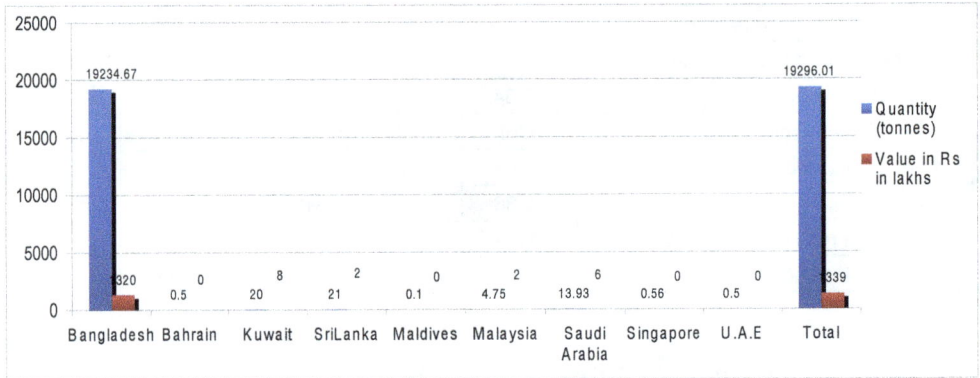

Figure 1.8: Country-wise Export of Apple from India during 2001-02.

Source: **Directorate of Horticulture, Govt. of J&K.**

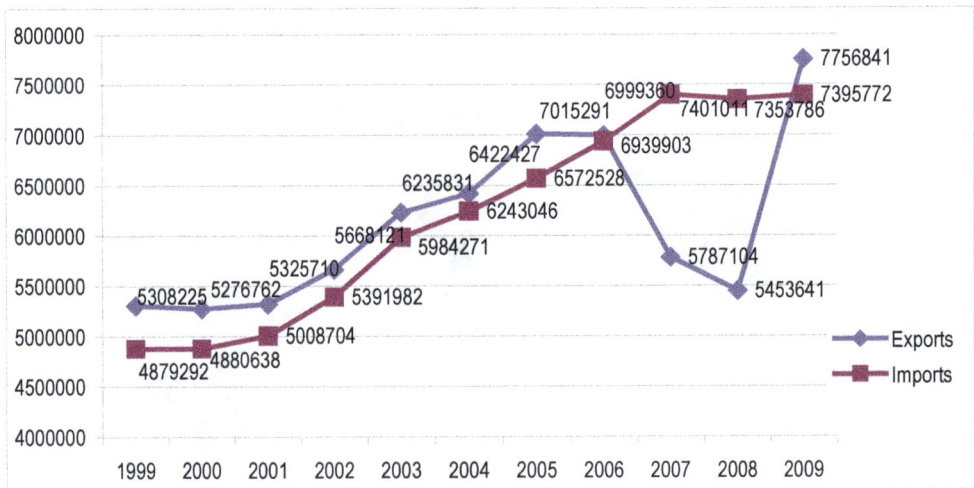

Figure 1.9a: Global Import and Export of Apple (in Metric Tonnes).

Source: **FAO STAT.**

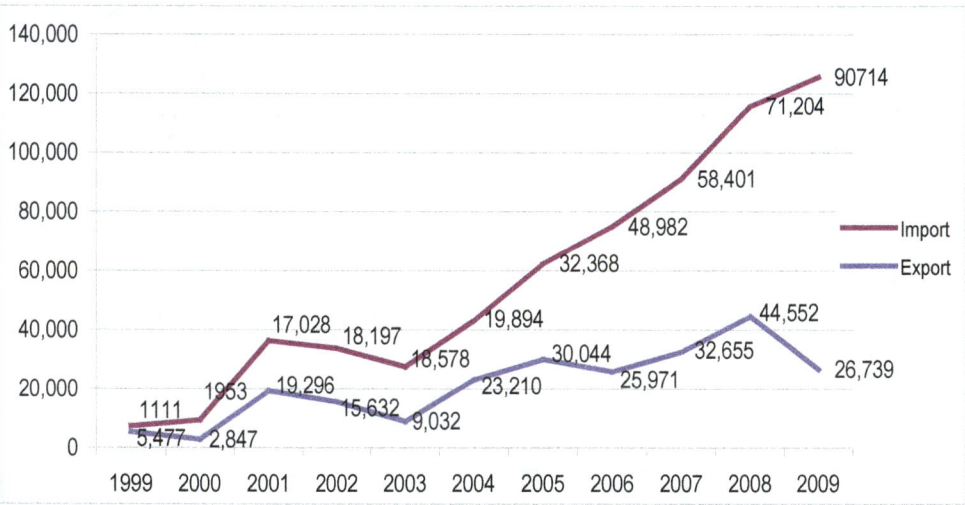

Figure 1.9b: India's Share in Global Trade of Apple (metric tonnes).

Source: **FAO STAT.**

The apple has been traded worldwide and it shows continuous increase from last ten years. The global export of apple was 5308225 metric tonnes in the year 1999, since than it grows continuously and it was 7756841 metric tonnes in the year 2009 show an increase of 40 per cent. Similarly the global import of apple was 4879292 metric tons in the year 1999, since than it grows continuously and it was 7,395,772 metric tons in the year 2009 shows an increase of 51 per cent. The export of apple by India shows a continuous increase from 1999 to 2009. In 2009 the exports were 5477 metric tons and it grew up to 26739 metric tonnes shows an increase of 4.88 times.

Similarly the import of apple by India shows a continuous increase from the year 1999 to 2009. It was 1953 metric tonnes in 1990 and it rose up to 90714 metric tonnes in 2009 -2010 showed an increase of 50 times. However recent data of Directorate General of Commercial Intelligence and Statistics (DGCIS), Kolkata depicted a compound annual growth rate of 28 per cent during last two decades in the imports mainly from Washington and China. Overall, in recent year export of apple to other countries is almost negligible because of stiff competition from other countries. However, some meagre quantity of apple is traded with Pakistan in Barter system through LOC trade agreement between two countries.

(a) Apple Imports in India

Apple imports in India grew nearly with a 28 per cent growth rate between 1999 and 2015. India imported approximately 155410 tons of apple during 2012 and the import price of Washington apple was 83 INR per kg. Apples accounted for major chunk of fruit imports despite the high import tariff rate of Rs 56 and India's customs duty on apples is high. Delicious group constitutes approximately 80 per cent of the apples imported into India. The remaining portion included varieties such as Fuji, Royal Gala, and Granny Smith. Generally imported apples are priced approximately double than domestically produced apples. Higher import tariff and transportation cost along with higher margins charged by importers were important contributing factors. Important qualities by which imported apple fetch premium price and give a tough competition to domestic apple include more tempting with their bright red, gleaming surface and they taste the same all year round. To improve quality and yields to better compete with imported apples overall, Indian growers have made little effort in the value chain (Tables 1.7 and 1.8)

Table 1.7: Status of Apple Imports

Year	Import in MT	Year	Import in MT
1999	1111	2008	64955
2000	1953	2009	62348
2001	6586	2010	90714
2002	12524	2011	122265
2003	20093	2012	155410
2004	22051	2013	198924*
2005	15845	2014	254622*
2006	27856	2015	325916*
2007	35832	CAGR	28 per cent

Source: Directorate General of Commercial Intelligence and Statistics (DGCIS), Kolkata.

* Projected figures calculated at 28 per cent growth rate.

Table 1.8: Price Analysis of Imported Apple in India

Particulars	Price in US$ per 20 Kg box	Price in INR per 20 Kg box	Price in INR per Kg
Import price of Washington apples	25	1650	83
Expenses incurred by importer	17.1	1128	56
Importer's margin	3.3	217	11
Price at whole sale market	45.4	2996	150
Expenses incurred by trader	1.1	72	3.5
Trader's margin	2.2	145	7.0
Retailer's purchase price	48.7	3214	160
Retailers expenses	4.4	290	14.5
Retailer's margin	8.9	587	29
Consumer price	62	4092	200

Source: NABCONS-2014, $= 66 INR.

(b) Agri Export Zones (AEZ)

With the primary objective of boosting agricultural exports from India, in March 2001, Government of India announced a policy of setting up of Agri Export Zones (AEZs) across the country. The Central Government has sanctioned 60 AEZs comprising about 40 agricultural commodities.AEZs is spread across 20 states in the country.

The objective of setting up AEZs is to converge the efforts made, hitherto, by various central and state government departments for increasing exports of agricultural commodities from India. The AEZ takes a comprehensive view of a particular produce/product located in a geographically contiguous area for the purpose of developing and sourcing raw materials, their processing/packaging, and leading to final exports.

Table 1.10: Identified Agri-export Zones (AEZs) for Apple

Sl.No.	Name of the State	Name of AEZ	No of AEZ
1	J and K	Srinagar, Budgam, Pulwama, Baramullah, Kupwara and Anantanag	6
2	Himachal Pradesh	Kinnaur, Shimla Sirmor, Kulu, Chamba and Mandi	6

Source: APEDA Agri. Exchange

Chapter 2

An Overview of Apple Value Chain

S. A. Wani and F. Naqash

School of Agricultural Economics and Horti-Business Management,
Sher-e-Kashmir University of Agricultural Sciences and
Technology of Kashmir, Srinagar
E-mail: dr.shabirwani@rediffmail.com

1. Background

Apple (*Malus Borkh*) is one of the oldest fruits known to man. It is a deciduous fruit (fruit which shed their leaves). The deciduous fruits are divided into pome, soft and stone fruits. The apple is among the pome fruits. Apple tree is small and deciduous reaching 3 to 12 meter (9.0 to 39 feet) tall with broad often densely twiggy crown blossoms are produced in spring, simultaneously with budding of leaves.The fruit matures in autumn and is typically 5 to 9 cm (2 to 3.5 inches) in diameter. It has been found wild in most temperate parts of the world and cooler higher hills of sub-tropical areas. It was, probably, first domesticated in the Caucasus, but fast spread all over Europe, even in pre-historical times. From Europe, apple spread to USA, Australia and South America. Apples are found wild in the hills of North India, but nothing much is known about the cultivated apple. The earliest plantations must have been established in Kashmir by the turn of the sixteenth century. The apple is the pomaceous fruit of the apple tree, species *Malus domestica* in the rose family (Rosaceae). The tree originated in Central Asia, where its wild ancestor, *Malus sieversii*, is still found today. There are more than 7,500 known cultivars of apples, resulting in a range of desired characteristics. Different cultivars are bred for various tastes and uses, including cooking, fresh eating and cider production.

Apple is considered as one of the most important and widely grown temperate fruit of the world with regard to its acreage, production, economic returns, high nutritive value and popularity. Apples are full of healthy antioxidants, fiber, vitamins and minerals. One medium sized apple contains 95 calories and 4.4 g of dietary fiber. In addition, an apple is a good source of potassium, phosphorus, calcium, manganese, magnesium, iron and zinc. Apples also contain vitamins A, B1, B2, B6, C, E, K, folate, and niacin. Apples come in different shapes and sizes, so the amount of calories and vitamins in apple varies. Best of all, apples contain no fat, sodium or cholesterol (Wani *et al*, 2015).

Apple is the fourth widely produced fruit in the world after banana, orange and grapes with a growing demand. About 76 million tonnes of apples were grown worldwide in 2012, with China producing almost half of this total (49 per cent). The United States, with more than 5 per cent of world production is the second-leading producer. Other important global players are Turkey (3.8 per cent),Poland (3.8 per cent), and India (2.9 per cent). The largest exporters of apples in 2009 were China, the U.S., Turkey, Poland, Italy, Iran, and India while the biggest importers in the same year were Russia, Germany, the UK and the Netherlands (FAO, 2014).

India's share in the total world apple production is merely 2.9 per cent. The average productivity of apple in India is nearly 6-8 tonnes per hectare, which is much lower than that of countries like Belgium (46.22t/ha), Denmark (41.87 t/ha), and Netherlands (40.40 t/ha). Substantial progress has been recorded in previous plan periods in apple fruit cultivation in terms of area coverage, production and productivity and still a vast potential exists for both vertical and horizontal expansion. Yield of apple has shown an increase from 4.12 to 13.07 M.T. per ha (1975-2010), yet it is far below the level achieved by advanced countries. Jammu and Kashmir, Himachal Pradesh, Uttarakhand and Arunachal Pradesh are the major apple producing states of India. The two important states namely J&K and Himachal Pradesh accounts for 92 per cent of the total production and about 85 per cent of the total area under apple in India. As far as productivity of apple is concerned J&K has the highest productivity (8.6 tonnes/hectare) followed by Himachal Pradesh (3.9 tonnes/hectare) and Uttarakhand (2.16 tonnes/hectare)(NHB, 2014).

Apple is the principle fruit crop of Jammu and Kashmir and accounts for 51 per cent of total area of 2.72 lakh hectare under all temperate fruits grown in this state. Average yield of commercially important apple cultivars per unit area is the highest in the country ranging between 10-12 tonnes/ha, but it compares poorly to the yields of 20-40 tonnes/ha in horticultural advanced countries of the world. Climatic and other agro-ecological factors of Kashmir are ideally suited to the cultivation of many apple varieties. Apple is produced by large number of small farmers scattered around the valley whereas, the consumers are located throughout the country. Apple is cultivated in almost all the ten districts of Kashmir region, with Baramulla, Shopian, Kulgam and Pulwama being the highest producers. The harvesting of fruit begins from August for Early Maturing Cultivars and continues till November with peak activity in September and October. Small produce, lack of time ability, knowledge of marketing system and liquidity potential *etc.*, prevent them to undertake direct marketing of apple. The marketing system for apple is highly complex and is composed of different marketing channels for distribution of apple in different markets. In each channel varying number of functionaries are involved and numerous specialized business activities called marketing functions are to be performed by them (DES, 2014).

Only around 1.6 per cent of the country's production gets exported. Presently, a small quantity of apple produced in India is exported, mainly to Bangladesh and Sri Lanka. Export of fruit from outside J&K State, has occupied a prominent place in trade of the State but it is showing fluctuating trend over the years. The total quantity exported ending Nov. 2012-13 is 7.21 lakh MTs, against 10.30 lakh MTs in the year 2011-12. Although, the production under horticulture sector is increasing year by year but there is no significant growth in the export of horticulture produce outside the State. The reason for low growth in export of fruits outside the State is introduction of Market Intervention Scheme (MIS) under which "C" grade apples are procured at a support price of Rs.6 per kg for processing into juice concentrates in the locally established juice processing units. Secondly, India is also importing fruits from foreign countries as free trade policy is in force at the country level. The Government is making all efforts to promote exports from Jammu and Kashmir State (Naqash Farheen, 2015).

J&K leads in terms of production and provides the maximum marketable surplus. About 30 per cent of A grade, 40 per cent of B grade and 30 per cent of C grade of pre-falls and culled apples account for substantial quantum of more than 3 lac tonnes which needs to be exploited as raw material for processing industry. The increased production yielded some good results and our export worth Rs. 4500.00 crore is expected during 2014-15 as against Rs. 5000.00 crore during 2013. Food processing industry offers tremendous opportunity for commercial exploitation of horticulture of the State but commercial processing is around 1 per cent only due to lack of post harvesting and processing facilities as well as unscientific packaging. Therefore, opportunities are open for exploiting the potential under processing, with individual, joint venture and Government efforts (Economic survey, 2015).

Agri-food systems are undergoing rapid transformations characterized by the emergence of private standards and different systems of vertical value chain governance. The emergence of integrated supply chains is one of the most visible market phenomena in India. Increasing concentration on processing, trading, marketing and retailing is being observed in all the segments of supply chains. Consequently, production, processing and distribution systems are adapting to such changes.

Production, value addition and marketing of apple are capital intensive and have been a major employment provider among horticultural crops. While the produce moves from one chain actor to another chain actor, it gains value in the form of price mark up. The chain actors, who actually transact a particular product as it moves through the value chain, include input dealers (*e.g.*, nursery suppliers), farmers, traders, processors, transporters, commission agents, post harvest contractors, wholesalers, retailers and final consumer.

Weak production and supply chain along with poor marketing strategies, low transparency in the marketing system have together completely eroded incentive for producers to improve quality and productivity of apple. The low quality of apple is linked with mono-culture of a few old cultivars; faulty pruning and training practices; use of seedling rootstock of unknown performance; deficiency of suitable pollinizers; ineffective control of pests and diseases; lack of institutional credit and efficient factor inputs are some other bottlenecks which have turned the terms of trade against producers. The improvement in the production is quite important, but marketing has also an equal role to give a crop commercial orientation. There have been multi-dimensional efforts to increase the production of apple in the state but market regulation has not received proper attention (Shaheen and Gupta, 2002). Apple marketing being complex phenomena requires special treatment and utmost care in the Kashmir Valley. Present marketing system in the state has an inherent tendency to shift more benefits to intermediaries at the cost of apple growers. The present marketing structure is such that 87 per cent of the marketing functions are solely performed by these powerful intermediaries (Ahmed, 2008 and Bhat, 2010).

The improvement of systems of harvest and post harvest handling must occur at the critical juncture in the value chain between the large farmers and the large wholesalers/importers. The key to making this happen will be the development of closer and more effective coordination between these two categories of value chain actors. This is necessary so that once good quality apples are produced, the needed post harvest procedures and infrastructure are in place so that the product delivered to the cold storage unit will actually meet the agreed quality standards and be properly treated for prolonged storage. Overall, the apple value chain shows considerable dynamic growth potential. Actors in it, at both the farm level and at the key large wholesaler/importer level are making significant new investments in upgrading production and storage capacities.

(a) Rationale

The value chain framework has been used as a powerful analytic tool for strategic planning and is useful in identifying and understanding crucial aspects to achieve competitive strengths and core competencies in the market place. The value chain can be used to diagnose and create competitive advantages on both cost and differentiation. The momentum for the increased production of apple cannot be sustained unless simultaneous efforts are made to improve the marketing of apple through the development of effective marketing system and marketing agencies. The farmers will get the remunerative prices for their surplus produce only when the effective and efficient marketing system is in place. The value chain analysis helps to map the value chain of a specific product involving various value chain actors, which may use qualitative or quantitative approach. While the produce moves from one chain actor to another chain actor, it gains value in the form of price mark up. The current study enables in understanding the flow of apple produce, its treatment at each level, price mark up, roles being played by each value chain player as well as factors affecting the value chain. This study will try to highlight the production and marketing efficiency of entire value chain of apple in the J&K state.

2. The Issues

(a) Lack of Irrigation Facilities

The majority of the farmers (above 80 per cent) faced the problem of proper irrigation facilities especially during the spray season and fruit maturity period (July/August). Although, Kashmir is having abundant water resources, still the growers are facing water scarcity. It is the weak water resource management and improper channelization which is responsible for this important problem.

(b) Lack of Resources

Lack of resources generally faced by marginal farmers, results in lower investmentfor better production technologies. There is a need for creation of durable resources through contract/co-operative/corporate farming to cater such needs of the growers.

(c) Shortage of Effective Fungicides and Pesticides

In present scenario, farmers face the problems of disease and pests which are threat to industry and with this menace growers harvest less than thirty per cent grade A apple as against more than 50 per cent in HP and 80 per cent in Europe. A study conducted in Jammu and Kashmir, revealed that the fungicides and insecticides available in the market are not effective to rectify the problems. Other common glitches are poor adoption of spray schedule, advent of unregistered agencies/Spurious fungicides, loan market linked with pesticide trade and resistance and resurgence.

(d) Introduction of High Density Planting

High productivity can be achieved through shifting from conventional method of planting to High/Ultra High density planting. Most of the developed countries have already shifted to High/Ultra High density planting and have achieved yields of about 40-50 MT/ha and USA about 100 MT/ha through high/ultra- high density planting.

(e) Deficiency of Labour

The shortage of skilled labour is another problem perceived by apple growers. This problem becomes more acute at the harvesting stage of apple when it overlaps with the paddy harvesting operations. Consequently, the farmers have to pay higher wages in order to complete the work in time.

(f) Deficiency of Latest Technical know-how

There is lack of latest technical know-how among the growers. Field functionaries do not make proper and sincere efforts to disseminatethe latest technical know-how to the growers which is considered to be one important cause for low production and productivity in apple.

(g) Replanting/Rejuvenation Problems

About 40 to 50 per cent of the area under apple in J&K State needs replanting as they are more than 50 years old. Replanting will increase the production and productivity level of old orchards. Replanting involves introduction of new plants in between two rows of existing old trees and gradual removal of old trees once the new plants start giving yield. Rejuvenation may be done relatively on a mid-age trees, whose root portion is good but the scion portion is damaged or left unattended for many years and on those trees which have a lower quality and yield.

(h) Planting Material

A general problem is that scarcity of quality planting material and there is huge gap between the demand and supply of certified planting material, which needs to be addressed. It is also necessary to discourage the farmers to purchase the planting material of unknown quality from unregistered nurseries.

(i) Improved Varieties

The most common cultivated variety in the valley is that of Delicious group of apples (70 per cent is that of Red Delicious and Royal Delicious). The Delicious varieties are alternate bearers, susceptible to scab disease, sensitive to weather change and late bearing cultivars that cannot attract the off-season premium of early cultivars. There is a need to diversify in to other varieties of apple and overcome some of the problems associated with Delicious group of varieties. While introducing new varieties, the requirements of early varieties, spur bearing self-pollinating varieties especially for high density plantations, better types of pollinator varieties for interplanting should be considered.

(j) Insurance Cover for Risk Mitigation

Apple faces various kinds of risks during production and is there is no risk mitigation mechanism in place. The possible risks associated with apple production include hail storm during fruit development; snow fall or frost occurrence during flowering and fruit bud development during spring; sudden drop/fluctuations in temperature/inclement weather condition during fruit development which may lead to less fruit set and thereby affecting the yield drastically. The State Government should work with the Insurance companies to put in place some kind of insurance cover for the apple crop.

(k) Lack of Equipment and Machinery

It was observed that 65 per cent of the farmers are in lack of adequate farm machinery and equipment's like power tiller, power sprayer, pumps, *etc.* Most of the growers in the state are marginal and small orchardists which are not financially sound. It is suggested that the government should make available these equipment's on subsidized rates. The financial institutions should also provide loans to growers at low rate of interest in order to purchase these equipment's.

(l) Smallholder Apple Growers

The most of the apple growers belong to the smallholder category in the state of Jammu and Kashmir. In absence of contract/corporate and co-operative farming small holding size of the farm culminates into decline in the productivity and quality of the output.

(m) Market Malpractices

The major destination of Kashmir apples is the Azadpur *Mandi*, which is a buyers' market and designed to be so. Manipulation of prices by traders in the *Mandi* is resorted through a) stopping apple trucks at the border of entry in to Delhi b) use of cold/CA stores of Kundli, Industrial Growth Centres of Haryana *etc.* to alter supply of apples in the *mandi* c) keeping away small buyers with artificially high price quotes and later reducing prices to low levels to benefit preferred buyers and use of proxies in auctions. However, the markets within the state are comparatively better in price determination and transparency

(n) APMC Law

APMC law stands passed in the State, but has not been implemented. The law needs amendments in line with the changes suggested by the Central Government and adopted by several states. This has led to the proliferation of unlicensed traders, agents acting in the market with non-transparent auction procedures. The APMCs do not earn revenues through collection of *mandi* tax and are unable to invest to improve the conditions in the *mandi*.

(o) Lack of Initiative for Direct Marketing

Direct marketing provides better price realization to the growers. Already such arrangements have been in place in Himachal Pradesh (HP) in apple and strawberry. Mother Dairy, NAFED, SurAgro Fresh, Bharti, Mahindra, Futures' Fresh, Big Bazaar *etc.* are procuring top quality apple varieties from orchards of HP. Adani Group also already started direct procurement and buying or through franchise (hub points) *etc.* These initiatives which benefit growers are missing in J&K. Agriculture is a state subject, it is up to the State to adopt innovations and reform their laws.

(p) Dearth of Financial Availability

Most of the apple growers in the state are marginal and small orchardists. The rising cost of inputs for maintenance of orchards has made the cultivation of the crop away from their reach. They have no capacity to invest in better production technology. Their access to financial institution has been beset with innumerable problems like insufficient availability of timely credit, exorbitant rate of interest, *etc.*

(q) Problem Associated with Apple Plucking

Plucking is the primary activity and most delicate process dealing with post-harvesting operations. Apple, being a perishable crop, needs utmost care during the plucking so as to give a safeguard to the apple and make the effective possibility of transfer of apple to the consumer with best form, taste and high nutrient value at desired time and place with maximum consumer satisfaction and producers benefits. Thus, efficient, talented skilful manpower is needed to pluck the fruit. But in the state at present still more than 80 percent fruit is plucked in its traditional form. No, innovative method is used so as to increase the shelf life of the produce in the state. Moreover, the Development Departments are not coming in front to address the problem.

(r) Grading Problems

There is existing still traditional, informal and non-registered grading system in the state of Jammu and Kashmir. Mixing of C grade apples with A and B grade apples is a common practice The resultant unreliable quality leads to a discount in the prices quoted, as traders want to make allowances for possible low grade apples. The absence of equipment to sort, grade and pack is clearly felt. Himachal Pradesh with a clear sorting, grading and packing protocol is able to fetch higher prices at the farm gate for even lower grade fruits than Kashmir.

(s) Highly Indebted Pre-harvest Contractors

The commission agents have made the pre-harvest contractors (PHC) highly indebted by providing time to time and need base financial support to PHC for purchasing fertilizers, pesticides, packaging material, and labours and for other pre-harvesting activities subject to the conditions that produce must be marketed through the commission agent. Finally smallholder growers had to share the indebtedness of PHCs.

(t) Communication Gap

There is a lack of communication between the scientists and apple growers in the state of Jammu and Kashmir. More than 90 per cent of growers don't rely on the suggestions and advises of the experts. Moreover, the institutional service failure was also recorded. The experts are not delivering their services up to the expectations of the growers. There is technological gap because of this reason as the extension wing of government departments are not making proper and sincere efforts to disseminate the technical know-how from research stations to the farmers. Therefore, the communication gap is responsible for the low quality and less quantity of produce.

(u) Problem of Marketing Credit and Commission Agent's Monopoly

The marketing of Kashmir apple is still dominated by the private commission agents and have some how a monopoly power in dealing and handling the entire horticulture produce in general and of apple in particular. Apple growers are lacking the availability of finance so as to make the arrangement for picking, packing, transporting and marketing.

(v) Lack of Organised and Regulated Markets

The dominance of commission agents and hegemony of intermediaries which becomes hurdle not allowing the markets to be organized. Moreover, the apple markets are not properly regulated by the intuitional frame work so that the concept of "minimum price support programme" gets implemented and the fruit growers will be benefited and market risk will be minimized.

(w) Lack of Apni Mandi

Most prominent problem faced by the fruit growers in general and of apple growers in particular is the non-availability of *"Apni Mandi"* as is in Punjab.

3. Value Chain Map and Value Chain Actors

A value chain is the full range of activities including design, production, marketing and distribution; businesses go through to bring a product or service from conception to delivery. For companies that produce goods, the value chain starts with the raw materials used to make their products, and consists of everything that is added to it before it is sold to consumers. The value chain is the core business process in an organization that created and delivered a product or service from concept through development and manufacturing into a market for consumption and the integrated supply chain consisted of suppliers and organization (USAID, 2008; DoA, 2011). This segment of this chapter gives an overview of the whole value chain of the Kashmiri apple in a value chain map with the main focus on the primary value chain actors, which make this value chain functional.

(a) Value Chain Map

Figure 2.1 illustrates the value chain map of various actors involved in the apple value chain, from producers at the bottom all the way to the ultimate consumers. On the left, various functions are shown of the primary value chain actors, while on the right-side the various service providers or support organizations can be seen. Especially, the government support organizations have a strong role in determining the enabling environment through policies, subsidies, *etc.*

(b) Value Chain Actors

The apple value chain map shown in Figure 2.1 is based on the Kashmir region, which is the main centre for commercial apple production (68 per cent) and trade in India. Other areas of production, include Himachal Pradesh and Uttarakhand,

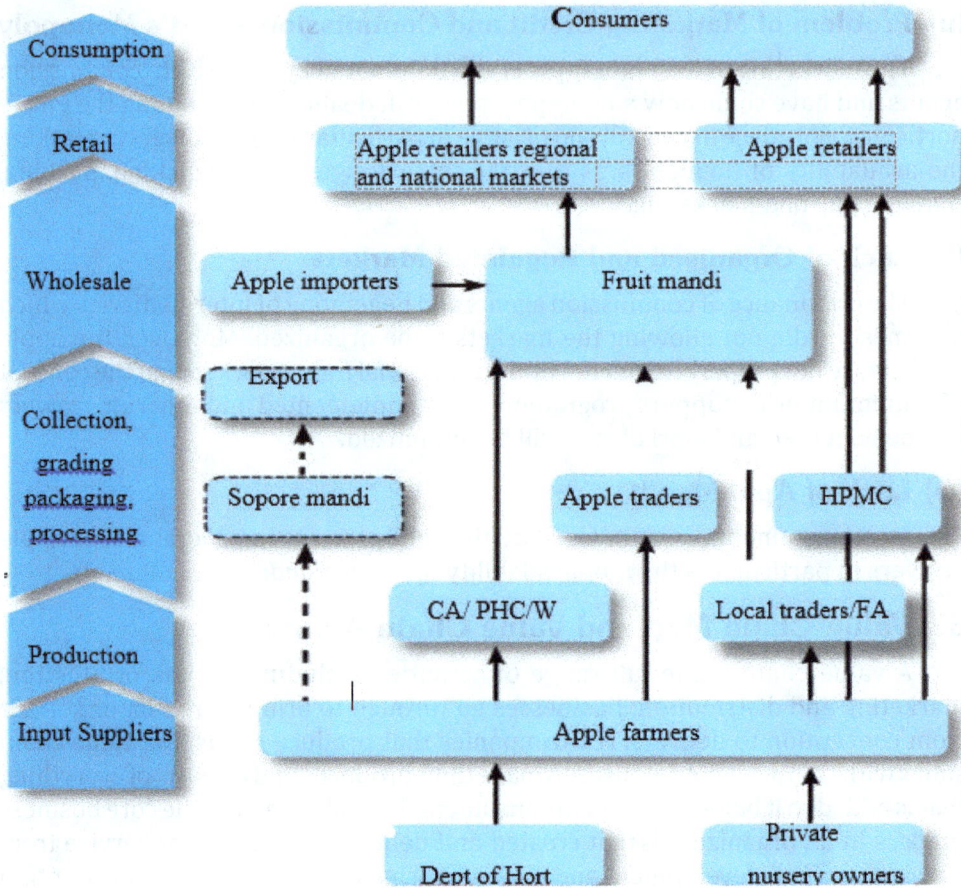

Figure 2.1: Value Chain Map.

and have similar production marketing structures. The main actors involved in production and trade are described below.

(I)   Production

(i)   Small Farmers

Small farmers with apple surfaces of between 0.1 ha and 0.7 ha, for whom apple production constitutes a major component of household revenue but is unlikely to be the most important source of income. These farmers will have a volume of production generally under 1000 boxes of apple that justifies some effort spent at harvesting and marketing, but they do not treat apple production as a commercial activity with a rationalized system of production that seeks to maximize returns to land or labour. They have limited receptivity to improved technologies and little ability to make new investments in apple production. Yields among farmers in this category are estimated to be in the 18 thousand kg per hectare range. The small size of most orchard plots is a major constraint for apple orchard development because it limits capital investment possibilities and, in the absence of any collective marketing, makes for a multiplicity of farm level points of sale that add to collection costs.

(ii) Large Farmers

Large farmers cultivating surfaces of over 0.8 ha (and upto 7 ha in one outlying case) that operate as true commercial apple orchards. Such farmers will invest in certified saplings of good genetic quality, prepare the soil on an annual basis, apply fertilizer and use gravity-fed flood irrigation. Mostly apply pesticides, although recommended dosages and spraying schedules are rarely respected. These farms generally produce over 2000 to 3000 boxes of apples in a season and many more if areas cultivated exceed one hectare. Yields may be similar to small farmers if no improved production technologies are used, but most farmers in this category can achieve yields in the 25 to 40 thousand kg per hectare range. Apple production for this category of farmer appears to be fairly profitable.

(II) Pre-Harvest Contractor

Pre – harvest contractors are the persons specialized in performing various marketing functions. They are efficient marketers of fruits. They overcome the difficulty of small produce by way of contracting more than one orchard at a time and perform most of the marketing functions themselves.

(III) Forwarding Agents

These are specialized persons operating in the apple producing areas. Their main business is to arrange the transportation of produce of their clients to different markets. For performing this function they charge commission on per box basis. These agents also supply packaging and other material to the orchardists.

(IV) Commission Agents (Arhatias)

A commission agent is a person operating in the wholesale market who acts as a representative of either a seller or a buyer. Orchardists consign their produce to commission agents in a particular market and take over physical handling of the produce and make arrangements for its sale, collect the money from the buyer, deduct his expenses and commission and remits balance to the seller. They sometimes provide advance money to the farmers for arranging packing material *etc.* with the stipulation that produce will be consigned to him for sale.

(V) Wholesale Trade

The wholesale markets in major cities are the critical link in the apple value chain that is characterized by a diversity of relatively fluid and unspecialized actors. An estimated 90 per cent of all apples (both domestic and imported) pass through the wholesale markets of Kashmir. The market provides the physical space where farmers, traders, and retailers come together in the greatest volumes. Although some actors bypass the wholesale market, it is clear that the main tendencies of the Kashmir apple market are determined within the confines of the wholesale markets—and particularly the one in Sopore. The major types of actors at this critical level are:

(i) Small Wholesalers

These players possess their own transport vehicles, most often a load carrier or a troller. They are the major buyer at the farm level and the source of the largest volumes arriving in wholesale markets. These are, for the most part, small traders with no fixed warehouse or depot, and who travel each day to *mandi*, as long as apples are available, to bring them to their home base, where they either sell into the wholesale market or make deliveries directly to market retailers. Their principal market function is to serve as the main locus for assembly and transport of apples from the farm level to the wholesale markets. They are in competition with farmers who may perform the same function if they have access to transport.

(ii) Large Wholesalers/Importers

These are generally incorporated registered businesses who have a fixed warehouse inside or outside the wholesale market. Most are registered as importers, as they import apples and other products. These actors buy domestic apples from individual larger farmers and use their own transport vehicles, generally trucks that have a capacity of over five tons, to evacuate the product. In general, actors in this category are small or medium enterprises with diversified activities and other sources of income that may include agribusiness activities, construction and real estate. They sell to distributors in the wholesale market, although they may also sell occasionally further downstream to retailers when they are dealing with domestic apples in smaller quantities.

(iii) Major Apple Importers

This category of actor is practically the only true specialist in the entire apple value chain; they limit their activities to the import and wholesaling of imported food products, including apples. They do not sell in less than pallet sized loads, whereas all other actors deal in 18-to-20 Kg crates of domestic apples. They sell mainly to distributors in the wholesale market. Like the large wholesalers/importers, they are officially registered enterprises. Apple importers often also have other lines of business activity.

(iv) Distributors

These traders, who are largely unincorporated "physical persons," assemble a variety of fresh produce from different sources, including imported and local products, and offer a basket of different fruits and vegetables that meets the desired needs of retailers. They generally rent smaller depots inside the wholesale market around the parking area where small wholesaler trucks park when they arrive to sell in the market. Others are located near vicinity of the market. They purchase pallets of imported products, which they stack and combine with purchases of domestic products of all types from small wholesalers arriving in the market (or apples from larger wholesalers/importers). They usually purchase domestic produce in crates. While the mix of both Kashmiri and imported products changes during the seasons, distributors will seek to constitute a relatively constant array of the main fruits and vegetables so that they can attract a regular clientele of retailers who appreciate the convenience of rapid shopping and not having to conduct too many transactions with different small wholesalers. Thus the ability to offer key products that are high in demand throughout the year, including apples, is a key factor in the success of individual distributors. There also seems to be some differentiation among distributors by quality levels, with the distributors located inside the main warehouses making an effort to offer a higher quality of produce at a higher price than those in the small wholesaler transactions zone of the market.

(VI) Retail Trade

No data on the breakdown of fruit sales volumes between the different retail market segments was found during the research for this study. Thus it is extremely difficult to assess trends in changing market shares of fruit and vegetables in general, or apples, in particular. Anecdotal observation and the reported opinion of market actors tend to support the view that the retail sector remains largely dominated by small neighbourhood retail grocery stores (including stores with a range of products and fruit and vegetable specialty retailers) and, especially, by the green (retail) market vendors. All categories of retailers buy local and imported apples mainly from distributors and occasionally from other sellers in the wholesale markets. No clear quality or product differentiation is practiced at the retail level, although market observers agree that some retailers in wealthier areas tend to buy and sell produce that is of better and more standard quality than retailers in lower income areas. The retailer earns his income from the difference in the sale and purchase price.

(VII) HPMC

The Horticultural Produce Marketing and Processing Corporation Ltd. was established in the year 1974 in Jammu and Kashmir. The main function of this corporation is to modernize apple marketing system by developing infrastructure for the post harvest handling of apples on scientific lines. This HPMC has set up many mechanical grading and packing houses, cold storage facilities in producing areas and consuming markets, transportation facilities both ordinary and refrigerated and fruit processing plants. In addition, they have to make arrangements for the sale of fresh fruits and processed products in various markets, supply of packing material to orchardists, etc. these facilities are offered to apple growers in the form of integrated marketing system to make use of them for better returns, through value addition.

(VII) Market Information Agencies

Market information ensures the smooth and efficient working of the marketing operation. Accurate, adequate and timely availability of market information facilitates, decision about when and where to market the fruit. The Horticulture Planning and Marketing collect prices of apple for different varieties and grades from different markets of the country and disseminate the same through All India Radio/Doordarshan,Srinagar and also publish in monthly bulletins for the benefits of the producers, traders and consumers.

(c) Value Chain Governance

The apple value chain has a long history and has been functional despite the problems and inefficiencies. While the changes in some parts of the value chain are incremental, other changes are radical especially having to deal with mind-set changes. The important issues to be dealt with in the value chain are to improve the quality and productivity of fruit on-farm, to ensure that quality fruits are segregated and offered to realise the premium prices in the markets, prevent glut in the market through store and controlled release of fruits (with the help of cold stores) for marketing and transform market practice.

References

Ahmad, N. 2008. *Problems and prospects of temperate fruits and nut Production scenario in India vis-à-vis international scenario*, Central Institute of Temperate Horticulture, Srinagar.

Bhat, J. 2010. Problem of apple marketing in Kashmir. *Abhinav Journal of Research in Commerce and Management* 1 (6): 105-111.

Chengappa, P.G. 2004. Emerging trends in agro processing in India. *Indian Journal of Agricultural Economics* 59(1): 55-74.

Department of Agriculture (DOA) 2011. *A value chain analysis of apple from Jumla*. Ministry of Agriculture and Co-operation, Lalitpur, Netherlands Development Organization (SNV), pp49.

DES (Directorate of Economics and Statistics) 2014.*Digest of Statistics, Planning and Development Department*, Jammu and Kashmir Government, Srinagar.

Economic Survey, 2015. *Directorate of Economics and Statistics*, Jammu and Kashmir Government, Srinagar 876 pp.

Food and Agriculture Organization (FAO) 2014. *Production Year Book*. Food and Agricultural Organization of the United Nations. Rome.

Naqash, Farheen. 2015. A value chain analysis of apple in Jammu and Kashmir. *MSc. Thesis submitted to Sher-e-Kashmir University of Agricultural Sciences and Technology of Kashmir, Shalimar*, pp. 94.

NHB, 2014. *Indian Horticulture Database-2014*, National Horticulture Board (NHB), Ministry of Agriculture, Government of India.

Shaheen, F.A. and Gupta, S.P. 2002. Economics of apple marketing in Kashmir province —problems and prospects. *Agricultural Marketing* XLV (2): 5-13.

USAID, 2008. *The Albanian apple value chain:* FSKG case study, Micro Report 120, pp59

Wani, S.A., Wani, M.H., Bazaz, N.H. and Mir, M.M. 2015. *Commodity profile of apple*, Network Project on Market Intelligence, Sher-e-Kashmir University of Agricultural Sciences and Technology of Kashmir, Shalimar, pp 42.

Chapter 3

Apple Farming in Himachal Pradesh: An Overview

Ravinder Sharma, Subhash Sharma
and Chandresh Guleria

Department of Social Sciences, College of Forestry
Dr. Y S Parmar University of Horticulture and Forestry, Nauni, Solan
E-mail: rsharmauhf@yahoo.co.in

1. Introduction

The beginning of the apple production in India was quite quixotic. At the beginning of the twentieth century, Samuel Evan Stokes landed in India from Philadelphia, USA, to join the Leprosy Mission of India. He settled in the area of Kotgarh and Thanedar in the present day state of Himachal Pradesh, and was re-christened as Satyanand. He was convinced that apple cultivation could end the ills of the farmers in Himachal Pradesh. In 1916, he brought in the saplings of the Stark Brothers' Delicious variety of apple from USA. It is from these first few saplings of the Delicious apples that the Indian state like Himachal Pradesh has become famous in India for apple cultivation.

In other hill regions of India, however, apple was introduced by the Europeans in the nineteenth century. It was introduced in the Nilgris through the botanical gardens in Ootacamund as early as the 1850s. The credit for their commercial plantation must go to Capt. R.C. Lee, a retired British soldier who planted the first apple garden at Bundrole in the Kullu valley in 1870. The European varieties, adapted to the colder climate and lower sunlight intensity, were greenish - yellow in colour and sub-acidic in taste. However, these varieties did not appeal to the Indians who produced a wide choice of sweet fruits.

Commercial apple production was given an impetus by SatyaNand Stokes, an American missionary turned Hindu, who imported the Red Delicious apple from the USA along with American technology, in the first quarter of this century. It was with the import of the red-coloured, sweet-flavoured Red Delicious that apple received its home coming in India. The Red Delicious quickly flourished in the bright sunshine and warm summers of the North Indians hills. Practically all the Delicious apples in India today, are the progeny of the plants imported by Stokes. Intensive research in production techniques and introduction of new cultivars has led to high productivity of quality fruit in Europe and USA and production has reached a level of 2,500 to 7,500 boxes per hectare as compared to500 to 1250 boxes per hectare in India.

2. Trends in Area, Production and Productivity of Apple

(a) National Scenario

In India, the production of apple is confined to Jammu and Kashmir, Himachal Pradesh, Uttarakhand, Arunachal Pradesh, Nagaland and Sikkim. However, Jammu and Kashmir and Himachal Pradesh are the most important states together accounting for 85 per cent of the total area and 95 per cent of the production in the country (Table 3.1).

Table 3.1: Area and Production of Apple in different States of India (2013-14)

States	Apple	
	Area ('000 Ha.)	Production ('000, Tonne)
Arunachal Pradesh	14.28 (4.56)	31.87 (1.28)
Himachal Pradesh	107.69 (34.40)	738.72 (29.58)
Jammu and Kashmir	160.87 (51.39)	1647.69 (65.97)
Sikkim	0.02	0.03
Tamil Nadu	0.01	0.03
Uttarakhand	29.97 (9.57)	77.45 (3.10)
Total	**313.04 (100)**	**2497.68 (100)**

Source: Horticulture Statistics Division, DAC and FW.

The area under apple crop registered a growth of 3.40 per cent during 1991-92 to 2013-14. The growth rate in production was recorded 5.40 per cent per annum. However, the productivity registered a growth of 2 per cent per annum which was lower as compared to growth in area and production (Table 3.2).

(b) State Scenario

Apple cultivation has made tremendous growth in the state during the last four

Table 3.2: Area, Production and Productivity of Apple in India (1991-92-2013-14)

Year	Area(000' ha)	Production (000' tonne)	Productivity (Tonne/ha)
1991-92	194.5	1147.7	5.9
2000-01	239.8	1226.6	5.1
2001-02	241.6	1158.4	4.8
2002-03	193.1	1348.4	7.0
2003-04	201.2	1521.6	7.6
2004-05	230.7	1739.0	7.5
2005-06	226.6	1814.0	8.0
2006-07	252.0	1624.0	6.4
2007-08	264.0	2001.0	7.6
2008-09	274.0	1985.0	7.2
2009-10	282.9	1777.2	6.3
2010-11	289.1	2891.0	10.0
2011-12	321.9	2203.4	6.8
2012-13	311.50	1915.40	6.10
2013-14	313.04	2497.68	8.0
CGR (per cent/annum)	3.40	5.40	2.0

Source: Horticulture Statistics Division, DAC and FW.

Table 3.3: Compound Growth Rates of Apple in Various Districts of Himachal Pradesh during 1973-74 to 2013-14

Districts	Area	Production	Productivity
Chamba	7.50* (0.094)	4.50* (0.454)	- 3.00* (0.421)
Kangra	0.50** (0.131)	- 1.10 (0.601)	- 1.60 (0.684)
Kinnaur	6.10* (0.092)	8.20* (0.283)	2.10* (0.222)
Kullu	3.00* (0.045)	2.80* (0.402)	- 0.20 (0.391)
Lahaul and Spiti	9.40* (0.239)	9.90* (0.527)	- 3.70* (0.795)
Mandi	3.30* (0.045)	2.80* (0.402)	- 0.40 (4.325)
Shimla	2.20* (0.044)	3.60* (0.360)	1.40* (0.396)
Solan	- 3.00* (0.379)	- 6.60* (0.366)	- 3.60* (0.545)
Sirmaur	0.80* (0.088)	- 4.60* (0.498)	- 5.40* (0.494)
Himachal Pradesh	3.27* (0.045)	3.50* (0.315)	0.30 (0.305)

* Significant at 1 percent level; ** Significant at 5 percent level, of significance.

decades in respect of area and production. Area under apple cultivation increased at a rate of 3.27 per cent, whereas, production recorded a growth of 3.50 per cent per annum, which was slightly higher than the growth in area. However, growth in productivity was found to be 0.30 per cent per annum which is a cause of concern (Table 3.3). Further district-wise highest growth in area (9.40 per cent) was found in Lahaul and Spiti district followed by Chamba (7.50 per cent) and Kinnaur (6.10 per cent). Highest growth rate in production (9.90 per cent) was found in Lahaul and Spiti followed by Kinnaur (8.20 per cent) which forms the major tribal belt of the state.It is further observed that the growth rate in area was higher as compared to production highlighting the fact the increase in production is mainly because of area expansion.

The area growth was quite impressive and significant in Lahaul and Spiti, Chamba and Kinnaur districts. These districts constitute backward tribal districts, which received added attention in the recent past by the state government for augmenting growth promotion in these untapped non-traditional areas. Moreover, climatic conditions are more favourable in these districts for the successful cultivation of apples. Shimla, Mandi and Kullu districts, are traditional areas for apple production. District Kangra and Sirmaur, which have lesser comparative advantage of growing apple, also experienced positive and significant growth of 0.50 and 0.80 per cent per annum respectively. It is interesting to note that, district Solan, exhibited a declining trend in the area expansion during the last 39 years. The area under apple crop in the state has shown increasing trends (Figure 3.1).

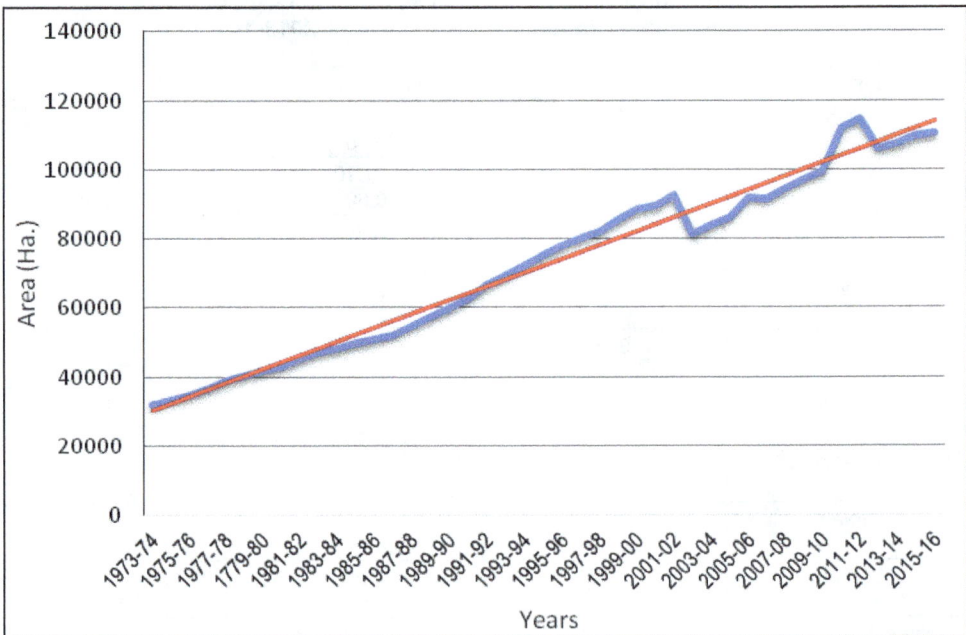

Figure 3.1: Trends in Area under Apple in Himachal Pradesh (ha), 1973-74 to 2015-16.

Himachal Pradesh has undergone a revolution in the apple production during last few decades. The area under apple has increased by more than six-folds since 1966-67. During 1966-67, the area under apples constituted nearly 58 per cent of the total fruit area in the state. In the later years, there has been relatively more emphasis of planting of other fruit trees in the state as a consequence of which the proportionate share of apple area has come down to nearly 48.30 per cent in the year 2011-12. More than 2 lakh farm families are engaged in apple cultivation out of which nearly 90 per cent are small and marginal with an average holdings of less than 0.6 hectares. Apple farming is the fastest growing economic activity of the state and is being grown in 9 out of 12 districts. Due to varied agro-climatic conditions across the districts, there exists large variation in the area and output growth of this crop activity.

Amongst all the fruits grown in the state, apple occupies the premier position in terms of production accounting for nearly 74 per cent of total fruit production. The results of the district wise analysis show that apple production has taken new stride, and has increased at a rate of 3.50 per cent per annum during 1973-74 to 2013-14 (Figure 3.2). The perusal of the table revealed that, the tribal districts *viz.,* Lahaul and Spiti, Kinnaur and Chamba has recorded a significant increase in the production, whereas, low altitude districts of Kangra, Solan and Sirmaur recorded a negative growth in the production which may be attributed to change in climatic conditions adversely affecting the productivity of the crop.

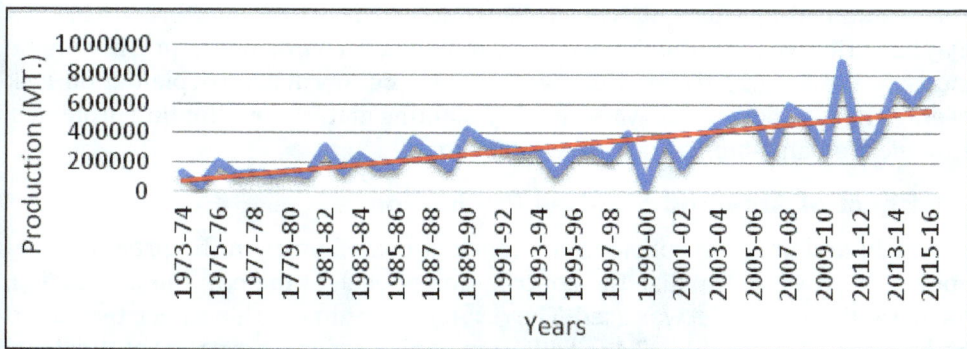

Figure 3.2: Trends in Production of Apple in Himachal Pradesh.

(c) Changes in Productivity

In Himachal Pradesh, the overall productivity of apple hovers around 4 tonnes per hectare as compared to 8 tonnes per hectare of all India level and is much below the international level of 40 tonnes per hectare.

It is very disheartening to note that except Kinnaur and Shimla districts, all the districts have exhibited the very waning trends in growth in productivity. The declining growth in productivity was significant for the districts of Solan, Sirmaur, Chamba and Lahaul and Spiti. Significant growth in productivity was observed

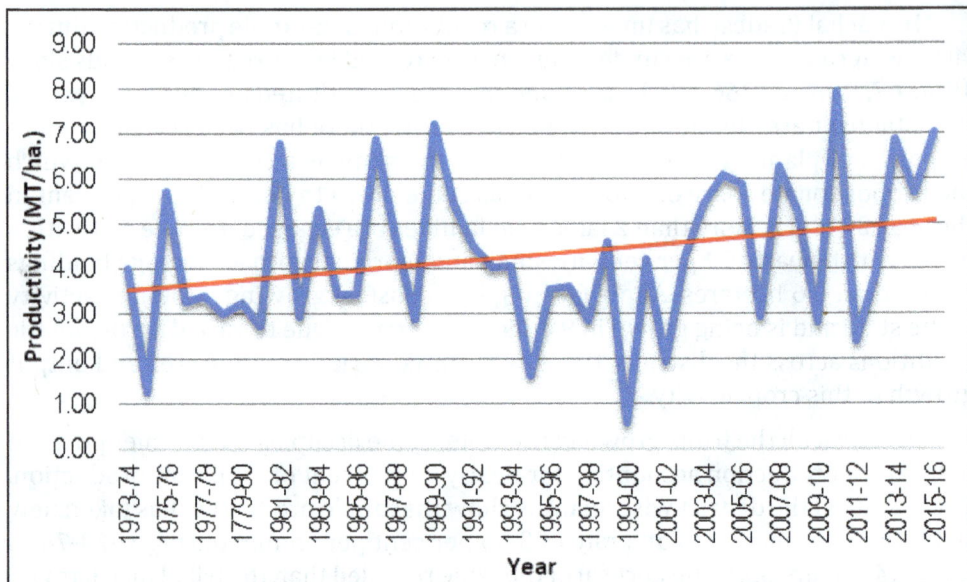

Figure 3.3: Trends in Productivity of Apple in Himachal Pradesh (MT/ha).

in only two districts namely Shimla and Kinnaur. The scenario of continued deceleration in apple productivity is a cause of concern. This dismal growth in yield may be attributed to predominance of old and senile orchards, development of apple industry under rain-fed conditions, global warming, low density of plantation, lack of efficient use of irrigation water, quality planting material, pollination problems, site selection and imbalanced use of resources.

(d) Effect of Area and Yield on Production

Yield and area are considered important contributors in the production of apples. In order to visualize the contribution of each of them in the production, Narula and Sagar model was used. The results of the analysis have been presented in the Figure 3.4 and Table 3.4 for different apple producing districts of Himachal Pradesh. The results of decomposition analysis through Narula and Sagar model as applied to 39 years data revealed that for the state as a whole increase in apple production was mainly due to area expansion (85.67 per cent) followed by yield effect which works out to 14.33 per cent, the relatively stronger area effect over the yield effect implies that whatever growth has taken place in the apple output it is mainly attributed to extensive cultivation rather than intensive cultivation. This also implies that yield increasing innovative technologies have not reached the farmers door to an expected extent.

3. Varieties of Apple

Over 700 accessions of apple, introduced from USA, Russia, U.K., Canada,

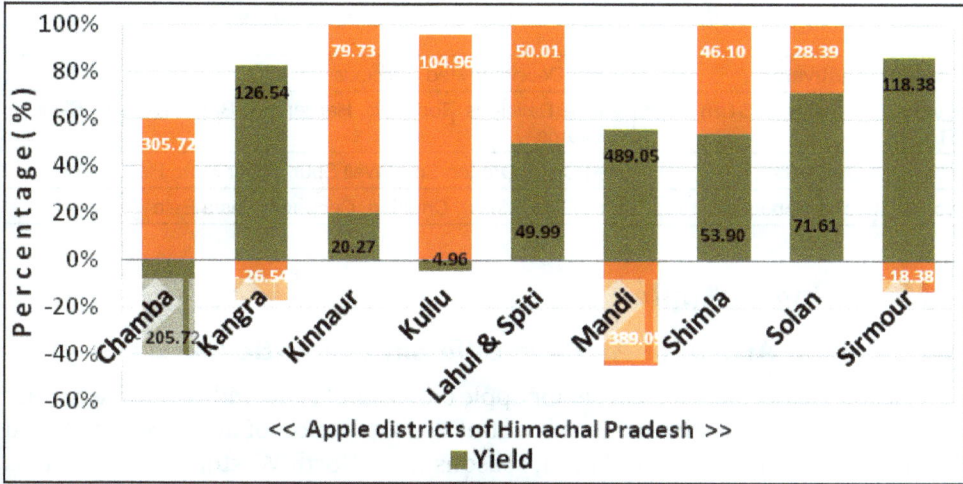

Figure 3.4: Percentage Contribution Due to Yield and Area in Production.

Table 3.4: Contribution of Yield and Area in the Production of Apples in different Districts of Himachal Pradesh, 1973-74 to 2011-12

District	Percentage Contribution due to	
	Yield	Area
Chamba	-205.72	305.72
Kangra	126.54	-26.54
Kinnaur	20.27	79.73
Kullu	-4.96	104.96
Lahaul and Spiti	49.99	50.01
Mandi	489.05	-389.05
Shimla	53.90	46.10
Solan	71.61	28.39
Sirmaur	118.38	-18.38
Himachal Pradesh	14.33	85.67

Germany, Israel, Netherlands, Australia, Switzerland, Italy and Denmark have been tried and tested during the last 50 years. The delicious group of cultivars is predominated in the apple market. The area covered under Delicious cultivars is 83 per cent of the area under apple in H.P., 45per cent in J&K and 30 per cent in Uttarakhand. In more recent times improved spur types and standard colour mutants with 20-50 per cent higher yield potential are favoured. The important cultivars recommended for cultivation are presented in Table 3.5.

Table 3.5: Recommended Cultivars of Apple

Standard Cultivars	Royal Delicious, Red Delicious
Early Colouring Standard Strains	Vance Delicious, Top Red, Hardeman, Bright-N-Early, Skyline Supreme
Spurtype Cultivars	Red Chief, Oregon Spur, Well Spur, Silver Spur, Red Spur
Non-Delicious non spur	Fuji, Gala, Mutsu, Criterian, Cooper-4, Vesta Bela

4. Marketing of Apple

(a) Trends in Arrival of Apple in different Domestic Markets

The biggest wholesale market for apple crop is the "Fruit and Vegetable market at Azadpur", in Delhi. About 70 per cent of the total trade of apple is distributed through this market. The marketing seasons from North Western States such as Himachal Pradesh is from July to October with a peak in August- September, whereas

Figure 3.5a: Different Varieties of Apple—Early Colouring Strains.

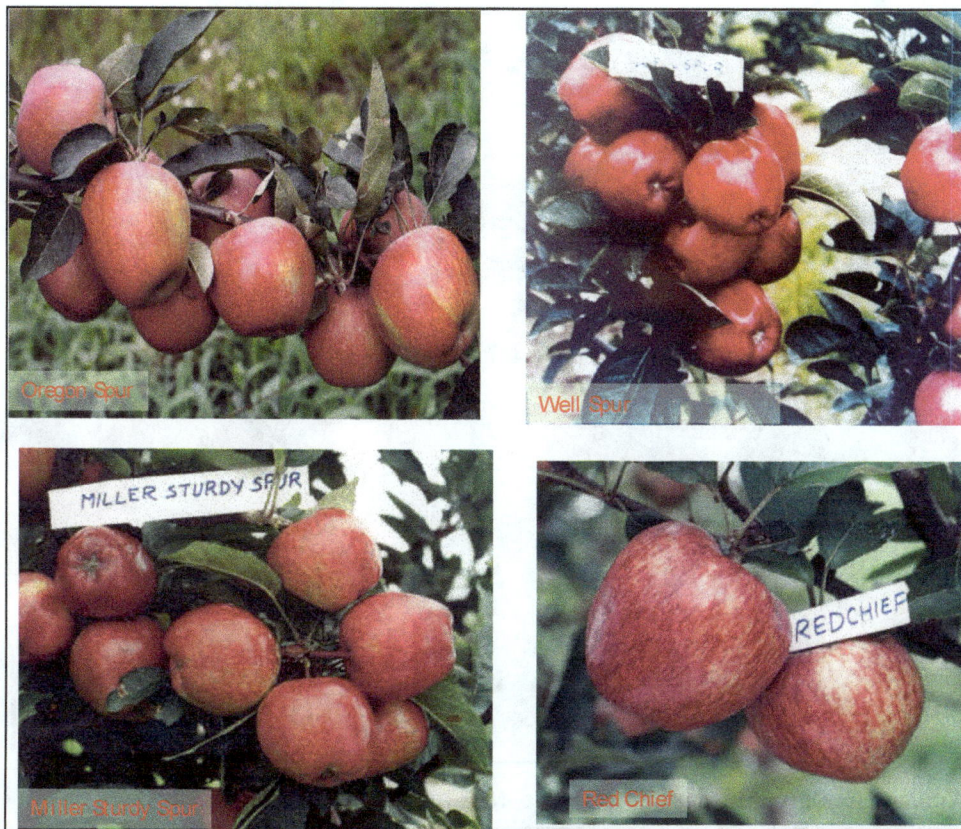

Figure 3.5b: Different Varieties of Apple–Spur Type Varieties.

in Jammu and Kashmir it is from August to November with a peak in September to October. Similarly from Uttarakhand crop comes into the market during June to October with a peak in July to mid-September. Analysis of market arrival data from Chandigarh, Delhi and Shimla markets revealed that still 98 per cent of produce is marketed in Delhi market.According to an estimate about 4.01 per cent of apple is lost during the post harvest operations.It was further found that about 50 per cent losses occur during harvesting and assembling stage followed by 14.89 per cent during grading and packing, 13.76 per cent in transportation, 11.67 per cent in the markets and 7.20 per cent during the storage. There exist widespread fluctuations in the prices of apple produce overtime and space which introduce an element of uncertainty in income level of the growers.

(b) Existing System of Marketing of Apple

Analysis of data on number of holdings according to size class revealed that majority of holdings are marginal and small and about 50 per cent of the area is

Figure 3.5c: Different Varieties of Apple–Non Delicious Varieties.

under these two classes (Table 3.6). Marketable produce also depends on the size of holdings. This holds true for the apple orchardists too. According to horticultural census (1989), 86 per cent of apple orchardists fall in the category of marginal growers. Only about one per cent growers come under medium to large category. This implies that marketable surplus in scattered and poses a problem in the marketing.

Table 3.6: Distribution of Operational Holdings in HP

Size Class	1995-96		2005-06	
	No of Holding (per cent)	Area (per cent)	No of Holding (per cent)	Area (per cent)
Marginal (< 1ha)	64.4	23.10	68.21	26.67
Small (1-2ha)	20.10	24.10	18.82	25.27
Medium (2-10ha)	15.0	45.0	12.60	41.86
Large (10ha)	0.50	7.80	0.38	6.20

Source: Statistical outline of Himachal Pradesh, Directorate of Economic and Statistics.

It was observed that of the total production of apple which was approximately 583000 MT of which 524000 MT reached in various *Mandis* during 2014. It can be observed that 48 per cent of Himachal apple was traded in APMCs situated in Himachal Pradesh and 45 per cent was traded in the outside markets of state (Delhi, Chandigarh, Bombay, Chennai, and Kolkata). About 4.80 per cent of Himachal apple was procured by the agro commodities trading houses. HPMC and HIMFED procured 14000 MT of culled apple from the Himachal Pradesh. The trend implies that with the development of APMCs in the state, orchardists who are mostly marginal and small prefer to sell their produce in these markets (Table 3.7).

Table 3.7: Production and Marketing of Apple in HP in 2014

Particular	Quantity (MT)
Total Production	583000 app.
Fruit dispatched to the market	524000 app.
Percentage of the apple produce sold in HP markets	48 per cent
Percentage of the apple produce sold in outside HP markets	45 per cent
Procured by private companies	4.80 per cent
HPMC (culled fruit)+HIMFED	14000

Source: Author's estimate.

(c) HP Business Models for the Procurement and Distribution of Apple

At present, three different business models are practiced for the procurement and distribution of apples in Himachal Pradesh which are Commission agents in traditional APMC markets, Semi-direct company buyers and direct company buyers.

(I) Commission Agents in Traditional APMC Markets

After harvesting the farmers pack their apples in cardboard boxes and transport them by small trucks to the *Mandis*, travelling on an average of about 20 kilometres from their farm. A commission agent in the *Mandi* works with the farmers by acting as a liaison between the farmers and buyers. There are major inefficiencies in this supply chain model. From the grower's perspective the major disadvantage is that he or she does not know before hand the prevailing price of apples at the *Mandi*. Word-of-mouth and/or cell phone communication are the only means of price discovery for the farmer. This information is often unreliable and insufficient for determining where, when, and at what price to sell the product. Once the farmer arrives at the *Mandi* with the produce, he or she discovers the price. In most cases, the farmer must sell at whatever price the apples get at auction by the commission agent. Farmers are left with a few options for two principal reasons firstly; shortage of storage facilities in the absence of cold chain infrastructure thus forcing the farmer's to sell their fruit immediately following harvest. Secondly, farmers lack financial training and do not understand that transporting their apples to the *Mandi* and incurring the transportation costs generally puts them at the mercy of whatever price the commission agent offers. Farmers simply cannot afford to pay the cost of transportation more than once. Before the APMC Act reforms, farmers were not only dependent on commission agents to sell their apples but also to get loans in the absence of a formal credit mechanism. Reliance on a commission agent makes the entire transaction very asymmetric where the farmer has very little power relative to the commission agent. With little power in the hands of farmers, cheating in the weighing of the apples has become standard practice and farmers are not in a position to demand otherwise.

(II) Semi-direct Company Buyers

Retail companies such as Reliance, Mahindra and Mahindra, and Spencer's have hired their own agents in the state. These agents buy from farmers on behalf of their company and compete aggressively in the field for larger volumes of good quality apples. Since most of the growers produce small quantities of apples, the purchasing company needs many buyers to handle the large volume of purchases. Moreover, since the packing and grading of apples is not standardized, a great deal of time is spent finalizing the deal with farmers. This makes it difficult to monitor and control the entire operation of apple procurement. Semi-direct company buyers purchase approximately 30 per cent of overall apple production in the state.

(III) Direct Company Buyers

At present, the direct company buyers include Adani Agri-fresh and Fresh and Healthy Enterprise Limited of the Container Corporation of India (CONCOR), Ministry of Railways. Unlike commission agents and semi-direct company buyers, direct company buyers work throughout the year to train farmers in scientific cultivation practices and post harvest management. These training sessions are organized by company personnel, who send experts hired by them periodically to visit villages and invite farmers to participate in training sessions free of charge. Both Adani Agri-fresh and Fresh and Healthy own CAS facilities, which are technically far superior to conventional cold storage, as the former controls the entire atmosphere and not just the temperature.

Table 3.8: Procurement of Apples by different Firms in Himachal Pradesh

Name of Firm	Year	Quantity Purchased (MT)	Average Procurement Price (Rs. per Kg)
M/S Adani Agri Fresh Limited	2006	4766.560	31.30 to 32.50
	2007	15409.990	24.30
	2008	19704.482	25.30
	2009	8783.940	41.75
	2010	16317.156	26.19
	2011	6000.00	N.A.
	2012	16268.000	N.A.
	2013	21068.057	N.A.

Name of Firm	Year	Quantity Purchased (MT)	Average Procurement Price (Rs. per Kg)
M/S Fresh and Healthy Enterprises Limited	2006	1060.000	29.62
	2007	10940.000	27.87
	2008	7720.000	35.50
	2009	2720.000	44.12
	2010	8400.000	39.29
	2011	6904.000	N.A.
	2012	1603.580	N.A.
	2013	10205.000	41.57
M/S Dev Bhumi Cold Chain Private Limited	2006	-	-
	2007	690.080	25.00 to 45.00
	2008	723.240	35.80
	2009	285.520	33.82
	2010	139.875	25.89
	2011	6280.000	N.A.
	2012	7000.000	N.A.
	2013	2871.588	36.62

Some information has been collected and presented in Table 3.8. The analysis of data showed that procurement through this channel is picking up. A linear growth rate of 6.35 per cent per annum was recorded. Price received by the farmers varied between Rs 25 to 45 per Kilogram. It was further observed that prices of different company buyers also varied depending upon their quality parameters set for the crop procurement.

(d) Distribution and Marketing Channels

Distribution comprises movement of apples from producer to ultimate consumer. In this process the fruits have to pass through more than one hand, except when it is directly sold to consumer by the producer, a rare phenomenon. In this chain various agencies like growers, pre-harvest contractors, wholesalers, and retailers are engaged. The chain of intermediaries/functionaries are called the marketing channels. The following channels are identified as important channels through out Himachal Pradesh for marketing of their produce.

A. Producer ⟶ Pre-harvest contractor ⟶ Commission agent/wholesaler ⟶ Retailer ⟶ Consumer

B. Producer ⟶ Forwarding agent ⟶ Commission agent/wholesaler ⟶ Retailer ⟶ Consumer

C. Producer ⟶ Commission agent/wholesaler ⟶ Retailer ⟶ Consumer

D. Producer ⟶ Producers Cooperative Society ⟶ Commission agent/ wholesaler ⟶ Consumer

E. Producer ⟶ HPMC ⟶ Processing unit

F. Producer ⟶ Agro commodities trading houses ⟶ Consumer

It can be observed from the Table 3.9 that more than half of the apple (56.04 per cent) in the selected farms are traded through the marketing channel C *i.e.* Producer -Commission agent/wholesaler-Retailer –Consumer followed by marketing channel D *i.e.* Producer –Producers Cooperative Society- Commission agent/wholesaler- Consumer (19.29 per cent). 5.60 per cent apple was traded through the Agro commodities trading houses. HPMC procures just 2.77 per cent of the apple in the study area.

Table 3.9: Average Quantity of Apple Marketed through different Channels in Study Areas of Himachal Pradesh

Marketing channel	Quantity Sold (per cent)
Producer-Pre-harvest contractor - Commission agent/wholesaler -Retailer – Consumer	7.02
Producer- Forwarding agent - Commission agent/wholesaler-Retailer –Consumer	9.28
Producer - Commission agent/wholesaler-Retailer –Consumer	56.04
Producer –Producers Cooperative Society – Commission agent/wholesaler-Consumer	19.29
Producer-HPMC- Processing unit	2.77
Producer-Agro commodities trading houses –Consumer	5.60
Total	100

Economics of marketing through different agencies/business models was estimated and the results have been presented in Table 3.10. Analysis of data revealed that though the maximum price was realized in the Delhi market, but the average price was found higher in the markets situated within the state. The reason for this was that the produce in the Delhi market is sold on the grade basis, whereas in the local markets it is sold on average basis. The proportion of quality produce is less in the total production and only large growers preferred to sell their produce in the Delhi market. It is interesting to know that only one per cent of the total growers fall in the category of large growers. The cost per kilogram of the produce varied between Rs 9.50 to Rs 11.50 in the local and Delhi market. It was found that in the local market average price per kilogram was higher as the produce is sold on wholesale grade basis. With the introduction of private players, growers are selling their produce to them also. But these players are buying produce of a specified quality and grade. Different players have different parameters for buying the produce. Almost all the players are buying produce from the apple growing areas above 2000 masl. Major advantage to the growers is that they do not incur marketing cost or if there is any it is very negligible. However, they cannot sell their whole produce since it does not comply the standards fixed by these companies. However, high quality produce is being sold through this channel at premium price. Moreover, these companies are also providing consultancies to the farmers free of cost to improve the quality of produce.

Table 3.10: Comparative Economics of Marketing of Apple through different Agencies

Market/ Producers Company	Price(Rs/kg)			Marketing Cost Rs/kg	Returns (Rs/Kg)			Remarks
	Av.	Mini	Max.		Av.	Mini	Max.	
Outside state (Delhi)	53	32	74	11.50	41.50	20.50	63.50	Produce is sold acc. to grades.
Within state (APMC Parwanoo)	58	36	71	9.50	48.50	26.50	61.50	Produce is sold on average basis
M/S Adani Agri Fresh Limited	56	42	70	-	56.00	42.0	70.0	Company provides crates and also charges for carriage
Container Corporation of India	58	31	72	2.50	55.50	28.50	69.50	cost of box and assembling Expenditure is charged from the growers
All Fresh Supply Mgt. Pvt Ltd.	50.0	37	64	-	50.00	37	64.0	Company has installed grading and packaging machines in the apple producing areas.

A recent trend has been observed that buyers are camping in the production areas to buy the produce locally or in the APMCs of the state, resulting into buyer's market for the apple in the state.

5. Constraint Analysis

Production of apples has increased many-fold during last four decades which has created many problems in the field of production and marketing. The problems faced by the orchardists in the area of production and marketing of apple have been discussed under two heads,viz., production and marketing.

(a) Production Constraints

(I) Shortage of Skilled Labour

Shortage of skilled labour for training and pruning of trees and plant protection measures is major problem in the state. Non-availability at peak periods and lack of technical know-how also affected the production of the crop.

(II) Chemical Fertilizers

The cost and timely availability of chemical fertilizers due to transportation problems affected the adoption of package of practices by the growers which adversely affected the production.

(III) Plant Protection Chemicals

Plant protection chemicals constitute an important critical input in apple production. High prices of chemicals, non-availability in time, and availability of spurious chemicals are the main problems faced by orchardists in the state.

(IV) Plant Material, Farm Yard Manure and Irrigation Problems

Healthy plant material is key to quality production of apple. Non-availability of healthy and genetically improved spur varieties of apple posed the problem in the expansion of apple cultivation. Similarly, use of FYM is vital for health and production of fruit plants. Actual need of FYM is rarely met in the state. Irrigation is one of the critical inputs which directly effects the productivity of apples as majority of orchards are rainfed.

(b) Marketing Constraints

Marketing of apple is as important as production. Lack of markets and improved marketing practices contribute to the complicated nature of the marketing of apple in the hills. Some of the major problems are:

(I) Grading and Packing Problems

Grading and packing are basic marketing activities to be completed by the growers. In the absence of sale by grade, the returns remain lower. Moreover, grading and packing is generally done manually. Hence higher wage rates and non- availability of skilled labour for apple grading becomes the main problems

(II) Packing Material

Apple being fragile in nature needs good packaging which may ensure least damage to fruits during transportation. The deterioration of quality of fruits may result into non-remunerative prices. The shortage of boxes, high prices, non-availability of credit, and unavailability in time and desired place becomes major problem to the growers.

(III) Storage Problems

Apple produce being perishable, require immediate disposal. Due to lack of cool chain system, huge losses are borne by the participants of marketing process. Farmers of the state do not have enough scientific storage facilities for apple. Storage is normally carried out in some improvised or ill ventilated sheds at home. The inappropriate storage facility normally, increases the quantitative and qualitative losses. Marketing seasons of Himachal Pradesh apples extend from July to end of October which coincides with that of Kashmir apple from August to December, forcing the producers to resort to distress sales.

(IV) Transportation Problems

Transportation is one of the important marking functions required in apple marketing because consumers are situated at longer distances from producing areas. Transportation involves bringing produce from orchards to road head and then road head to consumers. Often family and hired labour is used for carrying the produce from orchards to assembling points. After doing packing at the assembling point, the produce is carried to road-head either on human backs or on the mules. From the road-head, after doing appropriate marketing, motorized transport is hired for taking the produce to local or terminal markets. The major concern is high transport cost which in the absence of all-weather roads aggravates the problem further.

(V) Market Intelligence

Market intelligence plays a significant role in the marketing of perishables. The information regarding the market demand, arrival and prices prevailing in the market are very important as the same can affect the income of the growers. However, in the absence of efficient market intelligence the growers suffer a lot.

(VI) Malpractices

The regulation of markets was done with a view to watch the interests of the sellers and buyers by curtailing the malpractices prevailing in the markets. The apple growers report that, the intermediaries are still charging miscellaneous marketing charges from the sale proceeds, while these charges are expected to be paid by them as per the market regulations.

References

Abdul, Rauf. 2009. Production and marketing of apple in Himachal Pradesh and Jammu and Kashmir – a comparative study. *PhD Thesis, Dr. Y. S. Parmar University of Horticulture and Forestry, Nauni Solan, Himachal Pradesh, India.*

DES, 2013. *Statistical outline of Himachal Pradesh,* Directorate of Economic and Statistics.

GoI, 2014. *Hand Book on Horticulture Statistics*, Department of Agriculture and Cooperation, Ministry of Agriculture, Government of India(GoI), New Delhi.

FAO. 2013. Crop Production Statistics.

Kashyap Rachit and Guleria Amit. 2015. Socio-economic and marketing analysis of apple growers in Mandi district of Himachal Pradesh. *Journal of Hill Agriculture.* Vol. 6 (2): 202-206

Kireeti, K. 2013. Productivity analysis of apple orchards in Shimla district of Himachal Pradesh. *M.Sc. Thesis- Dr. Y. S. Parmar University of Horticulture and Forestry, Nauni Solan Himachal Pradesh, India.*

Sharma, Isha. 2016. Economic analysis of apple cultivation in Kullu district of Himachal Pradesh.*M.Sc. Thesis- Dr. Y. S. Parmar University of Horticulture and Forestry, Nauni Solan Himachal Pradesh, India.*

Chapter 4

Envisioning Apple Value Chain: J&K Banks Initiative towards Rural Prosperity

Sajjad Bazaz and Rashid Ud Din

Apple Project J&K Bank, Corporate Headquarters,
M.A. Road, Srinagar (Kashmir)
E-mail :rashid.din@jkbmail.com

1. Prologue

An independent study conducted in 2011 by Enterprise Solutions to Poverty (ESP) - an innovation group headed by Ms. Nancy M Barry, who is recognized as a global leader in building finance and enterprise systems, revealed that over 75 per cent of small farmers get their financing from moneylenders, with only 10 per cent from banks. Even fixed investment for irrigation or tree replenishment is rare. The study suggested that small farmers will need access to money, technology and procurement options if the high cost of trader finance and procurement is to be avoided, and if farmer productivity and incomes are to increase.Interestingly, according to the study, farmers prefer moneylenders to banks because traders provide a market as well as money, and are willing to finance for emergencies as well as for crop requirements.The study revealed that 75 per cent of apple growers get their financing from traders and are often unaware that its effective interest cost of financing from moneylenders are 36 per cent to 54 per cent.Farmers interviewed during the study perceived that charges by the traders are similar to what a bank would charge – which averages 12.7 per cent. The farmers were not aware of bank financing opportunities, but think that there are high documentation requirements and hassles to get a bank loan.

Precisely, the study findings revealed that apple farmers showed that they have to rely on moneylender for advances. During the study it was found that majority of the farmers haven't any experience with bank loans. The larger farmers rely on savings to finance their operations. Many farmers are slowly replacing their orchard stock, meaning that superior trees could be planted, and that their investment loans are desired. Irrigation is not used, but would have a substantial impact on productivity.

Potential for horticulture lending is huge and agriculture involves about 70 per cent of the population and contributes 23 per cent to the GDP (Economic Survey, 2015). Apple accounts for 86 per cent by value of the state's horticulture output. J&K state produces 65-68 per cent of India's total apple output (Horticulture Data Base, 2014) and controls 49 per cent of land dedicated to apple cultivation (Department of Horticulture, 2015). Apple being an important principal crop of J&K state, but it has been observed that one of the reasons adversely affecting the production is lack of proper finance at proper time. This has created disinterest among growers, as the returns are not matching the kind of time and labour they put in the production of various varieties of the fruit. Overall, the private financers were predominantly controlling the produce, as they have been lending growers at exorbitant interest rates. They even enter into futures contract with the growers to buy the entire produce at the rates, which were lower than the market scenario. This practice had drastically eroded the margins of the apple growers and simultaneously lowered their enthusiasm.

While realizing the financial difficulties of the growers, J&K Bank had already devised a scheme to finance the fruit growers. But the response of the growers to this formal channel of financing was not to the desired level. To bring enthusiasm among the growers and also as a significant step towards actualization of the Bank's strategy of increased investment in the economic development of J&K State, the J&K Bank modified the scheme and re-introduced it under the name of "Apple Advance Scheme". The scheme was tailored to make it more relevant, need based and hassle free even to a common grower for development of orchards, production, and marketing of fruit crop in a free and fair manner without financial intervention/support from non-institutional agencies like money lenders, outside *arthias* and commission agents. The revised Apple Advance Scheme of the bank aims at providing adequate and timely credit for comprehensive credit requirements of the grower under single window, with flexible and simplified procedure, adopting whole farm approach, including the production credit needs and a reasonable component for consumption/subsistence needs. Earlier the scheme was not having any credit support to the grower for his day to day financial/subsistence needs and the grower was always dependent on and at the mercy of local traders/money lenders for advance payments against the fruit crop for his day to day financial requirements. So, all fruit growers owning orchards in fruit bearing stage were brought under the ambit of the scheme. Besides, all local traders/*arthias*/ commission and forwarding agents having a market standing of at least one year too were extended the benefits under the scheme.

The scale of finance under the scheme is assessed on entire fruit bearing orchard/land holding owned by the grower or any of his family members for a period of three years. For loans up to Rs.1.00 lakh, no revenue records will be insisted on. However, the grower shall have to furnish an affidavit duly attested by an executive magistrate, as a declaration of his orchard land holding. The loan limit will be to the extent of 90 per cent of the total financial requirements assessed for his orchard land at Rs 3.20 lakh per acre. A margin of 10 per cent is to be contributed by the grower in the shape of costs incurred on fertilizer/pesticide/fungicide application and post harvest maintenance of the orchards called 'sweat equity'. The scale of finance for local traders/*arthias* will be assessed on the basis of number of fruit boxes marketed/forwarded during the previous year with a reasonable increase of 25 per cent (maximum) on average growth during the last three years. However, the fresh borrower under the scheme should be having minimum market standing of one year. Precisely, the revised Apple Advance Scheme of the bank has been tailored as per the actual requirements of the producers and provides an opportunity for them to enhance their returns. Furthermore, the growers while having access to the required finances at proper time, can raise the production to the optimum level.

2. Intervention by J&K Bank

(a) J&K Bank Apple Project: A Brief

In the backdrop of the study conducted in 2011 by Enterprise Solutions to Poverty (ESP) - an innovation group headed by Ms. Nancy M Barry, J&K Bank adopted a mission mode method to realign its Apple Advance Scheme entirely to the benefit of the apple growers' community. It's here 'J&K Bank Apple Project' was launched in February 2012 to rescue the apple growers from exploitative noose of usurious moneylenders and empower them sufficiently in terms of finance and guidance to realize the true potential of apple economy. Hassle free access to adequate finance with minimum documentation remains hallmark of the 'Apple Project'. In essence, the project envisaged complete institutional financing of the apple crop to empower particularly the grass root grower to ensure that the state proceeds ahead towards becoming a rich horticulture economy. Over the past five years, the project has successfully aimed at transforming the state into an agriculturally rich economy with major contribution from apple industry. During all these years the bank conducted the project in a way that a common grower today remains at the center of state's apple economy.

In its pilot-study of the project, the bank initially focused on apple rich districts of Baramulla and Shopian with gradual and time-bound extension of other objectives within project implementation period of five years. Under the project special teams of Apple Project Managers headed by Zonal Coordinators were constituted, who worked on a door-to-door awareness method. An extensive awareness programme was launched to apprise the growers about the schemes. These teams comprising of 3 to 4 officers literally camped in the villages of the two districts and prepared the profile of growers at their door-steps in these areas. Further, these teams counseled the grass root growers individually, in groups and advised them in the matters of finance and other related issues to bolster their confidence and efforts towards self-reliance and sense of ownership of their crop. Moreover, timely camps were organized in which some experts on horticulture imparted free training and awareness to the growers on the use of fertilizers, pesticides, methods of pruning, irrigation, control of rats and rodents, art of picking of apple and packing *etc.* The growers were also informed about scientific methods of cultivation and use of technology, which go a long way to enable them to grow more and save much. The project envisaged to cover 2.88 lac apple growers in the valley as per the data base created by the bank. In order to tap all the apple growers yearly targets for number of growers as well as for financial assistance were set.

While conducting the project in districts of Shopian and Baramulla in the initial years of the project implementation, the bank has been able to tap as many as 49425 apple growers and disburse a huge amount of Rs 1095.13 crore among them. As on December 2016, almost 50 per cent of the overall, apple growers have been tapped in the valley across all its districts and the bank has created a core agriculture portfolio of Rs 2979 crore. Before the launch of the Apple Project, the bank used to have a meager exposure of Rs 400 to Rs 500 crore towards traders, commission agents *etc.* which used to qualify for agricultural advances as per the priority sector guidelines.Thus the project has been instrumental in creating a new business opportunity for the bank and put in initiatives for economic development of rural population and apple industry in the state.

During the first year of implementation for crop year 2012 when the project was implemented on a pilot basis in two districts of Shopian and Baramulla, 91per cent target in terms of number of growers and 106per cent targets in terms of loan amount was achieved. It's noteworthy that out of the profiled growers numbering 2.88 lac, around 25per cent of the apple growers had orchards in juvenile phase with little or no production. Besides, it was found that pending land mutations largely prevalent in rural areas acted as an impediment for achieving the envisaged targets under the scheme. During the implementation period of the scheme in the last five years several additions/measures were undertaken which include:

- ☆ Profiling of the apple growers throughout the valley and creation of in-house data base of apple growers with all relevant details including size of orchard, account and contact details.
- ☆ A tie up with SKUAST Kashmir for providing modern technological support and latest farming techniques on improving quality and quantity of apples, improving tree density, drip irrigation *etc.* to apple growers.

(b) Enablers/Support Structures

Besides financing the growers, the Bank has entered into a strategic tie-up with the Sher-e-Kashmir University of Agricultural Science and Technology and State Horticulture Department to guide growers in better farming practices. This has resulted in better yields. Besides, various advisory services through experts are being extended to the growers. Almost 3 lakh farmers are being enrolled for SMS services with strategic partners for daily inputs about weather, use of fertilizers, market situation *etc.* In order to have a reliable data base of apple growers, the bank has already completed the profiling exercise of apple growers which has been centralized for assessing further scope of the project. The Bank has also sped-up its financing to the critical support structure like CA stores, which extend shelf-life of apples to 10-12 months, to help and sustain the small apple growers in the process. Days are not far away when apple growers of the state shall be made available finance facility against their Warehouse Receipts as in vogue in other parts of the country.

(c) High Density Plantation

J&K Bank has been always looking for the opportunities to boost the production of apple in the state which include adoption of global best farm practices as the fruit is main contributor of the state exchequer. Recently, the government of J&K has come up with the new scheme of financial assistance to the growers for cultivation of apple under high density apple plantation. Under the scheme the Horticulture Department has taken the initiative of planting the exotic high yielding varieties of apple and the scheme shall enable even a smallest farmer to set up high density apple orchard owning just 5 *kanals* of Land. Under high density apple plantation scheme about 166 trees can be planted per *kanal* of land as against the 15 trees per *kanal* under the traditional method of cultivation. Likewise the production of apple on one hectare of land under high density apple plantation scheme shall be around 83 MT as against 45MT per hectare under traditional method of cultivation. The total revenue expected to be generated per hectare under high density plantation shall be around 25.50 lac as against Rs 20.25 lac per hectare from the crop under traditional method of cultivation. Plants under high density apple cultivation start commercial production just after second year of planting as against 8 or even 10 years in case of plants under traditional method of cultivation leading to the huge gap of revenue generated (about 17 lac per hectare) under high density plantation during the initial 5 years till the plants under traditional method start commercial production. These figures clearly depict the advantages of high density apple plantation scheme over the traditional methods of cultivation of apple to the growers of the valley who resort to the traditional methods of cultivation from years together.

Under this scheme Govt. shall provide the 50 per cent financial assistance upfront to the grower for establishment of the orchard in the shape of subsidy. Out of the remaining 50 per cent of the project cost 30 per cent is provided by our bank in the shape of JK Bank High Density Apple Plantation Finance scheme with nominal interest rates and the rest 20 per cent of the project cost is borne by the borrower from his own sources which mostly comprise of sweat equity (*i.e.* land development, pit digging *etc.*). There is a moratorium period of three years and the grower has not to pay any installment/interest for these three years. The grower has to pay interest of fourth years and from fifth years and onwards and he has to repay the loan in five equated yearly installments at a competitive rate of interest of only 9.5 per cent p.s with yearly rests. Under high density apple plantation finance scheme the cost to be financed shall be initial investment costs like cost of plant material, drip irrigation system, trellis system, *etc.* and operating cost for first two years and financial costs for three years till orchard reaches productive stage. Farmers owning and possessing land suitable for being used productively for high density apple orchard shall be eligible for financial assistance under the scheme. Growers interested in the scheme shall approach the concerned district Horticulture Department who after verifying the suitability of land and other parameters of the scheme shall forward the case to our bank for final disposal.

(d) CA Storage Facility

The total production of apple is around 18.00 lac metric tons in the valley. The A grade apples feasible for CA storage is 8.74 lac MT assuming 50 per cent of apples are B, C, and D not suited for storage. If the minimum requirement of CA store is envisaged to the tune of 25 per cent of A grade fruit production, then the total requirement of CA store is 2, 18,000 MT. Out of which bank has already financed 19 CA stores with capacity of 81,830 MT with credit exposure of Rs.540.10 crore and in principally agreed to finance about 07 CA stores with the capacity of 28,200 MT's with credit exposure of 148.56 crore. Thus the total capacity 1,10,030 Mt's has been covered under bank finance till date and there is an additional capacity creation of 1, 07,970 MT.

3. Project Possibilities

(a) Technological Outreach

(I) Farm Mechanization

Farmers have over the period of time developed a high sense of adoption of technology. Even the small orchardists now opt to keep some basic tools and implements *e.g.* grass cutter, weed cutter, power sprayer etc, which not only increase the productivity but saves his time as well. The bank has been on a holistic plan to enable farmers go for these farm implements at a very cheap interest rate and less loan formalities.

(II) Advisories

The project has conceived a mechanism to reach out to the apple growers with different kind of advisories. Under this mechanism a team from our bank and experts from the SKUAST will reach out to the farmers and share the ideas about the new marketing strategies, and advise them about the techniques of tackling the weather vagaries and other climatic changes so as to keep them aware about the possible future planning.

(III) Involvement of Corporate Sector for Non-Financial Services

This will enable farmers to get a double benefit of availing the services of corporate sector for non-financial/farm expertise under the roof of same institute.

(IV) Farmers Club

The Bank has conceived another idea of formation of Farmers' Clubs at least one for each rural branch in terms of the extant guidelines of NABARD where we could successfully put across them the concept of transfer of technology, financial inclusion and financial discipline. This shall be a maiden exercise of its kind in the state of Jammu and Kashmir where farmers shall be afforded a valid and effective platform to push their demands through to the bankers for coming to their financial and non-financial rescue.

(V) Agriclinics and Agribusiness Centres

The establishment of Agri-Clinics and Agribusiness Centres (ACABC) will supplement the efforts of government extension system and will make available supplementary sources of input supply and services to needy farmers. Under this idea bank intends to select horti-preneurs in consultation with agriculture universities/KVK's of the state for involvement of qualified youth of the state in the initiative. This is sure to change the entire ecosystem of farming practices in the state and create employment at a large scale.

(VI) Technological Support

The Bank is on way to bring about complete technological transformation by promoting mass use of the latest technology like point of sale machine, whereby, a farmer can buy the agri-inputs like seeds, fertilizers, pesticides *etc.* at ease and without hassle and also get updated information via easy sources like SMS alert.

(b) Interest Subvention

One of the major reliefs in the scheme was that the bank pleaded on behalf of growers at the central level and got a special sanction for the extension of interest subvention scheme for them through the bank. By virtue of this interest subvention scheme, the loans under apple finance scheme of the bank are granted at an interest rate of 7 per cent only. Further 3 per cent interest rebate is given to the growers if they repay the loan within the stipulated period. This means the loans to such prompt pay masters are available at 4 per cent, which is a huge leap towards empowerment of the growers.

(c) Proper Monitoring

As the project progresses ahead, a proper monitoring system has been set in place at the corporate level to assess and scrutinize the actual unfolding of the project on ground so that timely interventions are made instantly wherever the need arises.

(d) Impact

This relief has changed the fortunes of the growers and now the manipulative trend that was set earlier has been reversed to an extent whereby the traders themselves approach the grower for procuring the crop. Earlier, the grower was at the mercy of the traders and the moneylender's.As the bank engages in impact study of its pilot effort, it was revealed that significant strides have been made towards rescuing the farmers from the exploitative noose of usurious money lenders.The impact of J&K Bank Apple Project is that many farmers in the valley have shifted their agricultural practices to horticulture. Especially in south Kashmir many farmers have converted their agricultural land into orchards after they noticed changing living standard of their fellow farmers engaged in horticulture activities. The growers are of the opinion that the rate hike of the apple crop last year was the direct result of their strength to bargain with their counterparts outside state *Mandis* which could be made possible only by their sense of being empowered financially by the bank.

4. Conclusion

As the Bank has already extended the implementation of its 'Apple Project' to all the districts of the valley and the bank envisages tapping more. Out of a total 2.89 lac apple growers in the state, bank's finance has been extended to about 140000 growers so far. Tapping the remaining growers will generate credit of about Rs.6500-7000 crore more.

Chairman and CEO, J&K Bank, Parvez Ahmad who heralded the Apple Project in the initial years of its implementation as Incharge Executive President, while commenting on the role of J&K Bank in making the apple industry vibrant said, "Our horticulture sector in Kashmir region is mainly driven by apple industry. It's an underserviced and potential sector and next to personal and consumption loans the major attraction for the bank in the retail sector is the apple finance. Keeping in view in the major contribution of this sector in the State Domestic Product and its classification in the priority sector coupled with the relaxed asset classification norms, we see a lot of advantages for the bank to have a special focus on this sector. Bank has already customized its apple finance scheme to meet variant needs of the small apple growers. We have already set up plans for the growth and development of apple industry and realize its true potential. He further stated that the bank is very much on track to facilitate to a common grower the freedom of choosing buyers and negotiating prices for his hard-earned produce. "It is this choice that lies at the center of not only his financial empowerment but economic success of the whole apple industry," he said.

Chapter 5

Classification of Apple Growing Soils of Kashmir

Nayar Afaq Kirmani, J. A. Sofi,
Juvaria Jeelani and Khushboo Farooq

Division of Soil Science,
Sher-e-Kashmir University of Agricultural Sciences
and Technology of Kashmir, Srinagar
E-mail: afaqnayar@gmail.com

1. Introduction

Soil classification is the grouping of soils into well-defined groups or taxa on the basis of similarities and differences in their characteristics. Soils are classified in a systematic manner to organize their knowledge, so that the properties are clearly conceived and relationships easily understood. The scientific knowledge helps in the transfer of soil and land use technologies internationally not only by pedologists, but also by the other users of soil and land, such as, geologists, agronomists, botanists, farmers, ecologists, *etc.* for the benefit of mankind. United States Department of Agriculture in 1975 brought soil taxonomic system of classification, which was designed to meet the soil survey needs of many countries of the world including India. The publication of the second edition of *Soil Taxonomy: A Basic System of Soil Classification for Making and Interpreting Soil Surveys* (Soil survey staff 1999), followed by the several editions of *Keys to Soil Taxonomy* (Soil survey staff, *2014*) serves two purposes. It provides the taxonomic keys necessary for the classification of soils in a form that can be used easily in the field and also acquaints users of the taxonomic system with recent changes in the system.

Apple being the dominant fruit crop of the J&K state has emerged as the leading cash crop of the state. The area under apple was 60, 000 hectares in 1980-81 and has increased to 86,000 ha in 2000-2001 (Anonymous, 2014). The area under the crop almost doubled in last 15 years and swelled upto 1, 60,000 ha in 2014-15 (Figure 5.1). The second degree polynomial trend can be fitted with R^2 value of 0.988, which can predicts further increase in the area under apple cultivation in future decades. Its cultivation was previously confined to *Karewas* and *Kandi* areas, but due to high economic returns farmers have started growing apples on low-lands too which has resulted in the production of poor-quality fruits (Najar *et al.,* 2009). There has been increase in the production as well productivity upto 2000-2001 afterwards there has been decrease in overall productivity, but production followed the area trend, with some fluctuations after 2011-2015, with huge damage by floods in 2014.

The crop is being cultivated in almost all types of soils present in the valley which are, side and reposed slopes, fluvial broad and narrow valleys, flood plains, river terraces, *karewas,* upper and lower piedmont plains.

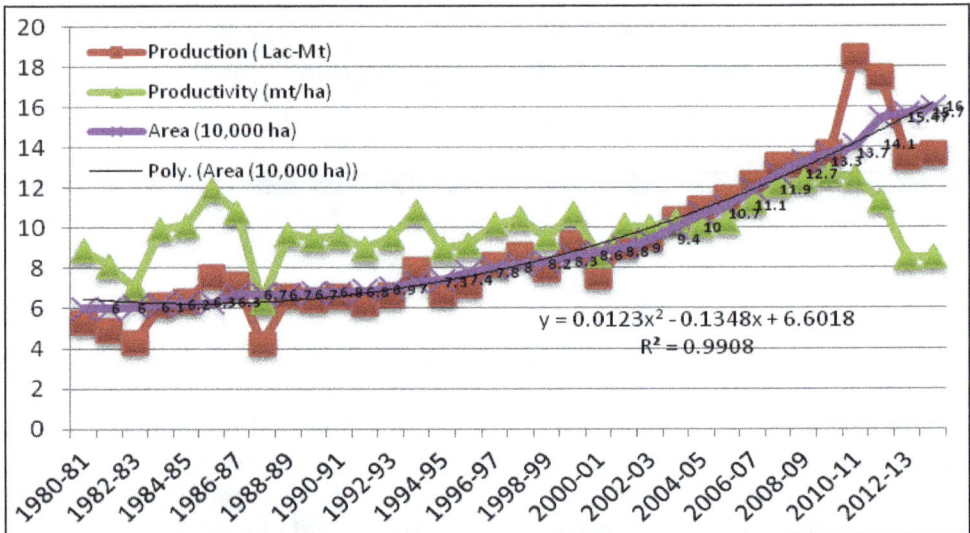

Figure 5.1: The Trends in Area, Production and Productivity of the Apple Cultivation in Jammu and Kashmir.

Source: **Department of Horticulture, 2016**

2. Methodology of Soil Classification

The methods in the Soil Survey Manual (Soil survey staff, 2003) and as per the "Proforma for soil-site description and soil characteristics" (Sehgal, 1994) is used for collection of field data. Soil colour was compared with Munsell's soil colour notation. The other morphological characteristics like, texture, structure, consistency and other special features are all described and noted for each profile. Soil samples are collected from different horizons of each soil profile, and are subjected to various laboratory analysis in terms of their physical, chemical and mineralogical makeup.The soils are then classified as per "Soil Taxonomy" (Soil Survey Staff, 1999) by following "Keys to Soil Taxonomy" (Soil Survey Staff, 1998, 2014) for identification and characterization of epi-pedons (surface) and endo-pedons (sub-surface).The mineralogical analyses are carried out for sand, silt and clay fractions using X-ray differactometer and other methods in order to authenticate the primary and secondary minerals present in the soil. The samples were dispersed after removal of all cementing agents like $CaCO_3$, organic matter *etc.*

3. Site and Morphological Characteristics of the Apple Growing Soils

The state of J&K occupies almost a central position in the Asian continent. Geographically valley lies on the North flank of Pir Panjal range and southern flank of Great Himalayan range is located at 33° 30´ N to 34° 30´ N and 74° 10´ E to 75° 03´ E (Kirmani *et al.,* 2013). The site characteristics of some profiles under apple cultivation are mentioned in Table 5.1. The sites are distributed in different districts of the valley with majority fall in the district Baramulla, known as the fruit bowl of Kashmir valley. The crop is grown in the large altitudinal variation of 1600 to 1800m, with almost flat slope gradient, although the terraced cultivation and on slope cultivation is not uncommon in the hilly regions of the valley.

Table 5.1: Site Characteristics of some Profiles Studied under Apple Growing Areas of Kashmir Valley (Kirmani, 2005, Najar *et al.,* 2009)

Location (District)	*Physiography*	*Altitude (m)*	*Aspect*	*Slope (per cent)*	*Natural Vegetation*
Trehgam (Kupwara)	Upland surrounded	1820	Northern	<1 per cent	Willow, poplar, apple tree, coniferous, shrunks, wild rose, indigo and shrubs
Shakipora	Upland surrounded	1830	Southern	<1 per cent	
Nowpora (Baramulla)	Karewa	1760	Northern	<1 per cent	
Mathen (Anantnag)	Karewa	1770	Southern	<1 per cent	
Kaimoh (Anantnag)	Alluval fan	1670	Southern	<1 per cent	
Sopore (Baramulla)	Alluval fan	1630	Northern	<1 per cent	

LOCATION (District)	Elevation (amsl) (m)	Topography	Slope gradient (%)	Slope Length (m)	Erosion/ Runoff	Drainage	Gr.Water Depth (m)	Flooding	Stone Size Diameter (cm)	Stoniness Sur. Cov. (%)	Rock out crops (Distance apart) (m)
Shirpora (Bla.)	1700	Flat	0-1	50-150	None/Very Slow	Well drained	> 10	No	< 2.5	< 3	No
Nihalpora (Bla.)	1700	Flat	0-1	50-150	None/Very Slow	Well drained	> 10	No	< 2.5	< 3	No
Bahrampora (Bla.)	1675	Flat	0-1	0-50	None/Very Slow	Well drained	> 10	No	< 2.5	< 3	No
Kreeri (Bla.)	1700	Flat	0-1	50-150	None/Very Slow	Well drained	> 10	No	< 2.5	< 3	No

The cultivation is very common in the table lands known as the *Karewas,* comprising of 450-500 m thick pile of sediments. The lower, middle and upper *karewa* sub-groups show fluvial; lacustrine; eolian and pedogenetic environments respectively (Pal and Srivastava, 1982). The slope gradient of the profiles studied ranged from 0 – 1 per cent with slope length of 0 – 50 m and 50 – 150 m with none to very slow erosion and runoff was also found very slow. All the soils were well drained with ground water depth greater than 10 m with no flooding and no rock out crops. Stone size was found less than 2.5 cm with stoniness less than 3.0 per cent of surface coverage.

Morphological characteristics comprise of soil colour, texture, structure, consistency, cutans and several other features of the horizons of soil profiles as can be perceived in the fields which signify the inherent characteristics of soil development during soil genesis. The morphological characteristics of representative soils profiles in terms of various soil characteristics are mentioned briefly (Table 5.2).

Table 5.2: Morphological Characteristics of some Soil Profiles of Apple growing Areas of Kashmir Valley (Kirmani, 2005)

Location	Horizon	Depth (cm)	Boundary Dist. Topo.	Diagnostic Horizon	Matrix Colour	Mottle Colour	Texture	Coarse Fragments
Shirpora (Bla.)	A	0-20	a s	A$_p$	10YR 6/3(D) Pale Brown: 10YR 5/3(M) Brown	Nil	l	Nil
	B	20-55	g i	B$_1$	10YR 3/4 DarkYellowish Brown	Nil	cl	Nil
	B	55-90	g w	B$_{21}$	10YR 3/4 DarkYellowish Brown	Nil	cl	Nil
	C	90-120+		C	7.5 YR 3/2 Dark brown	Nil	cl	Nil
Nihalpora (Bla.)	A	0-20	a s	A$_p$	10YR 7/3(D) veryPale Brown: 10YR 6/3(M) Pale Brown	Nil	l	Nil
	B	20-55	g i	B$_1$	10YR 3/4 DarkYellowish Brown	Nil	cl	Nil
	B	55-88	g w	B$_{21}$	10YR 3/4 DarkYellowish Brown	Nil	cl	Nil
	C	88-120+		C	7.5 YR 3/2 Dark brown	Nil	cl	Nil
Bahrampora (Bla.)	A	0-15	a s	A$_p$	10YR 6/3(D) Pale Brown: 10YR 5/3(M) Brown	Nil	l	Nil
	BA	15-30	g w	BA	10YR 4/3 Brown	Nil	cl	Nil
	B	30-60	g i	B$_1$	10YR 3/4 DarkYellowish Brown	Nil	cl	Nil
	B	60-90	g w	B$_{21}$	10YR 3/4 DarkYellowish Brown	Nil	cl	Nil
	C	90-120+		C	7.5 YR 3/2 Dark brown	Nil	cl	Nil
Kreeri (Bla.)	A	0-20	a s	A$_p$	10YR 6/3(D) Pale Brown: 10YR 4/3(M) Brown	Nil	l	Nil
	B	20-50	g w	B$_1$	10YR 5/3 Brown	Nil	cl	Nil
	B	50-85	g w	B$_{21}$	10 YR 3/3 Dark Brown	Nil	cl	Nil
	C	85-120+		C	10YR 3/4 DarkYellowish Brown	Nil	cl	Nil

c =clear; g =gradual; s =smooth; i=irregular; w =wavy; d =diffused : l =loam; cl =clay loam; sil =silty loam

Location	Diagnostic Horizon	Depth (cm)	Structure Size Grade Type			Consistence Dry Moist Wet			Porosity Size Qty.		Cutans Type Th.ness Qty.			Nodules	Roots Size Qty.		Effervescence (with dil. HCl)
Shirpora (Bla.)	A$_p$	0-20	m	1	sbk	dsh	mfr	ws wp	f-m	c	Nil			Nil	m-f-c	m	Nil
	B$_1$	20-55	m	2	abk	dh	mfi	wvs wvp	f-m	c	T	mThick Cont.		Nil	m-f-c	m	Nil
	B$_{21}$	55-90	c	2	abk	dh	mfi	wvs wvp	vf	m	T	Thick Cont.		Nil	m-f	c	Nil
	C	90-120+	c	2	abk	dvh	mvfi	wvs wvp	vf	m	T	Thick Cont.		Nil	f-vf	f	Nil
Nihalpora (Bla.)	A$_p$	0-20	m	1	sbk	dsh	mfr	ws wp	f-m	c	Nil			Nil	m-f-c	m	Nil
	B$_1$	20-55	m	2	abk	dh	mfi	wvs wvp	f-m	c	T	Thick Cont.		Nil	m-f-c	m	Nil
	B$_{21}$	55-88	c	2	abk	dh	mfi	wvs wvp	vf	m	T	Thick Cont.		Nil	m-f	c	Nil
	C	88-120+	c	2	abk	dvh	mvfi	wvs wvp	vf	m	T	Thick Cont.		Nil	f-vf	f	Nil
Bahrampora (Bla.)	A$_p$	0-15	m	1	sbk	dh	mfr	ws wp	f-m	c	Nil			Nil	m-f-c	m	Nil
	BA	15-30	m	2	abk	dvh	mfi	wvs wvp	f-m	c	T	mThick patchy.		Nil	m-f-c	m	Nil
	B$_1$	30-60	c	2	abk	dvh	mfi	wvs wvp	vf	m	T	mThick Cont.		Nil	m-f	c	Nil
	B$_{21}$	60-90	c	2	abk	dvh	mvfi	wvs wvp	vf	m	T	Thick Cont.		Nil	f-vf	f	Nil
	C	90-120+	c	2	abk	dvh	mvfi	wvs wvp	vf	m	T	Thick Cont.		Nil	vf	f	Nil
Kreeri (Bla.)	A$_p$	0-20	m	1	sbk	dsh	mfr	ws wp	f-m	c	Nil			Nil	m-f-c	m	Nil
	B$_1$	20-50	m	2	abk	dh	mfi	wvs wvp	f-m	c	T	mThick patchy.		Nil	m-f-c	m	Nil
	B$_{21}$	50-85	c	2	abk	dh	mfi	wvs wvp	vf	m	T	mThick Cont.		Nil	m-f	c	Nil
	C	85-120+	c	2	abk	dvh	mvfi	wvs wvp	vf	m	T	Thick Cont.		Nil	f-vf	f	Nil

m =medium; c =coarse; 1 =weak; 2 =moderate:: d =dry; m =moist; w =wet; h =hard; sh =slighty hard; fr =friable; fi =firm; s =sticky; p =plastic :
:S =size, f =fine; m =medium; c =coarse::Qty. f =few; m =many; c =common :: T =Argillans:: e =slight; es =strong ev =voilent

The diagnostic horizons in general were observed to be A_p – B_1 – B_{2t} – C with varying depths of the horizons. The colour of the soils in the surface horizon ranged from brown (10YR 4/3 m) to pale brown (10YR 6/3m). The sub-surface soil varied in colour from dark yellowish brown (10YR 4/3 m) to dark brown (7.5YR 3/2m). Shinde *et al.* (1984) studied some typical soils of lacustrine deposits of Jammu and Kashmir and reported that the colour of the soils was light yellowish brown (10 YR 6/4 D) to darker yellowish brown (10 YR 5/4 D). Mushki (1994), while studying the soils of Kashmir, found that the *Karewa* soils had colour value of 4 to 6, Chroma 2 to 4 and the hue of 10YR.

Verma *et al.* (1990) studied the soils under forests of Kashmir valley and observed that the surface soils (26-46 cm) have dark brown to very dark grey colour with 10 YR hue, Chroma 1-3 and value 3. Mahapatra *et al.* (2000), while working with the soils of Kashmir valley, observed that the altitude and relief have significant bearing on the soil properties. The high altitude soils have a colour value ranging from 3 to 4, chroma 3 to 4 with a hue of 10YR, whereas the *Karewa* soils had a value of 2 to 4, chroma 2 to 4. Katoo (2001) while studying the soils of Lethpora Command area reported that the soil colour varied from 10 YR 3/4 to 10YR 3/2 in surface horizon and 10 YR 4/3 to 10YR 4/4 in the B-horizon.

The soil structure in surface horizon was observed to be sub-angular blocky with weak grade and medium structure in all the profiles studied while as it was found as angular blocky with moderate grade of coarse size in sub-surface B_{2t} horizon (Table 5.2). The structure of kerawa soils was observed to be moderately developed angular to sub angular blocky (Shinde *et al.*, 1984, Katoo, 2001). The soil structure of some forest soils of valley have weak to moderate granular in the surface and moderate to strong sub-angular blocky in the sub-surface horizons (Verma *et al.*, 1990).

The consistency in the surface soil was found hard (dry), friable (moist), sticky and plastic (wet), while as in sub-surface horizons it was found to be very sticky and plastic. Thick clay cutans were observed in the sub-surface horizons. The root zone of these soils was observed upto the depth of 120 cm, with increase in the fineness of the roots with depth. Shinde and Talib (1984) while studying the saffron growing soils of Pampore and Kishtwar *Karewas*, reported the rooting zone of 20-75 cm of these soils.

4. Physico-chemical Properties

Slight acidic (6.52) to neutral (7.15) pH was observed in surface horizons and sub-surface horizons respectively, of these profiles (Table 5.3). Electrical Conductivity was normal and average organic carbon content was observed to be 8.90 g kg^{-1} and ranged from 6.63 to 11.70 g kg^{-1} in surface layers and 3.9 to 9.75 g kg^{-1} in sub surface layers. Average calcium carbonate content was found to be 0.20 per cent in surface horizons and 0.50 per cent in sub-surface horizons. The CEC ranged from 9.4 to 13.33 cmol$_c$ kg^{-1} in surface and 13.99 to 16.37 cmol$_c$ kg^{-1} in sub-surface layers with high base saturation. Calcium was found to be the dominant exchangeable cation followed by magnesium and potassium.

Table 5.3: Physico-Chemical Properties of some Soils under Apple Cultivation

Location	Depth (cm)	pH (1:2)	EC (dsm⁻¹)	O.C. (gm kg⁻¹)	CaCO₃ %	CEC cmol. kg⁻¹	Exchangeable Cations(cmole kg-1)* Ca⁺⁺	Mg⁺⁺	K⁺	% Base Saturation	C. Sand %	Fine Sand %	Silt %	Clay %
Shirpora (Bla.) P7	0-20	5.00	0.10	11.70	0.00	10.16	5.79	1.72	0.45	78.35	2.4	26.4	47.5	22.5
	20-55	6.23	0.02	5.85	0.00	13.99	7.98	2.24	0.29	75.07	2.5	24.1	48.2	32.8
	55-90	6.70	0.12	3.90	0.00	14.78	8.13	2.30	0.44	73.49	2.0	23.9	36.0	37.9
	90-120+	6.34	0.05	3.90	0.00	15.84	8.68	2.53	0.44	73.54	1.4	25.2	37.0	36.4
Nihalpora (Bla.) P8	0-20	6.24	0.03	6.63	0.20	13.33	7.40	2.22	0.58	76.51	3.0	25.2	46.2	23.8
	20-55	6.71	0.02	7.80	0.00	15.05	8.58	2.30	0.29	74.21	1.1	23.8	38.2	36.8
	55-88	6.85	0.02	6.63	0.00	15.44	8.76	2.40	0.44	75.08	0.8	25.0	36.0	38.2
	88-120+	6.67	0.06	5.85	0.00	15.71	8.65	2.51	0.87	76.61	0.4	23.9	36.5	39.1
Bahrampora (Bla.) P9	0-15	5.88	0.06	10.92	0.00	10.56	6.02	1.72	0.73	80.15	2.0	26.8	48.6	21.6
	15-30	6.16	0.02	9.75	0.00	14.78	8.22	2.37	0.44	74.54	1.3	23.4	42.5	32.6
	30-60	6.60	0.02	5.85	1.00	15.31	8.62	2.45	0.44	75.14	1.1	22.5	38.5	37.5
	60-90	6.26	0.03	3.90	1.20	15.18	8.61	2.38	0.44	75.26	0.8	23.6	36.3	39.0
	90-120+	6.76	0.03	3.90	1.80	16.37	8.99	2.62	0.44	73.58	0.3	24.2	36.8	39.6
Kreeri (Bla.) P10	0-20	6.16	0.02	10.53	0.00	9.37	5.34	1.60	0.44	78.71	2.4	26.5	48.0	22.5
	20-50	6.10	0.03	7.41	0.00	13.99	7.65	2.24	0.29	72.75	1.1	26.2	36.4	36.1
	50-85	6.60	0.02	5.46	1.00	15.84	8.85	2.53	0.44	74.62	0.9	25.3	34.3	39.3
	85-120+	7.87	0.03	5.46	1.20	14.52	8.28	2.32	0.44	76.00	0.8	22.8	36.6	39.1
Surface	Mean	6.52	0.14	8.90	0.20	12.32	6.85	2.03	0.67	77.75	2.45	26.3	47.28	22.6
	Range	5.0-6.24	0.02-0.1	6.63-11.70	0.0-0.20	9.4-13.33	5.12-8.73	1.58-2.5	0.44-0.96	76.51-80.15	2.00-3.00	25.2-26.8	46.2-48.00	21.6-23.8
Sub-Surface	Mean	7.15	0.07	4.80	0.50	15.08	8.4	2.4	0.58	75.58	1.44	25.25	41.972	36.44
	Range	6.1-7.9	0.02-0.12	3.9-9.75	0.0-1.8	13.99-16.37	6.23-9.86	1.86-2.85	0.29-1.04	72.75-76.61	0.30-2.5	22.5-26.2	34.3-48.20	32.6-39.6

The texture of surface soils was found to be loam, in the sub-surface horizon, clay loam texture was found through out the depth of profile (Table 5.3). Clay content increased on an average 22.6 to 36.44 per cent from surface horizon to sub-surface horizon, with decrease in silt and fine sand content. Lower pH in surface horizon may be attributed to higher amount of organic carbon and leaching of salts to lower horizons, which may be due to lower rates of mineralization in surface soils under temperate conditions. These properties may be associated with clay percentage in the profile, and indicates the presence of illitic clay in theses soils. The dominant soil texture of Kashmir soils ranged from clay loam (Peer, 1994) to silty clay loam (Handoo, 1983) and those of Lacustrine deposits was observed to be clay loam to silty clay loam (Shinde *et al.*,1984). The texture of almond orchard soils of Kashmir ranged from clay loam tosilty clay loam and the clay content was in the range of 25.0-32.0 and 24.5-39.5 per cent in surface and subsurface layers respectively, with an erratic distribution along depth (Mir, 1994). The Karewa soils were silt loam to clay loam in texture (Mushki, 1994) whileas the texture of cherry orchard soils of Srinagar district has been reported between loamy and silty clay loam (Dar,1996).

5. Miralogical make up of these Soils

The random powder x-ray diffractograms of sand fractions of these soils gave strong and sharp reflections of Primary quartz, Plagioclase, mica and orthoclase. The silt fractions of these soils also revealed the dominance of quartz, plagioclase and orthoclase. The strong peeks, between 9 to 10 A°, reveal the dominance of illite in clay fraction of these soils.

Salvation with glycerol allows the separation and positive identification of smectites between 17 and 18A° peaks. The qualitative estimates for clay minerals were found of the order of illite followed by mixed layer, vermiculite and then by chlorite (Kirmani, 2005; Kirmani *et al.*, 2103).

Table 5.4: Relative Abundance of Minerals in Clay Fractions (Qualitative) of Nihalpora Baramulla

Pedon	Depth (cm)	Illite	Vermiculite	Chlorite	Mixed Layer
Nihalpora	0-20	++++	+	++	+++
	20-55	++++	++	+	+++
	55-88	++++	+++	+	++
	88-120+	++++	+++		++

Source: Kirmani, 2005.

6. Soil Classification

Overall, the soils of J&K have been in general classified into four orders, Entisols, Inceptisols, Alfisols and Mollisols (Table 5.5). The soils of Side and Reposed Slopes fall under Entisols, Inceptisols and Mollisols order with Orthents, Ocherpts, Udolls and Udorthents suborder followed by Dystrocherpts and Hapludolls Great-Group followed by the Lithic Udorthents, Typic Udorthents, Typic Dystrocherpts and Typic Hapludolls sub-groups. Similarly the classification of other soil types *viz.* Fluvial Broadand Narrow Valleys, Flood Plains, River Terraces, Karewas, Upper and Lower Piedmont Plains are tabulated. The classification and development of lacustrine soils under apple cultivation of Kashmir valley have been studied (Kirmani, 2005: Kirmani *et al.*, 2013). The diagnostic horizons were found to be A_p – B_1 – B_{2t} – C in three soil profiles and A_p – BA – B_1 – B_{2t}– C in one profile. The surface horizon was found Ochric with well developed Argillic (B_e) horizon and were grouped as Typic Hapludalfs (Table 5.4).

Some karewa soils were classified as Vertic Hapludalfs and Typic Eutrochrepts, whereas the adjoining soils as Typic Hapludalfs and Fluventic Eutrochrepts (Shinde *et al.*, 1984). Sehgal *et al.* (1985) have characterized mountain soil on slopes formed on chloritic schist and valley soils from gneisses under Typic Hapludolls and Mollic Hapludalfs respectively in central Himalayas. Pal and Deshpande (1987) reported the Lithological discontinuity after 91 cm while there was an increase in clay content up to 71 cm, while studying two bench mark soils of Kashmir valley they also reported that these soils contained high content of silt-low sand indicating that parent material had loess origin and classified them as Mollic Hapludalf (Gogji Pather) and Mollic Haplaquepts (Wathora).

Gupta *et al.* (1988) classified the soils of Jammu and Kashmir as Ustorthents, Eutrochrepts/Haplumbrepts and Haplumbrept/Undorthents while as the soils developed on various physiographic zones of Kashmir valley *viz.* high altitude, *karewa* and valley basin have been classified as Agriudolls, Hapludalfs and Ochraqualf respectively (Jalali *et al.*, 1989) and the soils under forest as Typic Hapludoll, Lithic Hapludoll and Typic Argiudolls (Verma *et al.*, 1990).

Table 5.5: The Overall Classification of the Soils of Kashmir Valley

Sl.No.	Soils	Order	Sub-order	Great-Group	Sub-Group
1.	Side and Reposed Slopes	Entisols Inceptisols Mollisols	Orthents Ocherpts Udolls	Udorthents Dystrocherpts Hapludolls	Lithic Udorthents Typic Udorthents Typic Dystrocherpts Typic Hapludolls
2.	Fluvial Broad Valleys	Inceptisols Mollisols	Aquepts Ochrepts Udolls	Haplaquepts Eutrochrepts Hapludolls	Typic Haplaquepts Typic Eutrochrepts Dystric Eutrochrepts Fluventic Eutrochrepts Typic Hapludolls
3.	Fluvial Narrow Valleys	Entisols Inceptisols Mollisols	Orthents Aquents Ochrepts Udolls	Udorthents Fluvaquents Eutrochrepts Hapludolls	Typic Udorthents Aeric Fluvaquents Typic Eutrochrepts Dystric Eutrochrepts Fluventic Eutrochrepts Typic Hapludolls
4.	Flood Plains	Entisols Inceptisols	Aquents Aquepts	Fluvaquents Haplaquepts	Typic Fluvaquents Typic Haplaquepts
5.	River Terraces	Entisols Inceptisols	Fluvents Ochrepts	Udifluvents Eutrochrepts	Typic Udifluvents Fluventic Eutrochrepts Typic Eutrochrepts Vertic Eutrochrepts
6.	Karewas	Inceptisols Alfisols	Ochrepts Udalfs	Eutrochrepts Hapludalfs	Fluventic Eutrochrepts Typic Eutrochrepts Dystric Eutrochrepts Typic Hapludalfs
7.	Upper Piedmont Plains	Inceptisols	Ochrepts Aquepts	Eutrochrepts Haplaquepts	Fluventic Eutrochrepts Typic Eutrochrepts Dystric Eutrochrepts Aeric Haplaquepts
8.	Lower Piedmont Plains	Inceptisols Alfisols	Ochrepts Aquepts Udalfs	Eutrochrepts Haplaquepts Hapludalfs	Typic Eutrochrepts Dystric Eutrochrepts Typic Haplaquepts Aeric Haplaquepts Typic Hapludalfs AquicHapludalfs

Sources: Rana *et al.,* 2000; Ahmad, 2003; Kirmani, 2005; Kirmani *et al.,* 2013

The illuviation of clay with clay skins in most of the profiles of Lethpora command area have been classified as Hapludalfs and Eutrochrepts (Katoo, 2001). Dandroo (2001) classified the lower Munda soils into Mollisols, Entisols and Alfisols andNajar (2002) during his study on pedogenesis of apple growing soils of Kashmir valley classified the *Karewa* soils under orchards in Hapludalfs and Eutrochrepts. Ahmad (2003) classified some upland orchard soils of Baramulla district in Hapludalfs.

References

Anonymous, 2014. *Digest of statistics.* Directorate of Economics and Statistics, Government of Jammu and Kashmir. Pp-139.

Ahmad, Z. 2003. Characterizing and Nutrient indexing of apple (*Malusdomestica* Borkh.) orchard soils of Bangil area of Baramulla district.*Thesis submitted to Sher-e-Kashmir University of Agricultural Sciences and Technology of Kashmir, Shalimar,* pp. 1-104.

Dandroo, F.A. 2001. Characterization and classification of lower munda watershed soils in south Kashmir.*Thesis submitted to Sher-e-Kashmir University of Agricultural Sciences and Technology of Kashmir, Shalimar,* pp. 1- 94.

Dar, M. A. 1996. Nutrient status of cherry (*Prunusavium* L.) in orchards of Srinagar district.*Thesis submitted to Sher-e-Kashmir University of Agricultural Sciences and Technology, Shalimar,* pp. 1-140.

Gupta, R.D., Anand, R.R and Shardra, P.D. 1988. Characteristics and genesis of some alluvium-derived soils of Jammu and Kashmir. *Proceedings of Indian National Science Academy.*54, A (1): 120-130.

Handoo, G.M. 1983. Organic matter fractionation in some soil profiles of Jammu and Kashmir State developed under different bio and climosequences. *Ph.D. (Agri.) thesis submitted to Himachal Pradesh Krishi Vishva Vidhyalaya Palampur (HP).* India, pp. 1-128.

Jalali, V.K., Talib, A.R. and Takkar, P.N. 1989. Distribution of micro-nutrients in some bench mark soils of Kashmir at different altitudes. *Journal of the Indian Society of Soil Science.*37:465-469.

Katoo, N.A.2001. Characterization and classification of soils of Lethpora command area. *Thesis submitted to Sher-e-Kashmir University of Agricultural Sciences and Technology of Kashmir, Shalimar,* pp. 1-72.

Kirmani, N. A. 2005. Characterization, classification and development of Lacustrine Soils of Kashmir Valley.*Ph.D. Thesis submitted to Sher-e-Kashmir University of Agricultural Sciences and Technology of Kashmir, Shalimar,* pp. 1-96.

Kirmani, N.A., Mushtaq A. Wani and J.A. Sofi. 2013. Characterization and classification of Alfisols under lesser Himalayan temperate region. *Agropedology,* 23 (2):118-121.

Mahapatra, S.K, Walia, C.S., Sidhu, G.S. Rana, K.C. and Tarseem Lal. 2000. Characterization and classification of soils of different physiographic units in the sub-humid ecosystem of Kashmir region. *Journal of Indian Society of Soil Science,* 48:572-577.

Mir, G.A. 1994. Studies on almond (*Prunusdulcis* Mill) orchard soils of Kashmir. *Thesis submitted to Sher-e-Kashmir University of Agricultural Sciences and Technology, Shalimar,* pp. 1-115.

Mushki, G.M. 1994. Studies on apple (*Malusdomestica* Borkh) orchard soils of Kashmir.*Thesis submitted to Sher-e-Kashmir University of Agricultural Sciences and Technology, Shalimar,* pp. 1- 129.

Najar, G.R. 2002. Studies on Pedogenesis and nutrient indexing of apple (Red delicious) growing soils of Kashmir. *Thesis submitted to Sher-e-Kashmir University of Agricultural Sciences and Technology of Kashmir, Shalimar,* pp. 1- 204.

Najar, G.R., F. Akhtar, S.R. Singh and J.A. Wani. 2009. Characterization and classification of some apple growing soils of Kashmir. *Journal of the Indian Society of Soil Science.*57 (1) : 81-84.

Pal, D.K. and Deshpande, S.B. 1987. Parent material, mineralogy and genesis of two bench mark soils of Kashmir valley. *Journal of the Indian Society of Soil Science.*35:690-698.

Pal, Devendra and Srivastava, R.A.K. 1982. Land form configuration of the karewa floor and its implication on the quaternary sedimentation. Kashmir Himalaya. *In Himalaya: Land forms and processed, Prof V.K.Verma (Ed),* Today and Tomorrow publishers, New Delhi, pp. 47-48.

Peer, M.A.1994.Fractionation of potassium, its uptake and critical limits for rice. *Thesis submitted to Sher-e-Kashmir University of Agricultural Sciences and Technology, Shalimar,* pp 1-108.

Rana K. P. C.,Walia, C. S., Sidhu, G. S., Singh, S. P., Velaytham, M., Sehgal, J. 2000. Soils of Jammu and Kashmir: Their kinds, distribution, characterization and interpretation for optimum land use. *Soils of India Series, NBSS Publication No. 62.*pp 1-71.

Sehgal, J.L., Sys. C., Stoops. B and Tavernier R. 1985. Morphology, genesis and classification of two dominant soils of the warm temperate and humid region of the central Himalayas. *Journal of the Indian Society of Soil Science,* 33: 846-857.

Sehgal, J.L. 1994. Soil resource mapping of different states of India. *National Bureau of Soil Science and Land Use Planning, Nagpur, Soil Bulletin* No.23:39-40.

Shinde, D.A. and Talib, A.R. 1984. Studies on saffron growing soils of Jammu and Kashmir.*Journal of Indian Society of Soil Science.*32:777-780.

Shinde, D.A., Talib, A.R. and Gorantiwar, S.M. 1984. Composition and classification of some typical soils of saffron growing areas of Jammu and Kashmir.*Journal of the Indian Society of Soil Science*.32:473-477.

Soil Survey Staff. 1998. *Keys to Soil Taxonomy*. http//www.statlab.iastate.edu/soils/keytax/.

Soil Survey Staff. 2014. 12th edition of *Keys to Soil Taxonomy*. https://www.nrcs.usda.gov/wps/PA_NRCSConsumption/download?cid...

Soil Survey Staff. 1999. *Soil Taxonomy: A basic system of soil classification for making and interpreting soil Surveys.United States Department of Agricultural Handbook No. 436.* 2nd edition. United States Department of Agriculture and soil conservation, services, Washington.

Soil Survey Staff. 2003. *Soil Survey Manual*. http//www.statlab.iastate.edu/soils/ssm/.

Verma, K.S., Shyampura, R.L. and Jain, S.P. 1990. Characterization of soils under forest of Kashmir valley. *Journal of the Indian Society of Soil Science*. 38:107-115.

Chapter 6

Nursery Management and Propagation in Apple

Aarifa Jan[1], W. M. Wani[2] and M. A. Mir[3]

[1]ICAR-Central institute of Temperate Horticulture,
Rangreth, Srinagar – 190 007 (J&K)
[2]Sher-e-Kashmir University of Agricultural Sciences and Technology,
Kashmir, Srinagar
[3]Professor and Head, Department of Food Technology,
Islamic University of Science and Technology, Awantipora-J&K
E-mail: aarifa711@gmail.com

1. Introduction

The fruit production depends on topography, soil, climate, irrigation, rootstock and variety. The productivity of temperate fruits in India is 6.77 t/ha which is very low as compared to world average productivity 12-15 t/ha. Apple cultivation is mainly confined to Jammu and Kashmir, Himachal Pradesh, and Uttarakhand and partly in Arunachal Pradesh, Sikkim and Nagaland. The average productivity of Jammu and Kashmir is 10.2 t/ha, followed by Himachal Pradesh 6.9 t/ha and Uttarakhand 2.6 t/ha (2013-14) (NHB, 2014).The acute shortage of quality planting material of improved varieties is one of the major constraints limiting the production, productivity and faster development in fruit crops. As per the rough estimates annually 15-20 million fruit plants are required in temperate zone of India. For transforming the 30 per cent of traditional apple orchards of Jammu and Kashmir to high density system, 94 lakh plants are annually required (Srivastava *et al.,* 2014). Availability of quality planting material is a prerequisite to the success of horticulture development initiatives. The availability of quality seedlings at lower cost offers ample scope for large scale planting. Nursery management is an important tool for the success of such entrepreneurships which will help nursery men to run profitable businesses. Nursery is pre requisite for producing quality seedlings in lesser input and nursery management is a potential tool to execute the activity in successful means. Due to the diverse edaphic-climatic condition of India, the quality planting material requirements vary and also everlasting. Meanwhile the species or variety or genotypes suitable for cultivation in one region may or may not be remunerative in another region. Hence, development of location specific quality seedlings has the potential to increase the agriculture productivity. In order to facilitate the availability of quality planting material, the data on demand is must. In India, there are about 4409 fruit plant nurseries out of which 1575 fruit nurseries are under government sector and 2834 are under private sector, which have an annual target of producing 1387 million fruit plants. The main suppliers of perennial tree seedlings are the departmental/government and industrial nurseries. At present only up to 30-40 per cent demand for planting material is being met by the existing registered nurseries; the rest are met from the unorganized sectors, implying the need for establishing more nurseries in the organized sector.They are producing seedlings and vegetative propagules to meet their own seedling demand and also supply them to public to meet their raw material demand. As per survey conducted in Kashmir valley, most of the nurseries are not having well labeled mother plant orchards and most of the nurseries are supplying plants on seedling rootstocks. There is a wide gap of production and productivity levels among different apple growing states in India, while Kashmir is leading with productivity of 11 MT/ha, north eastern states are far behind with only 3 MT/ha. The major emphasis is to enhance the overall production and productivity of apple with an area expansion on large scale under superior cultivars followed by improved management system as lack of quality planting material has been identified as one of the major factors for poor performance of apple in India (Banday *et al.,* 2015). In order to supply the genuine planting material to farmers National Horticulture Board has initiated accreditation drive of private and government nurseries on the basis of their performance, basic facilities, annual production and availability of mother tree as well as rootstocks *etc.* In Jammu and Kashmir about 242 government and private apple nurseries have been registered by National Horticulture Board. Some of these nurseries and their production capacity are given in Table 6.1.

2. Nursery

A nursery is a managed site, designed to produce seedlings grown under favorable conditions. The fruit nursery is a place where fruit seed is sown to grow into seedlings (Plate 6.1). Branches (scion) or buds from good fruiting trees, which must also be healthy and disease resistant are joined onto these seedlings while still in the fruit nursery. All nurseries primarily aim to produce sufficient quantities

Table 6.1: Important Private and Government Apple Nurseries in J&K

Sl.No.	Nursery Name	Nursery Address	Name of Variety (s)	No. of Mother Plants	Production Capacity
1.	A. R. Nurseries	Mutalhama, Kulgam	Golden Delicious	07	3500
			Kullu Delicious	20	10000
			Red Delicious	50	22500
			Royal Delicious	20	10000
2.	Ahsaan Nursery	Chattergam, Budgam	Mollies Delicious	10	500
			Oregon Spur	10	2000
			Starkrimson	05	1000
3.	Al-Shajar Pvt. Fruit Plant Nursery	Kululhand, Doda	Red Delicious	05	5000
			Royal Delicious	05	1200
4.	Bakhtawar Seeds and Plants	Mir Bazar, Anantnag	Golden Delicious	11	4500
			Maharaji	10	4000
			Red Delicious	10	4000
			Royal Delicious	16	6500
5.	Bhat Kissan Nursery	Tankipora, Kulgam	Golden Delicious	10	4500
			Kullu Delicious	45	17000
			Red Delicious	40	16000
6.	Dar Nursery	Mutalhama, Kulgam	Golden Delicious	17	20000
			Red Delicious	58	20000
			Royal Delicious	25	20000
7.	Fruit Plant Nursery	Raj Bagh, Srinagar	Cooper IV	20	2000
			Lal Ambri	20	2000
			Mollies Delicious	78	8000
			Starkrimson Delici.	20	2000
			Vance Delicious	20	2000
8.	Fruit Plant Nursery	Batote, Ramban	American	53	2000
			Golden Delicious	40	3000
			Red Delicious	45	5000
			Royal Delicious	60	3500
9.	Fruit Plant Nursery	Bagh Bandipora	Gala Mast	30	1000
			Lal Ambri	30	1000
			Red Delicious	25	2000
			Royal Delicious	25	2000
10.	Fruit Plant Nursery, Chogal	Chogal, Kupwara	American	36	15000
			Golden Delicious	98	40000
			Red Delicious	150	75000

Sl.No.	Nursery Name	Nursery Address	Name of Variety (s)	No. of Mother Plants	Production Capacity
11.	Fruit Plant Nursery, Futlipora	Futlipora, Budgam	Red Delicious	130	50000
12.	Fruit Plant Nursery, Gool	Gool Ramban	Red Delicious	38	3000
13.	Fruit Plant Nursery, Guloora	Guloora, Kupwara	Red Delicious	42	8000
14.	CITH, Rangreth	Rangreth, Srinagar	Golden Delicious	10	2200
			Oregon Spur	10	2200
			Red Chief	10	2200
			Red Delicious	10	2200
			Red Gold	10	2200
			Royal Delicious	10	2200
			Starkrimson Delici	10	2200
15.	Fruit Plant Nursery, Haripora	Haripora, Shopian	Golden Delicious	30	2500
			Red Delicious	70	3000
			Starkrimson Delici	50	2000
16.	Fruit Plant Nursery, Harwan	Harwan Srinagar	Starkrimson Delicis	200	2000
17.	Fruit Plant Nursery, Khellani	Khellani, Doda	Golden Delicious	1	600
			Red Delicious	2	1500
18.	Fruit Plant Nursery, Sangaldan	Sangaldan, Ramban	Red Delicious	13	4000
19.	Fruit Plant Nursery, Siot	Siot, Rajouri	Lal Ambri	8	2000
			Red Chief	4	8000
20.	Fruit Plant Nursery, Tumlahal	Tumlahal, Pulwama	Red Delicious	16	2500
			Starkrimson Delici.	16	2500
21.	Fruit Plant Nursery, Zainapora	Zainapora, Shopian	Golden Delicious	22	200
			Lal Ambri	150	100
			Red Delicious	80	100
			Starkrimson Delici.	425	100
22.	Hi-tech Mother Fruit Plant Nursery (SKUAST-K)	Wadura Baramulla	Ambri	100	10000
			Oregon Spur	100	10000
			Red Chief	200	20000
			Red Delicious	200	20000
			Royal Delicious	200	20000

of high quality seedlings to satisfy the needs of users. Genuine and quality nursery plants are the foundation of successful fruit industry. To raise plants in the nursery seems to be easy but to maintain then in good state is very tedious. A seedling

Plate 6.1: Established Apple Nursery.

of high quality must meet the standards of performance on a particular site. For raising a healthy fruit nursery one should have complete knowledge of sowing time, depth of seeds, budding/grafting, transplanting of seedlings and their handling *etc.*

(a) Importance of Nursery

☆ Seedlings and grafts are produced in nursery from which the fruit orchards can be established with minimum care, cost and maintenance.

☆ The nursery planting materials are available at the beginning of the planting season. This saves the time, money and efforts of the farmers to raise seedlings.

☆ It assures the production of genetically improved quality planting material.

☆ It provides employment opportunities for technical, skilled, semi-skilled, unskilled labor.

(b) Problems in Nursery Plant Production

☆ Lack of mother plants of well adapted varieties and lack of infrastructure facilities.

☆ Lack of knowledge for establishment of separate mother block and seed orchard.

☆ Large gap in demand and supply of quality planting material.

☆ Lack of knowledge of nursery registration act.

☆ A wide variation in sale price of nursery plants in government and private sectors.

☆ Lack of technical know- how regarding standard of nursery plants.

☆ Lack of proper management of insects, pests and diseases.

(c) Components of a Good Nursery

The nursery site should be located in the well-drained fertilized soil, near to water source, free from soil pathogens and insects, availability of cheap and skilled labors and has good access to the main road for easy transportation. The site should be on gently sloping area and away from other tall crops for good drainage as well as to encourage air circulation. An appropriate site must be selected for the most effective, efficient, and economical design of a nursery. Careful observation of site conditions and an assessment of past and present climatic records are important. Standard nursery management aims at the most rapid production of healthy and quality planting materials.

(d) Nursery Management Practices

☆ Sourcing, collection and selection of seeds for propagation.

☆ Handling of seeds to hasten germination.

☆ Management practices of germination beds.

☆ Techniques of sowing the seeds in the germination beds.

☆ Pricking out and transplanting.

☆ Weed and pest control.

3. Nursey Establishemnt in Apple

(a) Bud Wood Band and Mother Tree Block

The bud wood scion bank can be established through introduction and selection of promising cultivars. Mother trees should be planted close (2 to 3 m) spacing depending upon the variety. The trees must be pruned hard regularly to develop hedges for production of scion wood in large quantities. Tree should be irrigated and fertilized with higher doses of nitrogen frequently to have more vegetative growth. The bud sticks/graft wood should be always taken from healthy and true to type tree progeny trees, free from viruses, diseases and insect pests. The past history of these trees must be known. The nursery men should have progeny trees of all commercial varieties of fruits that can be grown in that particular area. This is the most important step for production of quality nursery plants.

(b) Seed Orchard

Seed form Maharaji, Crab apples are preferred in apple for raising of seedling rootstocks. However, seeds of some commercial varieties like Golden Delicious, Granny Smith are also used. To meet the requirement of seeds, the rootstock of orchard of crab apple, Maharaji, Golden Delicious and Granny Smith cultivars are to be established.

(c) Clonal Rootstocks

Clonal rootstocks recommended in India are M-27 (ultra-dwarf), M-9 (dwarf), M-7 and MM-106 (semi dwarf), MM-111, MM-104 and MM-109 (semi vigorous). The demand for clonal rootstocks is increasing due to the advantage over seedling rootstocks. EMLA series of apple rootstocks are virus resistant, MM series of apple rootstocks are wooly apple aphid resistant. To meet the requirement of farmers, these rootstock stool beds need to be established and maintained in nursery for large scale multiplication. Some clones of M-9 rootstock of apple have been introduced from Netherlands by ICAR-CITH, Srinagar. These rootstocks have been planted in nursery for large scale multiplication (Plate 6.2).

(d) Site Selection

A good nursery location should provide the best possible conditions for seedling growth and ensure that the nursery is accessible, safe and comfortable for workers. Nursery should be located on levelled land near the road. It is important to choose the right place for a nursery. A site is needed where irrigation, mulching and composting and such daily maintenance will be easy. The nursery should be away from the approach of animals, brick kilns and stone crushers *etc.* Consider the following factors when selecting a nursery site. It is important to choose the right place for nursery raising.

(i) Water Supply

A reliable source of water is essential. This is especially true for nurseries located in areas with a distinct dry season. Ideally, position the nursery close to a near by spring, stream, pond, borehole or well. Consider installing water tanks to store water as a back-up for water shortages during dry periods.

(ii) Topography and Aspect

Nursery should be located some where flat, sheltered and well-drained. Avoid placing nursery in areas that are prone to flooding (*e.g.* at the bottom of a valley or in riverine areas), strong winds (*e.g.* at the top or a hill or in the middle of an exposed valley) or soil erosion (*e.g.* on a steep hill). Aspect (the direction your nursery faces) is also important. In summer months, strong afternoon sun (south facing in the northern hemisphere and north facing in the southern hemisphere) may damage seedlings. Therefore, it is generally best to position your nursery so that the seedlings face the morning sun (always rising from the east).

(iii) Soil

Ideally, suitable soil should be available near the nursery. This can be achieved by including sandy soil, river gravel or bark chippings in the growing medium, reducing the chance of the medium becoming water-logged.

Plate 6.2: Introduced Rootstocks of Apple in Nursery at CITH-Srinagar.

(iv) Access and Ownership

Nursery must be safe and accessible at all times for nursery workers. Ideally it should be located close to a road to help transport materials to the nursery and take seedlings from the nursery to your planting sites. Make sure that ownership of the land is clear before any construction begins. This may include consulting people on traditional use and ownership of the land in question and holding meetings with local land owners and land users that could be affected by its construction.

4. Propagation

Apple is commercially propagated through asexual method of propagation. But for raising seedling rootstocks sexual method is employed. The propagation of apple is classified into two groups: a) propagation of rootstocks and b) propagation of scion cultivars. Further, the rootstocks are propagated through seeds for raising seedling rootstocks and through layering, cutting and tissue culture for raising clonal rootstocks. The scion cultivars are propagated through grafting and budding techniques. In order to achieve success in apple orcharding, it is important to select a right type of rootstock for a particular cultivar and location and to raise the stock seedlings in proper manner so that there is good germination and the stock seedlings are ready for budding/grafting in shortest possible time. There should be adequate arrangement for supply of rootstock seeds of desired kinds and varieties. It will be desired to have rootstock trees in the nursery itself. Planting material of apple can be raised both by sexual (seed) and asexual (vegetative) methods. The description of methods for raising the planting material in apple is given below.

(a) Propagation of Rootstocks through Seeds

(i) Sowing of Seeds in the Nursery

The time of sowing of seeds of different plants depend on the nature of plant and varieties. Seedling rootstocks are used on which the scion varieties are grafted or budded. Seeds of self-pollinizing varieties like (Golden Delicious and Granny Smith), *Maharaji* and crab apples are used for raising seedling rootstocks. Bandana and Chandel (2015) while evaluating the seedling rootstocks of apple found that seedling rootstocks of crab apple, Red Gold and Granny Smith were better in terms of growth of rootstock and scion variety grafted on them. Seeds of apple Golden Delicious, Granny Smith, *Maharaji* and crab apples should be collected from well ripened fruits in September- October. Apple seeds will not germinate unless stratified. This involves keeping the seeds in moist conditions and subjecting them to a period of cold to allow after-ripening, during which embryo changes occur (Janick *et al.,* 1996). After-ripening will proceed at temperatures between 0 and 10°C, but the optimum is from 3 to 5°C. The time required may vary from 6 to 14 weeks and depends to some extent on the temperature. Abbott (1955) has shown that the temperature for after-ripening is critical. The seeds collected should be washed, dried and packed for sowing in November- December. The seeds are sown either in seed beds, polybags or *in situ*. The apple seeds should be sown in raised beds. Raised beds are created by forming a mound of soil on a well-drained area of the nursery floor around 10 cm high. The soil should be well pulverized mixed with FYM and sand in equal proportions. The seed bed should be sterilized to avoid contamination. The seeds should be sown at a depth of 3 to 5 cm in rows at spacing of 10-15 cm apart in seed beds (Chandel and Verma, 2015). After sowing the seeds should be covered with thin layer of sand and soil mixture. In case of dry spells after sowing of seeds, seed beds should be sprayed with water to facilitate the chilling treatment of apple seeds through stratification.

(ii) Care for Young Seedlings in the Nursery

After germination, young seedlings will require watering (again, avoid over-watering but never allow the growing medium to dry out). Monitor the health and condition of seedlings/saplings. Keep small and weak seedlings to one side. If some plants show signs of pest or disease, quickly prevent transmission to other plants. If plants show symptoms of nutrient deficiency, consider adding fertilizer to the soil. Be careful not to add too much as this may cause root burn. Before saplings are ready for planting they need to be hardened-off (a process that typically involves a gradual removal of shade and watering). This helps saplings to develop a woody stem and prepares them for the physiological stress. Drainage ditches should be cleared regularly to avoid water logging in the nursery. Fencing and shading materials should be inspected regularly and repaired and replaced as necessary.

(b) Production of Rootstocks through Vegetative Methods

The concept of vegetative propagation is that an exact copy of the genome of a mother plant is made and continued in new individuals. A piece of plant shoot, root, or leaf, can therefore, grow to form a new plant that contains the exact genetic information of its source plant. Whereas, sexual reproduction by seeds provides opportunity for variation and evolutionary advancement, vegetative propagation aims at the identical reproduction of plants with desirable features such as high productivity, superior quality, or high tolerance to biotic and/or abiotic stresses, and as such, plays a very important role in continuing preferred trait from one generation to the next. The most important vegetative propagation techniques for apple are the propagation by cuttings, layering, budding, grafting and micro propagation (Plate 6.3).

The most important reasons for vegetative propagation are:

a) Maintaining superior genotypes.
b) Problematic seed germination and storage.
c) Shortening time to flower and fruit.

Plate 6.3: (a) Apple Seedling Block for Budding at Zainapora Nursery; (b) Apple Rootstock Block for *in situ* Grafting at CITH-Srinagar.

d) Combining desirable characteristics of more than one genotype into single plant.

e) Controlling phases of development.

f) Uniformity of plantations.

Apple is mainly multiplied on crab seedlings of commercial varieties/clonal rootstocks. The clonal rootstocks are uniform in size and trees on these rootstocks are precocious in bearing and suitable in high density planting. The apple rootstocks recommended in India are given in (Table 6.2).

Table 6.2: Clonal Rootstocks of Apple

Sl.No.	Rootstock	Characteristics
1.	**Malling- 27** (*M.27*)	An extremely dwarfing rootstock, developed in East Malling Research Station, U.K. in 1929.It can be planted as close as 0.5 m x 1.5 m apart. It makes a tree of about 20 per cent the size of standard and half the size of M-9. It does not produce burr knots and root suckers. It is less susceptible to fire blight than M-9.
2.	**Malling- 9** (*M.9*)	It is most widely used dwarfing rootstock for apple and has originated as a chance seedling East Malling Research Station, U.K. It produces tree size of about 25-30 per cent of full size with most of the cultivars. The trees on this rootstock are precocious and tolerant to a wide range of soil and climatic conditions. The rootstock has poor anchorage, shallow root system and requires mechanical support to hold the tree firmly. This rootstock requires assured irrigation due to its shallow root system. Many sub clones of M-9 exist today. Clones vary with some degree of dwarfing imparted to the scion. The new clone M-9 337, a virus free clone of M- 9 is being encouraged in apple growing areas. It is slightly vigorous to M-9 but is productive also known MAC-9. Once crop matures, the tree on this clone stops vegetative growth and temporary staking is required.
3.	M.9 NIC 29	Developed in Belgium and is a selection from M-9. This rootstock is being recommended for cultivars that are less vigorous.
4.	**Geneva 65** (*G. 65*)	This is a patent rootstock from Cornell University, New York. It is very dwarfing produces trees smaller than M.9. It is precocious and productive. It is resistant to fire blight and collar rot. It has few burr knots and few suckers.
5.	**Vineland 3** (*V. 3*)	This is a new rootstock originated in Vineland, Ontario. It is slightly less vigorous than M. 9 but similar to M.9 clones M.9T337 and M.9 Flueren 56. Trees on V. 3 are as productive as M. 9 clones but are more yield efficient.
6.	**Budagov-sky 9** (*Bud 9*)	A dwarfing rootstock similar to M. 9 in size bred in Soviet Union, winter hardy and is resistant to collar rot. The tree requires support due to poor anchorage and starts bearing 2-3 years after planting.
7.	**Vineland 1** (*V. 1*)	The tree size is slightly larger than M. 26. Yield efficiency and fruit size is equal to or greater than M. 26. It is highly resistant to fire blight.
8.	**Ottawa 3** (*O. 3*)	This is most dwarfing rootstock to come out of cold hardy breeding programme at Ottawa. It is dwarfing than M. 26 but vigorous than M. 9. It is cold hardy, resistant to collar rot but susceptible to fire blight and woolly apple aphid. It produces few burr knots.
9.	**Malling 26** (*M. 26*)	A dwarfing rootstock introduced from East Malling in 1959. M. 26 is reported to be the most hardy of Malling series rootstocks. The tree is about 40 per cent of standard size, being larger and sturdier than M. 9 but smaller than MM. 106. Although its roots are not brittle, and anchorage is fair. It becomes self-supporting after about 5 to 8 years but support is recommended for early economic cropping.

5. Rootstocks of Apple

(a) Malling (M) Series

Developed at East Malling Research Station in England in 1913.The different categories of M series rootstocks are:

Dwarfing: M-8, M-9, M-26, M-27; Semi dwarfing: M-2, M-4, M-7; Semi vigorous: M-13; Vigorous: M-12, M-16.

Subclones of M-9 have been developed in different countries which are quite easy to propagate. Pajam 1, Pajam 2 by France; KI. 29 by Belgium; Burgmer series Germany and NAKB clones and Fleuron 56 in Holland.

(b) Malling Merton (MM) Series

Developed by crossing Malling rootstocks with Northern Spy to evolve wooly aphid resistant rootstocks.Semi dwarf: MM-10; Semi- vigorous: MM-109, MM-111; Vigorous: MM-109

(c) Merton Series

Merton 793 has proven as a very useful rootstock and is resistant to woolly aphid and collar rot. Other Merton rootstocks are Merton 778, 779, 789 and 793.

(d) Polish Series

Developed in Poland by cross of Antonovika and M-9, these rootstocks are winter hardy and resistant to crown rot. Poland series have five promising rootstocks (P1, P2, P16, P18 and P 22). All the rootstocks are dwarfing except P18.

5.1 Characteristics of an Ideal Rootstock

Apple tree grows as a composite tree (rootstock and scion). The scion qualities are controlled by rootstock. Ideal rootstock should fulfill the following criteria:

- ✰ It should be easily propagated.
- ✰ It should exert profound influence on vigour, precocity, productivity and quality of fruits of scion.
- ✰ It should have good root system to provide adequate anchorage to tree grafted on it.
- ✰ It must produce good, clean, upright stem so that grafting or budding can be easily done.
- ✰ It should have adaptability to soil and climatic conditions, and also resistant to different pests and diseases.
- ✰ It must be compatible with the scion varieties.
- ✰ It should be tolerant to high salts and should have winter hardiness.

5.2 Techniques for Raising Clonal Rootstocks

Apple trees are not grown on their own roots but propagated on rootstocks that control the tree growth. Clonal rootstocks of apple are multiplied by layering as well as by cuttings.

(i) Layering or Stooling

Layering has been found to be most successful method for raising of apple rootstocks. For raising the clonal rootstocks of apple through layering, the rooted clonal rootstocks are collected from registered nursery. The field should be prepared for establishing the mother stock. The clonal rootstocks are planted in winter in nursery beds at spacing of 1.0 x 1.0 m and allowed to grow for one season. At the end of next dormancy the mother stocks should be headed back leaving 10-15cm stub from ground level. Profuse shooting would be experienced in next growing season, the shoots are mounded with soil leaving the shoot tips open. Girdling the base of shoots by wiring or application of 2000 ppm IBA stimulates rooting. Khatik, (2010) reported that IBA 25,00 ppm was found best treatment for inducing better rooting in apple clonal rootstock M. 783. Proper irrigation should be done for profuse rooting. In January- February, rooted shoots are separated leaving the stub of mother stock there for producing the shoots for next year. On an average 15-20 rooted shoots are produced from a single stool.

(ii) Cutting

The hardwood cuttings are the common method of propagation clonal rootstocks of apple, which are prepared from fully mature tissues. The shoots of about one year old or more can easily be used for preparing hard wood cuttings. In case of apple the cuttings are made after pruning. Generally the cuttings of 15-20 cm length and having 3-5 buds are made. The lower cut is given in a slanting manner just below the bud to increase the absorption of nutrients. The upper cut is given at a right angle to reduce the size of the wound. After the cuttings are prepared they should be allowed to dry. These cuttings are usually tied in small bundles (20-25 cuttings) and buried in moist soil/sand for a certain period for healing of wounds, which is known as callusing. Planting time and growth regulators influence the rooting in apple rootstocks. Verma, (2013) while studying the effect of size of cuttings and concentration of growth regulators on rooting of hard wood cuttings of Merton 793 of apple found that 35cm long and 1.25-1.50 cm diameter cutting treated with 2,500 ppm IBA gave highest rooting of 65 per cent under shade net.

(iii) Tissue Culture

It refers to *in vitro* and aseptic culture of various plant parts like meristem tips, isolates of cell or protoplast in artificial growth medium. This technique can be used for rapid multiplication of clonal rootstocks in a limited space. During the recent years successful accomplishment of micro propagation through tissue culture technique has been made in clonal rootstocks of apple. In this technique plants are raised *in vitro* on artificial media under aseptic conditions. Well rooted plants are transferred to a mixture of sand and soil after 1-2 weeks to the nursery.

5.3 Propagation of Scion Cultivar

Apple scion cultivars are commercially propagated by grafting and budding. Some time cutting and micro propagation is also employed for nursery raising.

(i) T- Budding

Apple responds to chip budding and T budding. T budding is most successful technique in apple propagation. T budding is performed in July- September in Kashmir and June- July in Uttarakhand and Himachal Pradesh.It is also known as shield budding as the bud obtained from the bud stick resembles shield in shape and T budding as the two cuts made on the stock intersect such a way so as to form T shape. A perpendicular cut of 1 inch long is made on a smooth portion of stock and it is followed by a horizontal cut across the top at right angle to the first cut. The two cuts extend only through bark and should not go into the wood.

(ii) Chip Budding

This is performed in dormant season, fall or late winter, spring and late summer. In late spring it can be performed as bench grafting on uprooted rootstocks. The rootstock should have 13-25 mm thickness for successful combination of budding with wood. The scion is obtained by giving deep inward slanting cut (4-6 mm deep) just below the bud (5-10 mm below) at 30-45° angle. The second cut is made at 20-25 mm above the selected bud which goes download piercing the wood in such a way that it intersects the first cut. The chip of bud is obtained along with a piece of wood. A chip of same dimension is removed from the rootstock so as to fit the bud obtained from scion wood. The union is tied firmly with the alkathene tape.

(iii) Bench Grafting

It is very popular method of propagation of temperate fruits. It is done in dormancy season on uprooted apple rootstocks. The scion is collected and stored during pruning of apple. The bench grafting is performed either by tongue grafting or whip or cleft grafting. In tongue grafting equal thickness of scion and stock are taken. A slanting cut of 2.5-5.0 cm long is made by sharp grafting knife on both the stock and scion depending on their thickness and then reverse downward cut leaving ½ the length of slant cut, is given just below the tip of the cut surface on both scion and stock. The scion is inserted into the rootstock by sliding into the slanting cut surface. The graft union is tied firmly with the alkathene tape.

5.4 Operations after Budding and Grafting

(i) Removal of Alkathene Tape

The alkathane is removed after the bud has sprouted. In cases of budding it should be removed in April where as in grafting it should be removed in May- June. The alkathene tape should be carefully removed with the help of sharp blade. The alkathene tape in grafted plants should be removed after complete sprouting of graft scion and proper height and the union should be firm enough to hold the scion growth on the rootstock.

(ii) Deshooting

After grafting and budding, shoots arise from rootstock below the graft union. These shoots should be removed immediately after sprouting at regular intervals.

(iii) Staking

Staking of plants is must after the removal of alkathane tape from the graft union. The union is tender in summer and plants are blown off with higher wind velocities. Staking of grafted plants is usually done with wooden sticks.

(iv) Single Stem

In grafted and budded plants no scion shoots are allowed to grow and they must be pinched off from the growing scions.

6. Need for Quality Planting Material

The planting of poor quality low yielding apple varieties in orchards is mainly responsible for low quality fruits. The apple fruit production in India has shown an upward trend yet the potential has not been fully realized. The main reason for this is lack of quality planting material. At present the total number of nurseries available in India are 6724 (Table 6.3). Now a days we are shifting towards high density planting system which means planting of more number plants than optimum through manipulation of tree size for increasing productivity.The requirement of apple rootstocks for transforming into high density plantation is given in Table 6.4.

Table 6.3: Status of Fruit Nurseries in India

Sl.No.	Sector	No. of Nurseries
1	Public	1, 594
2	Private	4, 607
3	SAUs/ICAR	138
4	Model nurseries established under NHM	385
	Total	**6724**

Source: Singh *et al.*, 2008.

Table 6.4: Requirement of Plants for Conversion of 33 per cent Apple Orchards with High Density

Area (ha) 2014-15 under Kashmir Division	Area (ha) to be Converted to HDP (33 per cent of current area)	Change in Density with Use of HDP (trees/ha)	Trees Requires for Five Years (in lakhs)
1,44.733	48,000	1,000	480,000

Source: Division of Fruit Science, SKUAST-Kashmir.

7. Constraints in Nursery Production of Apple

The constraints faced while propagation and nursery raising in apple are given below:

a) Due to non-availability of sufficient land many nurseries are raised on same location for years, which results in increasing intensity towards infection of soil borne diseases.

b) The lack of bud wood or mother orchard which results in collection of scion wood/bud wood from different sources are not authentic.

c) Most of the bud wood is taken from trees with no pedigree as a result the quality of planting material gets affected.

d) The most of apple nurseries do not have separate rootstock banks for production of uniform planting material. The collection of seeds/ rootstocks from different sources results in poor quality of nursery plants.

e) Most of apple nurseries do not follow proper plant protection measures as a result the planting material is affected by pests and diseases.

f) The lack of trained/skilled manpower for propagation and other management practices.

g) There is improper implementation of nursery registration act.

References

Abbott, D. L. 1955. *Temperature and the dormancy of apple seeds.* p. 746-753. In: Rep. Int. Congr. Scheveningen.

Bandana and Chandel, J.S. 2015. Effect of varieties on seed germination, seedling growth and growth of budded plants of apple under protected conditions. *International Journal of Farm Science* 5: 74-82.

Banday F. A., Shrama, M. K. and Mir, M.S. 2015. *Status and strategies for meeting planting material requirement in temperate fruit and nuts.* In: K. L. Chadha, N. Ahmed, S. K. Singh and P. Kalia (Eds.) Temperate Fruits and Nuts. Daya Publishing House. New Delhi. pp. 284.

Chandel, J.J. and Verma, P. 2015. *Recent advances in propagation and nursery management of temperate fruits and nuts.* In: K. L. Chadha, N. Ahmed, S. K. Singh and P. Kalia (Eds.) Temperate Fruits and Nuts. Daya Publishing House. New Delhi. pp. 269-283.

Janick, J., Cummnis, J.N., Brown, S.H. and Hemmat, M. 1996. Apples. In: Janick, J. and Moore, J.N. (Eds.) *Fruit Breeding, Vol. 1, Tree and Tropical Fruits.* John Wiley and Sons, NewYork, pp. 1-77.

Khatik, P.C. 2010. Effect of auxin and rooting media on multiplication of apple clonal rootstock (Merton- 793). *M. Sc. Thesis, Dr. Y. S. Parmer University of Horticulture and Forestry, Nauni, Solan, Himachal Pradesh.*

Singh, A.K., Patel, V.B. and Singh, S.K. 2008. *Propagation of quality planting material in fruit crops for higher profitability.* In: income and livelihood security through horticultural development, J. N. L. Srivastava and Lallan Singh (Eds.). IFFCO Foundation, New Delhi, pp. 73-81.

Srivastava, K.K., Ahmed, N., Mir, J.I., Kumar, D. and Singh, S. R. 2014. Manual for quality planting material production, CITH, Srinagar.

Verma, P. 2013. Effect of pre-conditioning treatments, size of cuttings, plant growth promoting rhizobacteria and IBA on rooting in cuttings of apple (*Malusdomestica* Borkh.) clonal rootstock Merton 793. *M. Sc. thesis, Dr. Y.S Parmar University of Horticulture and Forestry, Nauni, Solan, Himachal Pradesh.*

Chapter 7

Description of Some Important Local and Exotic Apple Cultivars

W. M. Wani, Aarifa Jan and J. I. Mir

Division of Fruit Science,
Sher-e-Kashmir University of Agricultural Sciences and
Technology of Kashmir, Srinagar
E-mail: drwmwani@gmail.com

1. Introduction

Apple is cultivated worldwide as fruit tree and belongs to genus *Malus* and family Rosaceae. The genus has five sections including 122 species and subspecies (Chadda and Awasti, 2005). The cultivated apple is likely the result of interspecific hybridization and at present the binomial *Malus × domestica* has been generally accepted as the appropriate scientific name (Korban and Skirvin, 1994). Natural varieties of cultivated apple belong to *Malus pumila* Mill, while its hybrid varieties belong to *Malus domestica* Borkh. The main ancestor of apple is now considered to be *Malus sieversii,* which is wild from the Heavenly Mountains (Tien Shan) at the boundary between western China and the former Soviet Union, to the edge of the Caspian Sea (Morgan and Richards, 1993). The cultivated apple (*Malus domestica*) is reported to have originated in south western Asia in Caucasus region near Gilan in Turkestan. Apples have been grown for thousands of years in Asia and Europe, and were brought to North America by European colonists. It is not known when the apple was introduced in cooler parts of India, but evidence showed its presence in Agra in 1632. Baring variety *Ambri* which is indigenous to

Kashmir, all other varieties were introduced first by European settlers missionaries and later on by elite growers, nursery man and research introduction centers. Emperor Jahangir praised Kashmir apple in his book Turk-e-Jahangir which gives proof of their cultivation during 16th century. The first apple orchard in Himachal Pradesh was established at Bandrole in Kullu valley by Captain A. A. Lee, in 1870. The famous delicious varieties were introduced by Samuel Nicholar Strokes a residence of Philadelphia, USA in 1880 at Kotgarh in Shimla hills (Bal, 2002). The genetic variability found in the apple has allowed adapted types to be selected for different environments, and selection continues for new types to extend apple culture into both colder and warmer regions. The apple orchards are now found in Siberia and northern China where winter temperatures fall to –40°C and in high elevations in Colombia and Indonesia straddling the equator where two crops can be produced in a single year (Janick, 1974). Each country and area had its own local cultivars like Ambri in Kashmir. Lawrence (1895) in book "The Valley of Kashmir" wrote that the most popular apple in Kashmir is the *Amani* or *Amri*, which has apples a large round red and white sweet fruit, ripening in October and keeping its condition for a long time. This is the apple which is exported in large quantities, and it finds favour with the natives of India for its sweetness and its handsome appearance. The *kuddii sari* apple is said to have been introduced from Kabul. But in my opinion the best of the Kashmir apples, so far as flavour goes, is the little *Trel* which abounds in the neighbourhood of Sopur. The most widely grown cultivars by far are Golden Delicious and Delicious and its red sports, both chance seedlings of American origin. Golden Delicious has been widely and successfully used in breeding, and its seedlings, which make up a high proportion of the new cultivars, are rapidly changing the apple industry. Over 7,500 cultivars of the apple are known worldwide (Elzebroek and Wind, 2008) and breeders worldwide create more new selections annually, but only a few dozen types are widely produced commercially today (Janick *et al.*, 1996). The origin of controlled breeding of apples is attributed to Thomas Andrew Knight (1759-1838) who produced the first cultivars of known parentage. This technique continues to be the basis of all present day apple breeding programs. Until the latter half of the twentieth century most of the world's apple cultivars were chance seedlings selected by fruit growers. More than 10,000 cultivars are documented, yet only a few dozen are grown on a commercial scale worldwide (Way *et al.*, 1990). In 1983, the best known cultivars in the world were all chance seedlings found in the eighteenth or nineteenth centuries, Golden, Delicious, Cox's Orange Pippin, Rome Beauty, Belle de Boskoop, Granny Smith, Jonathan, McIntosh *etc.* (Janick *et al.*, 1996). In the past most small farms produced their own apples as fresh or preserved for use of local markets. But now a days the cost of modern apple production requires cultivars to have prolific, consistent yield of quality fruits which are amenable to handling, storage, and shipping and generate high consumer demand. Normally, apple production focuses on regular plantations established with a few highly productive cultivars of extraordinary quality. However, great quantities of apples are produced in small orchards, established with local, stress resistant cultivars, having good pomological qualities that are superior to cultivated cultivars and form a huge reservoir of variability (Mratinic and Aksic, 2012). Apple is the premier table fruit of the world and has been under cultivation since earliest times. Apple growing regions occur throughout the temperate zones of the world. In India, the major apple producing regions include Jammu and Kashmir, Himachal Pradesh, Uttarakhand,some parts of north-east and Nilgiri Hills. Jammu and Kashmir is the leading apple producing state in India with annual production 68 per cent of the total production in India. About 330 varieties of apple are known to have been under cultivation in Kashmir valley but only a dozen are propagated at present on commercial scale. J&K state has remained popular for its indigenous apple variety *Ambri* from remote past. This variety has been utilized in breeding programme extensively, as a result of which a few hybrids namely, *Lal Ambri* and *Sunehari* were released. Apple varieties are available in a great array of colours, sizes, and flavours. Apple cultivars vary in their cooking or fresh eating qualities, storage life, tree hardiness, pest susceptibility, and many other characteristics. Most adopted apple cultivars for commercial cultivation includes Oregon Spur, Red Chief, Well Spur, Vance Delicious, Gold Spur, Red Spur, Silver Spur, Top Red and Red Fuji (Verma, 2015).The description of some local and exotic cultivars of apple grown in India is given below:

2. Local and Exotic Cultivaris of Apple Gorwn in India

Sl.No.	Variety	Description	Image
1.	Akbar	Average fruit yield 245 MT/ hac. at 20 years of age. Fruit is medium to large in size, red coloured, matures in about 157 days after full bloom. Developed by SKUAST-K, from cross between (Ambri x Cox'x Orange Pippin) in 2001.	
2.	Ambri	Ambri is the most popular and is only indigenous cultivar of India. It is widely famous for its crispiness, aroma, flavor and attractiveness. Fruits are medium in size, conical in shape and blush red with stripes. It matures in the month of October and has long shelf life. The production of Ambri apples in Kashmir has increased in Shopian and Kulgam districts. In Jammu Ambri apple is grown in Batote region.	

Sl.No.	Variety	Description	Image
3.	American Apirouge	This kind of apple is small and round shaped and quite juicy. American Apirouge is very crispy and sweet in taste. The colour of fruit is solid red blush covering 75 to 100 per cent. It ripens in September and is out in market by mid- September. It offers good taste in desserts and comparatively much cheaper in price.	
4.	Anna	Developed in 1965 and is an early season apple variety from Israel. It has very low chilling requirement of less than 300 hours. Fruit colour is yellow with a red blush. This variety does not grow well in the cold and prefers heat and humidity.	
5.	Antonovka	Originated in Kursk (Russia) during 17th Century. A very old Russian variety with large, round, white fleshed apple of average quality. Pretty, pink-tinged flowers superb for cooking. It is a mid-season cultivar.	
6.	Belle de Boskoop	Originated in Netherlands. Fruit colour is greenish yellow. Fruit is large, flattened and regular in shape; excellent dessert apple; coarse, crisp, creamy flesh, with subacid flavor; cooks well and has good keeping quality.	

Sl.No.	Variety	Description	Image
7.	Benoni (Hazarat-bali apple)	Originated in Massachusetts in 1832. Fruits are yellowish-orange mostly covered with bright red and deep carmine striping, early high-quality dessert apple, fine-grained, crisp and juicy flesh. Apple is small to medium sized and its shapes may vary from round to slightly conical. This variety has early maturity and short shelf life. *Hazartbali* ripens in mid- July and is the oldest variety of apple available in Kashmir valley.	
8.	Black Ben Davis	Colour of fruit is bright yellow mottled with dark and bright red blushing. It is a mid-season cultivar. Noted for keeping well prior to refrigerated storage and is a low chilling variety. Developed in South-eastern US, in 1980s.	
9.	Breaburn	Originated in New Zealand 1952 as a chance seedling. The fruit is sold commercially in the UK. Fruit is medium sized and has a good keeping quality. The texture is crisp and juicy. The overall flavor is tangy with a good balance of sweetness and a hint of pear drops. It requires a long growing season.	
10.	Coe Red Fuji	It is precocious regular bearer, late maturing, high yielding cultivar, fruit shape is globose, moderate fruit ribbing, weak crowning at calyx end, solid flesh, aperture of locules on transverse section is moderately open.	

Sl.No.	Variety	Description	Image
11.	Cooper IV	It is mid maturing and regular bearing cultivar, fruit shape is cylindrical waisted. Fruit weight is (250-280g). Total soluble solids (13-15 °B) and fruit mature 15 days ahead of Red Delicious.	
12.	Cox's Orange Pippin (Kesari)	Discovered by Richard Cox in 1825, it was England's favorite apple for more than a century but it has recently lost ground to more modern varieties like Gala. One of the most celebrated apples in U.K, valued for its aromatic Orange colour and flavor. It is mainly grown in UK, Belgium and the Netherlands, but is also grown for export in New Zealand. Fruit colour is greenish yellow, red stripes over orange blush. Fruit is elliptical and oblate shape, firm and moderately juicy.	
13.	Early McIntosh	Cross of McIntosh X Yellow Transparent. Yellow, red striped, early-mid maturing, white, tender flesh with pleasant flavor. Good for home and local markets.	
14.	Fanny	Originated in Pennsylvania before 1869. Fruit colour yellow, mostly covered with crimson and darker red stripes. Early maturing variety. Medium to large and slightly ribbed; thin, smooth skin; yellowish-white flesh with red staining.	

Sl.No.	Variety	Description	Image
15.	Firdous	Developed from Golden Delicious × Rome Beauty × Prima in 1996 by SKUAST-K. Yield is (12-15 t/ha) at 20 years of age. Fruit medium in size, sweet with slight acidic blend, crisp, juicy having resistance to scab and moderate resistance to Alternaria and San jose scale.	
16.	Fuji	Bred in Japan in 1962 parents are Red Delicious × Ralls Genet. Dark red, conic shape, sweet, crisp, dense flesh is very mildly flavoured. Its keeping quality is well. One of the most widely grown apple varieties in the world.	
17.	Gala Mast	It is precocious, regular bearing mid maturing, high yielding sweet juicy with acidic ting. The fruit is conic shape, moderate fruit ribbing, red purple colour, flushed and mottled, aperture of locules on transverse section is closed or slightly open.	
18.	Gala	Gala was developed in New Zealand in the 1970's. Cross of world's best known apples (Kidd's Orange Red X Golden Delicious). Fruit colour is orange-red stripes over creamy yellow. Medium-sized, oval to round. Extremely firm flesh, very juicy, sweet and mildly aromatic. Mid season cultivar and one of the most widely available commercial fruit.	

Sl.No.	Variety	Description	Image
19.	Golden Delicious	Originated in Clay County, West Virginia, US, in 1914. This cultivar is a chance seedling possibly a hybrid of Grimes Golden and Golden Reinette. One of the most popular apples ever grown, and the parent of numerous crosses: Virginia Gold, Jonagold, Spigold, Gala, Honeygold, Mutsu, and many others. Conical fruit, medium to large; golden-yellow; with a fine, sweet flavor occasional russet patches. Due to its regular size, even colour and storage qualities the fruit is widely sold commercially. Golden Delicious is self-fruitful and an excellent pollinator tree. It is a late maturing variety of apple.	
20.	Granny Smith	Originated in Australia as a chance seedling in 1868. Dark to pale green, distinctive whitish dots on fruit - Popular light green apple noted for its late (November) ripening. Superior keeper, good for cooking and eating out of hand. A favourite variety, widely sold in the UK and also noted as common pie apple. Lime green colouring, a late maturing variety and is extremely tart.	
21.	Green Sleeves	Originated in Kent UK, in 1966. Cross of (Golden Delicious × James Grieve), good garden apple, with a pleasant but unexceptional flavor.	

Sl.No.	Variety	Description	Image
22.	Hardiman	It is early maturing and regular bearing variety of apple. Fruit shape is cylindrical waisted, red purple colour and fruit weight is (260-270g). The fruits are high in demand by consumers due to attractive shape and colour.	
23.	King Luscious	Originated in North Carolina in 1935, as a chance seedling. Greenish-yellow covered with deep red and overlaid with darker red striping, mid to late maturing.	
24.	Jonica	Fruit colour is bright red on a bright yellow green background. This variety has large fruit striped red over bright yellow. Fruit weight is (150-170g), fruit is firm, juicy and slightly tart. Finest dessert and eating quality having good cooking properties.	
25.	Lal Ambri	This was evolved by SKUAST–K Shalimar by crossing Red Delicious X *Ambri*. Fruit has excellent taste with *Ambri* aroma, medium to large in size, juicy firm texture, bright red colour covering 95 per cent of fruit surface and good shelf life. A better quality and high yielding apple, matures in last week of September.	

Sl.No.	Variety	Description	Image
26.	Laxton's Fortune	Laxton's Fortune (often known simply as Fortune) is a cross between Cox's Orange Pippin and a little-known American variety called Wealthy Laxton's Fortune. Apples have pale yellowish-green skin with red blushes, and some russeting.The flesh is sweet and juicy and crisp, though it becomes softer if left on the tree.	
27.	Liberty	Developed at Cornell in 1978 and is a cross between (Macoun X Perdue 54-12). Fruit colour is deep red. It is a late maturing variety, Juicy, fine-textured, white flesh with good flavor. Resistant to scab, fireblight and mildew.	
28.	Lord Lambourne	Developed in England (1921) and Parents are (James Grieve apple x Worcester Pearsrmain) The apple shape is broad globose conical and a mid-season apple. It has a distinctive orange blush mixed with a greenish yellow background and taste is sharp sweet.	
29.	Mayan	It is a very low chilling, early maturing variety, with fruits medium in size. Fruits are globose to slightly conical in shape with stripped red coloured over green yellow ground, fruit weight (115-150g).	

Sl.No.	Variety	Description	Image
30.	Maharaji Apple (White Dotted Apple)	Maharaji apple is a large sized apple with bright red color on a green base. It has some conspicuous dots. Skin of the fruit is crisp, juicy, and aromatic yet taste is bit acidic. This variety ripens in late October and it remains fresh for a longer period of time and gets sweeter with time. It is used mostly in confectionary.	
31.	Michal	It very low chilling, early maturing variety. Fruits are medium in size slightly conical in shape, with smooth calyx end, stripped red coloured skin over green yellow ground.	
32.	McIntosh	Originated in Ontario, Canada 1811. A popular cold tolerant eating apple cultivar in North America. Medium size, mostly red, with green coloring where shaded. The tough skin of the McIntosh makes it a good shipper. McIntosh is best suited to higher elevations. Introduced in 1870, presumed to be a cross of (Fameuse and Detroit Red). Offspring include Cortland, Empire, Macoun, Spartan.	
33.	Mollies Delicious	Developed in New Jersey, US in 1966. Lasts long in refrigeration. Good aftertaste - Large fruit, slightly conical, full red color—but not a Red Delicious cultivar. Has a snappy, high-quality flesh. Tree is vigorous and productive, will re-bloom if hit by a frost. A very heat tolerant variety.	

Sl.No.	Variety	Description	Image
34.	Oregon spur	Originated in US, 1968. It is a bud sport of Delicious with large number of spur bearing branches, conic fruit shape, strong crowning at calyx end, red purple group, dark intensity of over colour, solid flesh, medium depth of stalk cavity and large width of stalk cavity aperture of locules is fully open.	
35.	Pink Lady	Introduced by the Western Australian Department of Agriculture in 1979. It is a cross between (Golden Delicious X Lady Williams). Fruit has attractive pink blush over yellow undertone, medium to large in size, asymmetric or oblong; ribbing, or bumpy skin; crisp with dense flesh; firm, cream colored flesh resists browning; sweet-tart flavor. It is a late maturing variety.	
36.	Prima	It is medium sized apple. The 'Prima' apple is one of the modern disease resistant cultivars of domesticated apple which was bred by the PRI disease resistant apple breeding program in 1958 in USA. It is very resistant to apple scab and has some resistance against the other common diseases. It has a juicy flesh with a balanced mild sub-acid flavour, a red flushed skin over yellow background. It doesn't fall off the tree, and like most early harvest apple it doesn't keep well, even with refrigeration.	

Sl.No.	Variety	Description	Image
37.	Red Baron	It is a cross between (Golden Delicious X Daniels Red Duchess) and was introduced in 1970. Medium-size red and yellow apple with juicy flesh and a mild sweet flavor. Good for fresh eating with a storage life of 4 to 5 weeks. Tree is hardy and resistant to fire blight. Ripens in mid-September.	
38.	Red Delicious	Iowa, US. 1870. Original seedling known as "Hawkeye." Rights bought by Stark Brothers in 1893 unmistakable for its acutely conic shape, dark red colour and telltale bumps on bottom. Flavour is sweet and mild, bordering on bland. Poor choice for cooking or cider. First marketed as "Delicious" or "Stark's Delicious," name changed to "Red Delicious" in 1914 when Stark bought the rights to Mullin's Yellow Seedling, changing that apple's name to "Yellow Delicious". Red Delicious has many sports and ranks as the world's most prolific apple. Delicious or Red Delicious Apple is a world-popular and most widely grown variety of apple in Sopore region in Kashmir. Size of the fruit may vary from medium to large. The fruit can be found abundantly in local market immediately after it ripens in mid-September.	

Sl.No.	Variety	Description	Image
39.	Red Fuji	Originated in USA. This is very productive and annual bearer and late maturing variety, fruit shape in general globose. These crisp, juicy and aromatic apples are quickly replacing Red Delicious in orchards. Fruit keeps upto 12 months when refrigerated. It ripens in late October and is best pollinizer for Granny Smith.	
40.	Red Gold	It is used as pollinizer, a regular and heavy bearer, matures in mid-season. Fruit is globose, slightly oblong flesh colour is yellowish and fruit weight is (145-155g). Fruits are juicy, crispy and attractive in appearance.	
41.	Red Spur	Originated in US 1959. It is bud sport of Red Delicious from complete tree variation, fruits resembles to Rich-a-Red. It is a dwarf variety produces an impressive harvest of full-sized, sweetly flavoured fruits, similar in characteristics to those of the 'Red Delicious' family. Bred in the chilly climes of East Holland, the Red Spur is fully cold-hardy and has a columnar, upright habit making. It is the perfect feature plant for gardens or patios with limited space. It is also self-fertile.	
42.	Red Velox	Originated in Italy. Mutant of Red Delicious, tree is of moderate vigor, fruit is of medium size, dark red in colour, with light stripes, colour development is full and 100 per cent, shape typically of red delicious. Particular characteristics of the variety, is the early and homogenous colour on the whole tree.	

Sl.No.	Variety	Description	Image
43.	Rich a Red	It is a Sport of Delicious, originated from Monitor, United States in 1915. Fruit colour is greenish, red flavor is rich and complex with hint of coconut. Shape is boated compared to delicious with wide middle. It bears heavily and is self-sterile.	
44.	Rome Beauty	Originated in Rome, Ohio, United States in 19th century. Rounded, deep red, and very glossy. Crisp, juicy white flesh is mild as a dessert apple, but develops an extraordinary depth and richness when cooked. It has good keeping quality.	
45.	Silver Spur	USA, 1977. It is spur type, precocious regular bearer, late maturing and high yielding cultivar.	
46.	Spartan	Originated in Canada in 1926 cross between (McIntosh × Newtown Pippin). It is Late-mid to late variety available in October and November. The colour is dark purple and develops a bloom which does not affect taste. The flesh is white with crisp texture and juicy.	

Sl.No.	Variety	Description	Image
47.	Skyline Supreme	It is a bud mutant of Starking Delicious, fruit shape is cylindrical waisted, fruit colour is red purple, fruit weight is (195-230 g).	
48.	Stark Earliest	Originated in US in 1938. Fruits have red striping on light background. One of the earliest eating apples with fruits often being ready for harvesting as early as July. A really good flavoured, crisp apple - and very welcome for Summer fruit salads. It is self-fertile cultivar	
49.	Shireen	Developed from Lord Lambourna× R-12740- 7A in 1996. Average fruit yield 50-60 kg tree^{-1}. Fruit small to medium in size, sweet, juicy having good flavour and resistance to scab.	
50.	Starkri-mson	It is bud sport of Red Delicious. The tree is compact in size, mid maturing, precocious, regular and heavy bearing variety. Fruit shape is cylindrical waisted, red purple group and fruit weight is (230- 240g).	

Sl.No.	Variety	Description	Image
51.	Sunhari	Developed at SKUAST–K Shalimar, Deep golden colour fruit, small to medium in size firm melting texture with pleasant flavor.	
52.	Super Chief	Originated in US as a Sport of delicious in 1988. Tree is a super-spur and stays compact even on semi-dwarf roots. It is a consistent, annual bearer even when not thinned aggressively. This strain starts out as a stripe and fills in to a solid red ten days ahead of its parents.	
53.	Tydeman's Early Worcester	Originated in England in 1929 and is a cross of McIntosh x Worcester Pearmain. Crimson over yellow background colour. It is an early season cultivar ripens a month before McIntosh. It is medium sized apple juicy and crisp. It partially self-fertile, tip bearer and resistant to scab.	
54.	Vance Delicious	It is bud mutant of a Delicious, regular bearing, matures in mid-season, fruit shape is conical, weakly defined flush, strongly defined strips, purple red colour, strong crowning at calyx end, medium depth of stalk cavity, large width of stalk cavity.	

Sl.No.	Variety	Description	Image
55.	Vista Bella	Developed by Rutgers University, in US 1944. It is very early season variety (90-95 days after full bloom). A medium sized, glossy, very dark red apple of rather indifferent eating quality (it has a tendency to have water core). Fruit flesh is white offering tart flavor with hints of berries and crisp crunch.	
56.	Wealthy apple	Originated in Minnesota, US 1860 and is a cross between (Cherry Crab X Sops of Wine). Fruit colour is greenish yellow striped with bright red. It is a late maturing variety, good tasting when freshly picked and well-suited for sauce and other home processing. Tree stays small and is a heavy bearer.	
57.	Yellow Trans-parent	Russian variety. Fruit is yellow in colour and is early maturing variety, commonly known as June Apple. It comes into bearing very early and yields immense crop. Fruit is medium to large with green skin turning, yellow, or nearly white when fully ripe. Flesh is light, fine-grained, juicy, rich, sub acidic.	

References

Bal, J. S. 2002. *Apple: In: Fruit Growing.* Kalyani Publishers, Ludhiana. pp. 422.

Chadha, K. L. and Awasthi, R. P. 2005. *The Apple: Improvement, production and post harvest management.* Malhotra Publishing House. pp. 182-201.

Elzebroek, A.T.G. and Wind, K. 2008. Guide to cultivated plants. Wallingford: CABInternational.p. 27. ISBN 1- 4593- 356-7.

Janick, J. 1974. The apple in Java. *Hort Science* 9:13-15.

Janick, J., Cummnis, J.N., Brown, S.H. and Hemmat, M. 1996. Apples. In:*Fruit Breeding* (Eds. Janick, J. and Moore, J.N.) Vol. 1, Tree and Tropical Fruits. John Wiley and Sons, NewYork, pp. 1-77.

Korban, S. S. and Skirvin, R. M. 1994. Nomenclature of the cultivated apple. *Hort. Science* 9:177-180.

Lawerence, R. W. 1895. *The Valley of Kashmir*. Henry Frowde Oxford University Press Warehouse Amen Corner, E.C. pp. 353-355.

Morgan, J. and A. Richards. 1993. *The book of apples*. Ebury Press, London.

Mratinic, E. and Aksic M. F. 2012. Phenotypic diversity of apple (*Malus* sp.) germplasm in South Serbia. *Brazilian Archives Of Biology and Technology* 55 (3): 349-358.

Verma, M. K. 2015. Apple Production Technology. In book: *Training manual on teaching of post-graduate courses in horticulture (Fruit Science)*, Edition: 1st, Publisher: Post Graduate School, Indian Agricultural Research Institute, New Delhi. Editors: S.K. Singh, A.D.Munshi, K.V. Prasad, A.K.Sureja, pp.241-248.

Way. R. D., H. S. Aldwinckle., R. C. Lamb., A. Rejman., S. Sansavini., T. Shen., R. Watkins., M. M. Westwood and Y. Yoshida. 1990. Apples (*Malus*). P.1-62. In: *Genetic resources of temperate fruit and nut* (Eds. J. N. Moore and J. R. Ballington Jr.). *Int. Soc. Hort. Sci.*, Wageningen.

Chapter 8

Recent Advances in Apple Production in India

Khalid Mushtaq Bhat, Haseeb Ur Rehman and A. H. Pandit

Division of Fruit Science,
Sher-e-Kashmir University of Agricultural Sciences
and Technology of Kashmir, Srinagar
E-mail: haseebpom@gmail.com

1. Introduction

Apple is the most important temperate fruit of the North Western Himalayan region. It is predominantly grown in Jammu and Kashmir, Himachal Pradesh, Uttarakhand, accounting for more than 90 per cent of total production in India. The apple growing areas in India do not fall in the temperate zone of the world but the prevailing temperate climate of the region is primarily due to snow covered Himalayan ranges and high altitude which helps meet the chilling requirement during winter season extending from mid December to mid March.

Apple trees are large if grown from seed. Generally apple varieties are propagated by grafting onto rootstocks, which control the size of the resulting tree. There are more than 7,500 known cultivars of apples, resulting in a range of desired characteristics. Different cultivars are bred for various tastes and uses, including cooking, eating raw and cider production (Morgan and Richards, 2002). Trees and fruit are prone to a number of fungal, bacterial and pest problems, which can be controlled by a number of organic and non-organic means. In 2010, the fruit's genome was sequenced as part of research on disease control and selective breeding in apple production (FAO.org, 2016).

Worldwide production of apples in 2013 was 80.8 million tonnes (FAOSTAT, 2015) with China accounting for 40 per cent of the total (Table 8.1).

Table 8.1: Apple Production (2014-15) in World

Country	Production (MT)
China	39.7
USA	4.1
Turkey	3.1
Poland	3.1
Italy	2.3
India	2.2
Total	80.8

In India, J&K leads in apple production (19.7 lac MT) on an area of 161773 hectares with a productivity of 12.16 t/ha (Anon., 2016).

Apples originated in the Middle East more than 4000 years ago. Spreading across Europe to France, the fruit arrived in England at around the time of the Norman Conquest in 1066. The first trees to produce sweet, flavourful apples similar to those we enjoy today, were located many thousands of years ago near the modern city of Alma-Ata, Kazakhstan (Walkins, 1995).

Most historians believe the apple originated in the Dzungarian Alps, a mountain range separating Kazakhstan, Kyrgyzstan, and China, where wild apple trees still produce teensy apples the size and shape of the seedy and sour ancestors of the world's favourite tree fruit. Others insist the wild apple arose in the Caucasus Mountains between the Black and Caspian Seas. Whichever is true, as early humans migrated to other lands, they carried apples with them until apples became established throughout all of Asia, the Mediterranean region and the Middle East (Harris *et al.*, 2002 and Brown and Maloney, 2003).

The center of diversity of the genus *Malus* is in eastern present-day Turkey. The apple tree was perhaps the earliest tree to be cultivated, and its fruits have been improved through selection over thousands of years. Alexander the Great is credited with finding dwarfed apples in Kazakhstan in 328 BC those he brought back to Macedonia might have been the progenitors of dwarfing rootstocks. Winter apples, picked in late autumn and stored just above freezing, have been an important food in Asia and Europe for millennia (Webster and Wertheim, 2003).

Apples were introduced to North America by colonists in the 17th century, and the first apple orchard on the North American continent was planted in Boston by Reverend William Blaxton in 1625. The only apples native to North America are crab apples, which were once called "common apples". Apple varieties brought as seed from Europe were spread along Native American trade routes, as well as being cultivated on colonial farms. An 1845 United States apples nursery catalogue sold 350 of the "best" varieties, showing the proliferation of new North American varieties by the early 19th century. In the 20th century, irrigation projects in Eastern Washington began and allowed the development of the multi billion-dollar fruit industry, of which the apple is the leading product (Westwood, 1993).

Until the 20th century, farmers stored apples in frost proof cellars during the winter for their own use or for sale. Improved transportation of fresh apples by train and road replaced the necessity for storage. In the 21st century, long-term storage again came into popularity, as "controlled atmosphere" facilities were used to keep apples fresh year-round. Controlled atmosphere facilities use high humidity, low oxygen, and controlled carbon dioxide levels to maintain fruit freshness.

Apple was introduced in India by the British in the Kullu Valley of the Himalayan State of H.P. as far back as 1865, while the coloured 'Delicious' cultivars of apple were introduced to Shimla hills of the same State in 1917. The apple cultivar 'Ambri', is considered to be indigenous to Kashmir and had been grown long before Western introductions.

2. Varieties

Over 700 accessions of apple, introduced from USA, Russia, U.K., Canada, Germany, Israel, Netherlands, Australia, Switzerland, Italy and Denmark have been tried and tested during the last 50 years. The delicious group of cultivars predominates the apple market. The areas covered under Delicious cultivars are: 83 per cent of the area under apple in H.P., 45 per cent in J&K and 30 per cent in U.P. hills. In more recent times improved spur types and standard colour mutants with 20-50 per cent higher yield potential are favoured. The important selections are:

☆ Spur types - Red spur, Starkrimson, Golden spur, Red Chief and Oregon spur.

☆ Colour mutants - Vance Delicious, Top Red, Skyline Supreme.

☆ Low chilling cultivars - Michal, Schlomit, Anna, Dorsett Golden, Tropic Sweet, Tropical Beauty, Winter Banana, Vared, Tamma

☆ Early cultivars - Benoni, Irish Peach, Early Shanburry, Fanny

☆ Juice making cultivars - Lord Lambourne, Granny Smith, Allington Pippin.

In H.P. monoculture of a few cultivars such as Royal Delicious, Red Delicious and Rich-a-red have started showing negative impact on the apple industry. Serious problems of apple scab disease and outbreak of premature leaf fall and infestation of red spider mite are causing great concern. U.P. Hills, particularly the Kumaon hills division, have the unique advantage of early harvest of apple, mainly due to cultivation of early maturing varieties like Early Shanburry, Fanny and Benoni. The early maturing varieties are harvested 2-3 weeks before the arrival of fresh apple from H.P. and J&K, and hence fetch very remunerative prices (Laurens; Sharma and Kumar, 1993 and Sharma *et al.,* 2006).

Apple varieties should have climate adaptability, attractive fruit size, shape, colour, good dessert quality, long shelf life, resistance to pest and diseases, tolerance to drought conditions besides high productivity. The recommended apple varieties in Jammu and Kashmir are given in Table 8.2.

Table 8.2: Recommended Varieties of Apple in J&K

Season		
Early Season	*Mid Season*	*Late Season*
Irish Peach, Benoni, Ginger Gold, Early Shanburry, Vista Bella, Mollies Delicious	American Mother, Red Gold, Rome Beauty, Cox's Orange Pippin, Queen's Apple, Scarlet Siberian, Razakwari, Jonathan, Gala strains, Starkrimson	Yellow Newton, American Apirouge, Lal Ambri, Kerry Pippin, Chamura, Golden Delicious, Baldwin, Ambri, White Dotted Red, Red Delicious, King Pippin, Ambri, Granny Smith, Braeburn

(a) Pollinizing Varieties

Getting reliable and good yields from apple trees generally requires that more than one variety be present for proper cross-pollination (Pratt, 1988; Sedgley 1990 and Jana, 2001). The most important feature of pollinizing variety is its flowering habit should synchronise with the main variety. In addition to this, it should have abundant viable pollen, long duration of flowering, compatibility with main variety, self fruitfulness, regular bearing besides good commercial value (Singh and Mishra, 2007). The Delicious group of varieties are self incompatible and cross pollinated, whereas most of the English varieties are self pollinated and act as suitable pollinizers for Delicious group of varieties in the proper proportion of 11-30 per cent in main variety plantation depending on the situation of the orchard. The most suitable pollinizer varieties are Tydeman's Early, Red Gold, Golden Delicious, McIntosh, Lord Lambourne, Winter Banana, Granny Smith, Starkspur Golden, Golden Spur (Chadha, 2001 and Brittain, 1933). A combination of early, mid and late flowering pollinizers assured cross pollination for the main variety (FAO.org, 2016). Some important crab varieties like Red Flush, Crimson Gold, Yellow Drop, Manchurian, Snowdrift, Golden Hornet and *Malus floribunda* also acts as good pollinizers. Red Free and Liberty can also be used as pollinizers (Table 8.3).

In J&K, Allington Pippin and Tydeman's Early Worcester are used as efficient pollinizers for early varieties. Similarly, Granny Smith, Wealthy and Cortland are used for late maturing varieties. For Delicious group, commonly used pollinizers are Golden Delicious, Lord Lambourne, Summerred, Summer Queen, Red Gold and Crabs (Manchurian, Snow Drift and Golden Hornet).

Table 8.3: List of Various Pollinizers along with their Bloom Timing and Fruit Characteristics

	Cultivar	Characteristics			
		Ripening Period	Size/Shape	Colour	Taste/uses/Etc
Early Bloom (A)	Gravenstein	Early	Large round-flattened	Orange-yellow, red stripes	Old fashioned pie/cider/cooking apple
	Idared	Very Late	Large	Bright Red	Crisp, mildy acidic, used a lot for cooking
	Yellow Transparent	Very Early	Medium-Large	Transparent pale-yellow	Sweet, good for applesauce, cooking, eating
	Lodi	Very early	Large	Green	Cooking apple. Sweet-tart flavor
Early-Mid Bloom (B)	Braeburn	Very Late	Medium-large	Yellow, orange-red blush	High Quality, early bearing
	Liberty	Late	Medium-large	Orange, red blush covered	Disease resistant! Good for any use.
	Red Jonathan	Late	Medium, round	Bright red	Good eater/cooker, nice looking fruit
	Red Mcintosh	Mid	Medium-large	Bright dark red	Tart, spicy. Good for cider, fresh eating.
Mid Bloom (C)	Gala	Mid	Medium, oval-round	Golden, heavy scarlet striping	Semi sweet, very hardy, popular
	Golden Delicious L	Late	Large, conical	Golden Yellow	Mild, sweet. High quality
	Golden Russet	Very late	Medium	Gray-green to golden bronze	classic cider apple - very sugary juice, highly flavored
	Pink Lady L	Late	Medium-large, oblong	Green-yellow, pink blush	Great keeper, fine-grained white flesh
	Red Delicious	Late	Large round-oblong	Red	Most popular. Mild taste, good fresh.
	Redchief L	Late	Large round-oblong	Dark Red	Most popular. Mild taste, good fresh. An improved Red Delicious
	Yellow Newtown (Pippin)	Late	Medium	Greenish-yellow	Rich, aromatic, crisp, coarse, creamy flesh
	Jonagold	Late	Large	Bright yellow, red stripes	Slightly tart, good for fresh use, good keeper

	Cultivar	Characteristics			
		Ripening Period	Size/Shape	Colour	Taste/uses/Etc
Mid-Late Bloom (D)	Fuji	Late	Medium	Yellow green skin	Crispy, juicy, white flesh
	Granny Smith	Late	Large	Grass Green	Sweet popular for baking
	Honeycrisp L Medium Very	Late	Medium	Mottled red over yellow	Crisp flesh, excellent eating and keeping qualities
	Red Fuji	Late	Medium round-oblong	80 per cent red with a distinct stripe	Sweet, good keeper with great texture
	Spartan	Very late	Medium	Deep red pure white flesh	Great for fresh eating, heavy bearing
Late Bloom (E)	Cox's Orange Pippen	Late	Medium, conical	Orange w/red stripes	Excellent dessert apple, yellow flesh
	Red Rome Beauty	Late	Large, round	Bright red	Unique tart flavor. Good for baking
	Criterion	late	Red Delicious type	Yellow, reddish-pink blush	Mildly sweet, heavy producer, excellent keeper
	Red Spy (Red Northern Spy)	Very late	Large round-flattened	Green-yellow w/pink stripes	Tart, aromatic, good all purpose apple

3. High Density Planting in Apple

From a historical perspective, a high density orchard is defined as any orchard with more than 150-180 trees per acre. However, many highly productive commercial orchards today have 150-180 trees per acre and higher density could be anything over 180 trees per acre. Besides having an increased number of trees per acre, a high density orchard must come into bearing within 2-3 years after planting. To achieve this early production, it is essential to use a precocious dwarfing rootstock. Although it is possible to restrict the growth of trees on semi-dwarf rootstocks, they do not have the genetic capacity for early bearing (Mika and Krawiec, 1999; Herrere, 2002 and Dart, 2008).

The successful management of apple trees in any high-density planting system depends on maintaining a balance between vegetative growth and fruiting. If vigour is too low, excessive fruiting results, fruit size declines, biennial bearing increases and trees fail to fill their allotted space soon enough to make the orchard profitable. If vegetative vigour is excessive then flowering and fruiting are reduced and containment of the tree to the allotted space becomes problematic. The successful balance of vegetative vigour and fruiting results in 'calm' trees that produce heavy annual crops and require only a light annual pruning. Pruning and crop load management are the primary management tools along with fertilization and irrigation that are used to achieve a balance between vegetative growth and cropping throughout the orchards life. These management variables are affected by planting density, tree quality and tree training strategies (Haak, 2006).

High density orchards require trees propagated on dwarfing rootstocks. Presently, only three commercially available rootstock groups or types can be recommended to develop a high density orchard system. Having small trees is not enough; the trees must bear fruit early in the life of the orchard. The rootstocks that are commercially available to fit this niche are M.9, Bud.9, and M.26. No perfect rootstock exists, and the limitations and strengths of each rootstock must be evaluated to select the rootstock that performs best in a specific situation. Table 8.4 lists the major advantages and disadvantages of these rootstocks. Trees propagated on these rootstocks for high density systems need to be supported and irrigation is strongly encouraged. The Mark rootstock is no longer recommended because of decreased availability and drought sensitivity. Mark performed well in the Southeast when irrigated during the growing season beginning at orchard establishment.

Although only a few rootstocks are available for high density planting systems at present, several promising selections are under evaluation. The breeding program at Cornell University's Geneva Research Station has several selections that are in the advanced stages of evaluation and will be available in limited quantities in the next several years. The primary benefit of the rootstocks from New York is resistance to fire blight. However, advanced evaluation is still needed before large scale planting is recommended.

The size controlling Malling (M) and Malling Merton (MM) series clonal rootstocks were introduced in late 1960's at Fruit Research Station, Shalimar, J&K. Recently huge varieties of rootstocks has been imported by government, non government and other private nurseries in recent years. Typical characteristics of some important clonal rootstocks commonly employed in high density apple cultivation (Table 8.4).

Table 8.4: Rootstocks in High Density Planting Systems in Apple

Clonal Rootstock Series	Types/Characteristics etc.
Malling (M) series	Developed by Selection at East Malling Research Station in England in 1913, Dwarfing : M-8, M-9, M-26, M-27, Semi-dwarfing : M-2, M-4, M-7, Semi-vigorous : M-13, Vigorous : M-12, M-16, M-7 and M-13 are tolerant to excessive soil moisture, Most suitable for use as dwarfing interstock is M-27, M-9 is sensitive to injury from very low winter temperature.
	Pajam 1, Pajam 2, Kl.29, Burgmer Series NAKB clones and Fleuron 56 are other subclones of M-9 developed in different countries.
Merton series	Out of 4 Merton rootstocks (Merton 778, 779, 789 and 793), Merton 793 has proved a very useful rootstock. It produces trees little smaller than M-16, is resistant to wooly aphid and collar rot and is adaptable to a wide range of soil and induces early fruiting.

Clonal Rootstock Series	Types/Characteristics etc.
Malling-Merton (MM) series	Developed by crossing Malling rootstocks with Northern Spy to evolve wooly aphid resistant rootstocks, Semi-dwarf : MM-106, Semi vigorous : MM-104, MM-111, Vigorous : MM-109, MM-111 is resistant to drought while MM-104 is tolerant to wet soils. MM-106 is the most widely used rootstock.
Polish (P) series	Developed in Poland, Cross between Antonovka x M-9, Five rootstocks P1, P2, P16, P18 and P22 have been found promising, All these rootstocks are winter hardy and resistant to crown rot but susceptible to fire blight, P18 is semi-dwarfing while all others are dwarfing, The P2 and P22 show particular promise as dwarfing interstock.
Budagovsky (Bud) series	Bud series of rootstocks introduced by the Michurin College of Horticulture, Michurinsk, Russia. These rootstocks *viz.* Bud-9, Bud-490 and Bud-491 are of promise.Bud-9 and Bud-491 are dwarfing, while Bud-490 is semi-dwarfing. Bud-490 and Bud-491 are very winter hardy.Bud-9 and B-490 are resistant to crown rot
Ottawa series	There are two Ottawa series of apple rootstock, the Ottawa Hybrid Seedlings (OH) and the Ottawa Clonal series (O). 6 OH series rootstocks (OH1 to OH6): resistant to latent viruses.14 Ottawa Clonal series (O 1-O14): More productive and hard.
MAC series	MAC 1: Trees on this rootstock are approximately M-7 in size but do not sucker and are well anchored. MAC 9 (Mark): Trees on this rootstock are similar in size to M-9 or slightly larger and equivalent in productivity. MAC 24: Trees on this rootstock are semi-vigorous and in the MM-111 class. It is well anchored.
Geneva series	The joint Cornell University and United States Department of Agriculture-Agricultural Research Service (USDAARS) Apple Rootstock Breeding and Evaluation Program develops new rootstock cultivars with an emphasis on productivity, yield efficiency, ease of nursery propagation, fire blight resistance, tolerance to extreme temperatures, resistance to the soil pathogens of the sub-temperate regions of the US, and tolerance to apple replant disorder. In many trials in North America and other worldwide locations all of the released GENEVA® rootstocks have demonstrated a "per acre productivity" and "tree yield efficiency" similar or higher than current commercial standards M.9 and M.26.General GENEVA® Apple Rootstocks are resistant to Fire blight o Crown and root rots (Phytophthora), Replant disease complex, Woolly apple aphid resistant. Other characteristics include all are dwarf types that differ within dwarf sizes and Cold hardiness. *E.g.* G.16, G.41, G.213, G.214, G.814, G.935, G. 222, G.202, G.969, G. 30, G.210, G.890 etc.
Miscellaneous rootstocks	**Bemali**: First introduction from Balsgard, Sweden. Bemali produces trees about the size of M-26, is highly productive and resistant to wooly aphid. **Jork 9 (J9)**: This is an open pollinated M-9 seedling released by thre Jork research Station, West Germany, that has the advantages of more easy propagation with equal productivity over M-9. **Alnarp 2**: Introduction from Alnarp Fruit Tree Station, Sweden, is winter hardy, vigorous and induce early bearing and productive. **Robusta 5**: Originated in Canada, is vigorous, winter hardy, resistant to fire blight and easy to propagate.

The growth habits and characteristics of various varieties can influence high density plantings (Kumar and Sharma, 1992). Spur type apple varieties are strains that have compact growth habits with more spur and less shoot formation than standard strains of the same variety. Spur types frequently give trees about three-fourths the size of nonspurs of the same variety when both are grown under the same conditions. Tree vigour and ultimate sizes of various nonspur apple varieties can also differ when grown on the same rootstock under the same conditions (Soejima *et al.*, 2000). Vigorous varieties, such as Gravenstein and Yellow Newtown, must be spaced farther apart than the moderately vigorous Red Delicious. The less vigorous Rome Beauty, Jonathan, and Golden Delicious can be spaced closer than Red Delicious. Honeycrisp, Delicious, Braeburn, Empire, Jonamac, Macoun, Idared, Gala, NY674, Golden Delicious, McIntosh, Spartan, Fuji, Jonagold, Mutsu, *etc.* and tip bearing varieties such as Cortland, Rome, Granny Smith and Gingergold are mostly planted with high density system. Fuji Zhen Aztec, Gala Redlumb, Super Chief, Golden Delicious Reinders, Super Chief, Red velox, Oregon Spur, Well Spur and Silver Spur and Granny Smith has been introduced and evaluated by SKUAST-K, Shalimar for high density planting (Plate 8.1). The salient characteristics of these varieties/clones are as:

(i) Fuji Zhen Aztec

It is mutated Fuji, strong growing, flowering mid to late, alternate bearing, mid-large size fruit, strong red, shape is like Fuji, harvested in middle October, good storability, tastes good, highly crispy.

(ii) Gala Redlum

Moderately vigorous plant, medium size fruit, harvesting beginning in August, stable like Gala, colour is intensive red (100 per cent), crispy, highly tasty (TSS is 13-14 per cent), good storability.

(iii) Granny Smith

Granny Smith apples originated in Australia in 1868 when Maria Ann (Granny) Smith found a seedling growing by a creek on her property. Green with a slight pink blush if cool nights precede harvest, firm, medium grain, bright white flesh that resists browning when sliced, firm with strong tartness resembling that of a lemon; this famously green apple is bound to make your mouth water. Granny Smith is one of several apple cultivars that are high in antioxidant activity.

(iv) Super Chief

Super Chief like Scarlet Spur also belongs to Red Delicious, but differs in colour type. Super Chief was discovered by Charles Ray Sandidge in Washington State of USA. It is a "sport" of Red Chief and gets earlier coloration than its parent by 18 days. Super Chief is sometimes resembled with Scarlet Spur, but the difference is in their colour type. Scarlet Spur has blushed red colour while Super Chief has a stripped colour pattern. Mutation of Red Chief, mid-early bloom, fruit size 65-85 mm, typical spur type tree. It also helps in pollination, growth is weak to very weak, less vigour than Red Chief, pollinated by Granny Smith, Gala, Golden Delicious and

Plate 8.1: New Promising Varieties of Apple Grown in India.

Fuji Zhen Aztec

Gala Redlum

Granny Smith

Super Chief

Red Velox

Golden Delicious Reinders

Oregon Spur

Silver Spur

Well Spur

Firdous

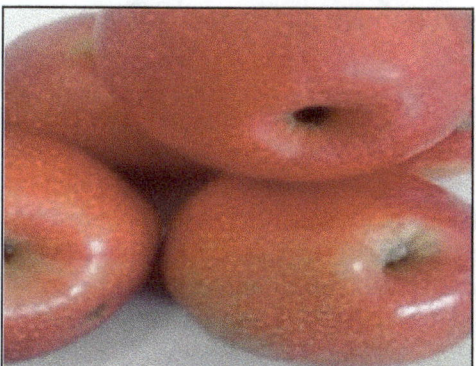

Coe Red Fuji

Shalimar 1

Shalimar 2

Shireen

Sunhari

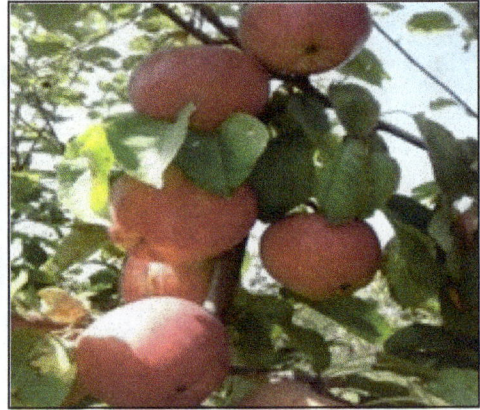

Lal ambri

Fuji, elongated fruit with prominent calyx lobes, intense red colour with stripes, fruit flesh is white, firm, crispy, juicy and good storage quality.

(v) Red Velox

Medium to large fruit size, dominant colour is dark red, covering colour percentage is 100, blushed, lightly striped covering colour, flattened conical shape, typical for Red Delicious, does not bruise easily, mutation of the Red Delicious Standard, variety owner is Griba Nursery, tree is of moderate growth, flowers a few days before Golden Delicious, yield is high and uniform, not susceptible to yield fluctuation, mid September, 1 week before Red Delicious Standard.

(vi) Golden Delicious Reinders

It is a clone of Golden Delicious. Its yield is high, regular and constant every year. It has an exceptional storage capacity. The fruit is of medium size and skin is yellow-green, turning more intensely yellow as the fruit ripens. The fruits are smooth with a nice, elongated shape. They are smoother than the Golden Delicious

Clone B and much more resistant to powdery mildew. The flesh is of medium firmness, very succulent with a wonderful taste and aroma.

(vii) Oregon Spur

It is also spur type developed by Trumbell in 1968 by mutation of Red King and is having stripped and dark coloured fruits.

(viii) Silver Spur

Spur type variety developed by Silvers in 1977 by mutation of McCormick and is having stripes with bright colour.

(ix) Well spur

Spur type (27 spur density), 42 cm^2 cross section area, 180 gm fruit weight, 27 kg per tree, 0.70 kg/cm^2 yield/TCSA.

4. Genetic Improvement of Apple in India

There is a considerable scope to bring about genetic improvement in this fruit crop and consequently selection of desired genotypes with good traits (Brown and Maloney, 2003 and Hancock *et al.,* 2008). Among the various apple cultivars grown in Kashmir, Ambri is indigenous to Kashmir and is known for its aroma, flavour and storability. But due to long gestation period, leathery texture of flesh, poor colour development and high susceptibility to scab this variety has fallen from favour and is not commercially grown now (A profile SKUAST-K and Banday, 2008). This cultivar is at the verge of its extinction and hence needs conservation. Already a lot of work has been done by SKUAST (K), Shalimar; YSPUH and F, Nauni (Mashobra), H.P; HETC, Raniketh, Uttarakhand; CITH Research Station, Mukteshwar, Uttarakhand and IARI Research Station, Shimla in this regard and the varieties released are discussed below (Table 8.5).

Table 8.5: Varieties Released by Various Institutes in India along with Important Characteristics

Variety	Characteristics etc.	Institute
Sunhari	Cross between Golden Delicious and Ambri, medium to large in size, flesh greenish white, sweet, juicy, matures in late September, yield potential 15-17 t/ha	SKUAST (K), Shalimar, J&K
Shireen	Cross of Lord Lambourne x Melba x R-12740-7A, Small to medium in size, dark red stripped, excellent taste, flesh crispy, matures in late August, yield potential is 15-17 t/ha.	SKUAST (K), Shalimar, J&K
Lal Ambri	Cross between Red Delicious and Ambri, red stripped fruit, white fleshed with firm texture, matures in late September, yield potential is 25-30 t/ha.	SKUAST (K), Shalimar, J&K
Firdous	Scab resistant, a cross of Golden delicious x Rome Beauty x malus floribunda, deep red fleshed fruit, yield potential of 12-15 t/ha.	SKUAST (K), Shalimar, J&K

Variety	Characteristics etc.	Institute
Akbar	Cross between Ambri and Cox's Orange Pippin, high yielding, medium to large size fruit, red stripped, flesh firm, crisp and whitish green, rich in juice, matures in late September, yield potential is 30-35 t/ha	SKUAST (K), Shalimar, J&K
Shalimar-1	Sunhari x Prima, developed and released in 2009, Early and Scab resistant, yields about 95 kg/tree (23.75 t ha⁻¹). The variety belongs to mid season group and has reddish pink, small to medium sized, crisp, juicy and sweet fruits.	SKUAST (K), Shalimar, J&K
Shalimar-2	Red Delicious x Ambri, developed and released in 2009, high yielding, average fruit yield of 106 kg/tree (26.50 t ha⁻¹) at 25 years on seedling rootstock. Moderately tolerant to scab and *Alternaria* leaf spot, the fruits are roundish, red mottled, juicy, crisp and sweet. The fruit has long shelf-life	SKUAST (K), Shalimar, J&K
Ambred	Red Delicious x Ambri, High yielding, good fruit quality	YSPUH and F, H.P
Ambstarking	Starking Delicious x Ambri, High yielding, good fruit quality	YSPUH and F, H.P
Amrich	Richard x Ambri, high yielding, good fruit quality, good fruit yield	YSPUH and F, H.P
Ambroya	Starking Delicious x Ambri, high yielding and good quality fruits	YSPH and F, H.P
Chaubattia Anupam	Early Shanburry x Red Delicious	HETC, Ranikhet, Uttarakhand
Chaubattia Princess	Early Shanburry x Red Delicious	HETC, Ranikhet, Uttarakhand
Swarnima	Benoni x Red Delicious	HETC, Ranikhet, Uttarakhand
CITH Lodh Apple 1	Bud sport of Red Delicious, high yielding, good fruit quality and improved over colour	CITH, Mukhteswar, Uttarakhand
Pusa Gold	Golden Delicious x Tydeman's Early Worcester, resistant to powdery mildew and apple scab	IARI,, Shimla
Pusa Amartara Pride	Resistant to powdery mildew and apple scab	IARI, Shimla

5. Pruning and Training Systems in High Density Apple

During the developmental years of an orchards life the trees have a juvenile character and the balance between vegetative growth and cropping is shifted toward vegetative growth. With high density systems such as the Tall Spindle the goal is to get the trees into cropping as soon as possible. This is best accomplished by minimizing pruning during the first 4 years. No heading cuts should be done to the leader or lateral branches at planting or for the next 4 years since the maximum growth and earliest cropping is achieved with no pruning. For the first 4 years pruning should be limited to the complete removal of unsuitable branches such as those lateral branches that are larger than 2/3 the diameter of the leader. Of much greater importance during the first 4 years is limb angle manipulation to change a young vigorous tree from a vegetative state into a reproductive state. With the Tall Spindle and Vertical Axis systems the artificial limb bending is limited to the 1st tier of branches while with the Solaxe even upper branches tied down. With the Tall Spindle and the Vertical Axis systems the bending of upper branches is achieved naturally by crop load (Robinson, 2001).

As the orchard reaches maturity, containment pruning of the canopy is essential to maintaining trees within the allotted space and to improve the light distribution to the lower portion of the tree. Good light distribution and good fruit quality can be maintained as trees age if the top of the tree is kept narrower than the bottom of the tree and if there is a good balance between vegetative growth and cropping. Pruning strategies based on shortening or stubbing back permanent branches that outgrow their allotted space generally are not as successful as limb renewal pruning strategies. This is partially because the most productive fruiting wood is cut off when a branch is shortened. In addition, stubbing cuts stimulate localized vigour on the affected branches which results in shading of the lower canopy. In our studies on how to manage the canopies of high-density systems, treatments where branches were shortened to maintain the conic shape to the tree, resulted in unacceptable yield reductions, a dense canopy resulting in interior canopy shading, excessive vigor compared to an unpruned control and poor fruit quality (Robinson *et al.,* 2013).

A more successful approach has been to annually remove 1-2 large upper branches completely and develop younger replacement branches. The removal of entire branches in the upper portion of high-density apple trees helps to open channels for light penetration which maintains fruit production and quality in the bottom of the canopy. This "limb renewal" pruning is the single most important pruning concept for mature high-density orchards to contain the canopy and maintain a conic tree shape. To assure the development of a replacement branch, the large branch should be removed with an angled or beveled cut so that a small stub of the lower portion of the branch remains. From this stub a flat weak replacement branch often grows. If these are left unheaded they will naturally bend down with crop. They are naturally shorter than the bottom branches thus maintaining the conic shape of the tree without stubbing cuts. This type of pruning does not stimulate vigorous regrowth. Our recommendation is to begin removing 1-2 whole limbs in the top of the tree once the tree is mature (year 6-7). This allows moderate pruning each year and a method to contain tree size. It also maintains good light distribution in the canopy without inducing excessive vigour. On trees with overgrown tops that need to be restructured, moderate renewal pruning (1-2 large upper branches annually) for a 4-5 year period can eliminate all of the large branches in the top of the tree (Robinson *et al.,* 2013).

Once branches have become horizontal or pendant under the weight of crop, they can be shortened by heading cuts without adverse effects since the terminal bud no longer exerts significant control over the branch. However, if the overall vigour of the mature tree remains high, leaving the pendant branches long will help increase cropping and reduce the vigour of the tree. After a number of years, if the pendant branches begin to shade the bottom half of the tree they should be removed with a renewal cut and a replacement branch developed. The natural bending of branches under the weight of fruit without heading can be used to great horticultural advantage in the tops of vigorous trees when it is desired to limit tree height. Often growers want to limit tree height by heading the leader in the top of the tree. If heading cuts are made on vertical shoots in the top of trees, vigorous regrowth results. If lateral shoots or limbs are manually bent horizontal or allowed to bend naturally under the weight of the crop they set heavy crops the next year. The crop will also act as a strong sink for resources, thereby, further reducing the vegetative vigour in the top of the tree. Once the top of the tree is fruitful and the leader has bent under the weight of the crop it can be shortened to a weak side branch without a vigorous response.

Many different training systems are being promoted for high density orchard management. There are four major systems currently in use, including the central leader, vertical axis, Hybrid Tree Cone (HYTEC), and the slender spindle (Robinson *et al.,* 2013). The major differences between the systems are tree height, density (spacing), and the way the leader is managed. Many trellis systems, such as the Ebro, Lincoln canopy, Y-trellis, vertical trellis, *etc.*, had been evaluated and standardized depending on location, rootstock employed and variety in question.

In apple commonly employed training systems include Bi-axis, Slender spindle, Super spindle/vertical axis, Tatura trellis or V system, Vertical Axis, Central leader (older)

(a) Central Leader

200-300 trees per acre. Usually used with semidwarfing rootstock. Leave only one trunk for the central leader. Remove branches with crotch angles less than 60 degrees. Remove all branches directly across from one another on the leader. Space lateral branches uniformly around the leader to prevent crowding as the limbs grow in diameter (Plate 8.2a).

(b) Slender Pyramid

Hybrid of central leader and vertical axis (limb renewal), 300-500 trees/acre, tree height 14-16'; Rootstocks: M.7, M.26, G.30, G.935, G.6210; Trees are not headed at planting; Bottom scaffolds are encouraged to fill in during first 2 year other lateral removed; Establish hierarchy of loose whorls around leader; Renewal pruning of higher tier laterals (Plate 8.2b).

(c) Vertical Axis

500-900 trees/acre, dwarfing rootstocks, narrow pyramid shape with dominant central leader, maximum height of about 12-14'. A vertical central leader (axis) is developed with relatively 'weak' fruiting branches arising around the leader. Tree density is between 1,000 to 2,500 trees/ha at a spacing of 4-5m x 1-2m and height can reach up to 3m. Maintaining apical dominance is important in the vertical axis system particularly during early stages of development to ensure weak fruiting branches - therefore no heading of the leader occurs. Branches are systematically renewed to prevent them from becoming permanent scaffolds. Support of two to three wire trellis is required. Similar to the spindle system, vertical axis systems are planted ideally using well feathered nursery trees (Plate 8.2c).

(d) Tall Spindle

1000-1500 trees/acre, dwarfing rootstock, highly feathered trees (10-15 feathers), early fruiting 2nd and 3rd leaf, No permanent wood. All scaffolds are renewed by complete removal when they become too big. Upper branches bent below horizontal to devigorate (Plate 8.2d).

(e) Spindle Bush

Spindle bush (or free spindle) are best suited to densities up to 2,000 trees/ha at 2-3m in height and 3-4m × 1-2m apart. In Europe, spindle systems are usually planted using well feathered two-year old nursery trees. At planting, a number of laterals are selected to form part of the permanent scaffolds in the bottom third of the tree. Competing laterals that develop at the end of the unpruned central leader have to be removed in a very early stage. As the leader grows, more scaffolds are selected and spaced equally. Leader dominance is important and if it is lost will result in a reduced tree canopy, whereas, if it becomes too strong lateral growth and development will be reduced. These systems can be free standing, however mostly utilise some form of support (either two to three wire trellis or individual supports) (Plate 8.2e).

(f) Super Spindle

The super spindle system is utilised for super high-density orchards on weaker rootstocks such as Quince C. Tree height is generally maintained at 2-3m. Generally there is a row spacing of <3m and a tree distance within the row of <0.8m, giving a density of more than 4,000 trees/ha. Some systems in Europe can reach 12,000 trees/ha when trained as super spindle (Plate 8.2k). This system has closely spaced compact trees with short fruiting wood or spurs evenly spaced along the central leader. Super spindles require a multi wire support system. The goal of this system is to achieve very high, early yields so new varieties can be introduced as quickly as possible to meet market demands. An additional benefit is low fruit production cost per hour of labour. Super spindle orchards usually use whips with a number of short feathers along the leader or even cheaper trees such as budded rootstocks (sleeping eye trees).

(g) Double Leader Systems

Double leader systems aim to achieve high leader densities whilst keeping tree numbers (and cost) down. Trees are planted at 3-4m × 1-1.2m spacing giving a tree density of around 3,000 trees/ha. However, the development of double

Plate 8.2a-k: Various Training Systems in Apple.

a. Central Leader System.

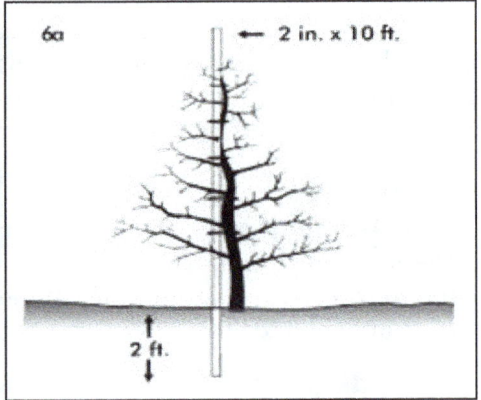

b. Slender Pyramid.

c. Vertical Axis.

d. Tall spindle.

e. Spindle Bush.

f. Double Leader.

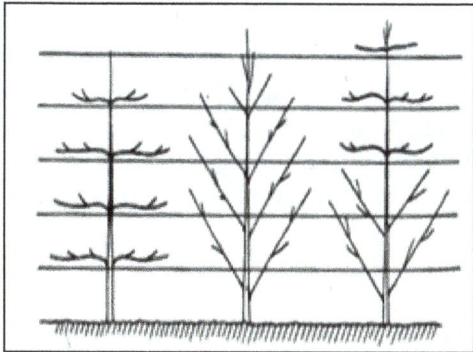

g. Palmette Systems.

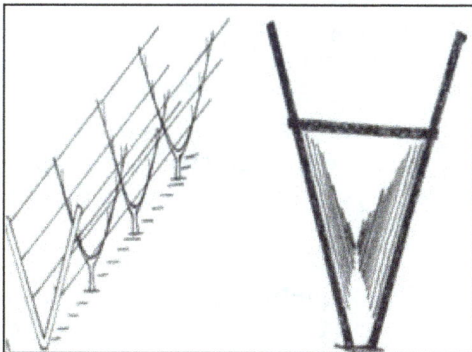

h. V/Y System.

j. Tatura Trellis.

i. V Hedge.

k. Super Spindle.

leaders mean that the leader density is 6,000 trees/ha. The Bibaum® system is a double leader system that was developed in Italy. This system involves planting Bibaum nursery trees, which are pre-formed with two axes in the nursery. Trees are planted at 3.3m × 1-1.25m spacing in a single row giving 3,000 trees/ha, with a leader density of 6,000/ha. Leaders are trained parallel to the row and are spaced approximately 50-60 cm apart. Pruning time was reduced significantly for the Bibaum system compared to both the spindle and candelabro (Plate 8.2f).

(h) Palmette

The palmette and its variations are generally limited to wide intra-row spacing's >2.0-2.5m with a tall tree which makes it best suited to planting densities of 700 to 1,500 trees/ha. There are a number of kinds of palmette training all with a central leader with scaffolds in the plane of the row only. Tiers of scaffolds are chosen each season and tied to wires to reduce vigour and promote spurring. These systems have been popular because the bending of branches on trellises controls growth and provides a balance of fruiting and vegetative growth (Plate 8.2g).

(i) V or Y Shaped Systems

Generally there are two basic shapes of canopies - Y shaped trees which have a vertical trunk and two opposing arms of the tree trained to either side of the trellis, and V shaped trees where the whole tree is leaned to one side of the trellis while the next tree in the row is leaned to the other side (Plate 8.2h).

(j) V Hedge

The V hedge system is widely used in the Netherlands and Belgium and is a variation of a Y shaped system. The planting distance in the V Hedge is 3.5 × 1.25 m which equals 2,057 trees/ha. These systems are planted using well feathered two year old nursery trees. Four feathers are kept as fruiting branches and considered as four central leaders on one stem. Tree height is maintained at 2 m with an opening of the V of 1.4 m. These systems can be planted more intensively, however light interception can be inhibited. There is no pruning at planting (Plate 8.2i).

(k) Open Tatura Trellis

The Open Tatura Trellis (OTT) system is a modification of the original Tatura Trellis (a Y shaped system), developed in the 1970s. In OTT there is a narrow strip of about ½ a meter that separates alternating diagonally planted trees within each row. OTT systems are planted 4-4.5m × 0.5-1m with 2,000 to 5,000 trees/ha. Trees can be trained in a number of different ways. Three of the most common are: single leader, double leaders (similar to a Bibaum® system) and, more recently, the cordon. Single leader OTT is similar to planting a slender spindle type system, with root systems about 0.5m apart and leaders 1m apart. Double leaders involve training two leaders on each tree (about 1m apart), to establish a high density of fruiting units at a lower tree cost. OTT with cordon allows for a moderately dense orchard of around 2,000 trees/ha with about 8,000 fruiting units growing up the wires. Nursery trees (usually whips) are bent over at planting and trained to the horizontal. Fruiting units are then encouraged at regular intervals along each cordon.OTT allows early production and high yields at maturity which means early returns on investment. It also reduces pruning and harvesting costs as the tree structure is simple and can be reached from the ground. However, establishment costs are often much higher than single row systems due to both the trellis construction and the early training of the trees (Plate 8.2j).

6. Conclusion

To improve apple productivity, breeders must develop varieties with multiple desirable traits. Development of new selections each year will require a well-coordinated evaluation plan using numerical data, molecular information and approaches, grower evaluations and consumer input for deciding the selection to be released. In order to identify new selections with high predictability of commercial utility as efficiently and quickly as possible, molecular tools and information contributing to marker assisted breeding and genomic selection approaches should be combined with classical breeding methods to maximize our likelihood of success and significantly increase the breeding efficiency as undesirable genotypes can be eliminated at very early seedling stage, particularly important for apple where in juvenile period is very long (Hemmet *et al.,* 1994). Apple breeding programs could be more efficient if more real scientific collaborations could be developed. Identified sources of various traits (fruit quality, storability and disease resistance) that have been identified at national/international centres need to be procured at an earliest by entering into MOU agreement with the co-operating centres. Apple breeding institutes must not only exchange bud, graft wood but also manage more efficiently the breeding programs and the diversity of sources of resistance. Latest advances in apple genetics offer opportunities for cultivar improvements as never before. High density planting employing use of spur type and new introductions on latest rootstocks systems resulting in higher productivity and overcoming biotic and abiotic stresses needs special attention. Best training systems specific to various varieties and agro-ecological conditions needs to be standardized to harness best results.

References

A Profile, Sher-e-Kashmir University of Agricultural Sciences and Technology, Shalimar. E-mail: skuast_k@rediffmail.com.

Anonymous, 2016. *District wise area and production of major horticultural crops in J&K state for the year 2015-16.* Directorate of Horticulture, Kashmir.

Banday, S. 2008.*The taste of Kashmir.* Greater Kashmir Nov 22, year 2008.

Brittain, W.H. 1933. Apple pollination studies in the Annapolis Valley, Nova Scotia. *Canadian Department of Agriculture Bulletin*, New Series 162: 1-198.

Brown, S.K. and Maloney, K.E. 2003. *Genetic improvement of apple: breeding, markers, mapping and biotechnology*, chapter Apples: botany, production and uses pp.31-59.

Brown, S.K. and Maloney, K.E. 2003. *The Genetic Improvement of Apple* Presented at the 46th Annual IDFTA Conference, February 17-19, 2003, Syracuse, New York.

Chadha, K.L. 2001. *Handbook of horticulture*. Apple chapter pp.119-131.

Dart, J. 2008. Intensive apple orchard systems. *District Horticulturist Intensive Industries Development,* NSW DPI Tumut Prime Fact 815.

FAO 2016. Book: *A manual of apple pollination* ISBN: 978-92-5-109171-5. Pp.11-21.

FAOSTAT. 2015. *Top production world* (available at: http://faostat.fao.org/ site/339/ default. aspx). Accessed 12 January 2017.

Haak, E. 2006. Growth intensity of apple-trees on clonal rootstocks before the beginning of fruit bearing. *Agronomijas Vçstis* (Latvian Journal of Agronomy), 9: 28-31.

Hancock, J.F., Luby, J.J., Brown, S.K. and Lobos, G.A. 2008. Apples. In: *Temperate fruit crop breeding: germplasm to genomics* (Ed. J.F. Hancock), pp. 1-37. Berlin, Germany, Springer Science and Business Media. 455 pp.

Harris, S.A., Robinson, J.P. and Juniper, B.E. 2002. Genetic clues to the origin of the apple. *Trends in Genetics* 18(8): 426-430.

Hemmat, M., Weeden, N.F., Manganaris, A.G. and Lawson, D.M. 1994. Molecular marker linkage map for apple. *J. of Heredity* 85:4-11.

Herrera, E. 2002. *Rootstocks for size control in apple trees.* Cooperative Extension Service College of Agriculture and Home Economics Guide H 307 pp. 1-4.

Jana, B.R. 2001. Effect of self and cross pollination on the fruit set behaviour of some promising apple genotypes. *J. of Appl. Hort.* 3(1): 51-52.

Kumar, J. and Sharma, R.L. 1992. Evaluation of some genotypes of apples under low altitude conditions of Kullu Valley of Himachal Pradesh. In: *Proc. Natl. Seminar on Emerging Trends in Temperate Fruit Production in India*, Sept. 4-5, 1992 at UHF, Solan, India. Abstr. No. 9: 6.

Laurens, F. 1999. Review of the current apple breeding programs in the world: Objectives for scion cultivar improvement. *Acta Hort.* 484:163-170.

Mika, A. and Krawiec, A. 1999. Planting density of apple trees as related to rootstock. In: *Proceedings of the International Seminar Apple Rootstocks for Intensive Orchards*. Warsaw-Ursynów, Poland. 77-78. p.

Morgan, J. and Richards, A. 2002. *The new book of apples.* London: Ebury Press.

Pratt, C. 1988. Apple flower and fruit: morphology and anatomy. *Horticultural Reviews* 10: 73-308.

Robinson, J. 2001. Taxonomy of the genus Malus Mill. (Rosaceae) with emphasis on the cultivated apple, *Malus domestica* Borkh. *Plant Syst. Evol.* 226: 35-38.

Robinson, T.L., Hoying, S., Sazo, M.M., DeMarree, A. and Dominguez, L. 2013. A vision for apple orchard systems of the future. *New York Fruit* 21(3): 11-16.

Sedgley, M. 1990. Flowering of deciduous perennial fruit crops. *Hort. Rev.* 12: 223-264.

Sharma, G., Anand, R. and Sharma, O.C. 2006. Floral biology and effect of pollination in apple (*Malus × domestica*). *Indian J. Agril. Sciences* 75(10): 667-669.

Sharma, R.L. and Kumar, K. 1993. *Temperate fruit crop improvement in India*. Chapter in Progress in Temperate Fruit Breeding 1: 149-156.

Singh, V.P. and Misra, K.K. 2007. Pollination management in apple for sustainable production. *Progressive Horticulture* 39(2): 139-148.

Soejima, J., Abe, K., Kotoda, N. and Kato, H. 2000. Recent progress of apple breeding at the Apple Research Center in Morioka. *Acta Hort.* 538:211-214.

Watkins R, 1995. Apple and pear. In: *Evolution of Crop Plants* (Eds. Smartt, J. and Simmonds, N.W.), pp. 418-422, Longman.

Webster, A.D. and Wertheim, S.J. 2003. Apple Rootstocks. In: *Apples Botany, Production and Uses* (Eds. Ferree, D.C. and Warrington, I.J.). CABI Publishing. 91-124. p.

Westwood M N, 1993. *Temperate-zone pomology, physiology and culture.* Timber press, Portland, Oregon, USA. 523. p.

Chapter 9

Research Developments for Increasing Productivity and Quality of Apple at ICAR-CITH, Srinagar

D. B. Singh, J. I. Mir, O. C. Sharma, S. Lal, A. Sharma, W. H. Raja and K. L. Kumawat

ICAR-Central Institute of Temperate Horticulture,
Old Air Field, Rangreth, Srinagar - 7
E-mail: javidiqbal1234@gmail.com

1. Introduction

Apple (*Malus domestica* Borkh), king of temperate fruits belongs to family Rosaceae. Among the total fruit production, apple alone covering 4.2 per cent of the total area and 2.4 per cent of total fruit production. Though there has been manifold increase in area, production and productivity, but the productivity has increased only 85-90 per cent over 1960-61 to 2013-14 in India. Apple cultivation is mainly confined to Jammu and Kashmir, Himachal Pradesh, and Uttarakhand and partly in Arunachal Pradesh, Sikkim and Nagaland. The average productivity of Jammu and Kashmir is 10.2 t/ha, followed by Himachal Pradesh 6.9 t/ha and Uttarakhand 2.6 t/ha (2013-14) (NHB; 2014). The Jammu and Kashmir has 170.6 thousand hectare area of apple and 1775.0 thousand metric tonnes which constitutes 66.0 per cent of the total area and 71 per cent of total apple fruit production. The China is the world highest producer contributing 2060 thousand hectares of area and 37000 Metric tonnes of production (FAO, 2015).The world highest productivity is 53.84t/ha in Austria which is too high as compared to India (8.0 t/ha). Further, there is great potential to reach the productivity upto 30-40 t/ha with the use of improved varieties and scientific production and protection technologies suitable to the region. Apple can be grown at an altitudes 1,500-2,700 m. above m.s.l. in the Himalayan range which experience 1,000-1,500 hours of chilling (the no. of hours during which temperature remains at or below 7° C during the winter season). The temperature during the growing season remains around 21-24° C. For optimum growth and fruiting, apple trees need 100-125 cm of annual rainfall, evenly distributed during the growing season. Excessive rains and fog near the fruit maturity period results in poor fruit quality with improper colour development and occurrance of fungal spots on its surface. Areas exposed to high velocity of winds are not desirable for apple cultivation. Loamy soils, rich in organic matter with pH 5.5 to 6.5 and having proper drainage and aeration are suitable for apple cultivation. In India major apple growing areas in different states are Srinagar, Budgam, Pulwama, Anatnag, Baramullah, Kupwara districts in Jammu and Kashmir; Shimla, Kullu, Sirmour, Mandi, Chamba, Kinnaur districts in Himachal Pradesh; Almora, Pithoragarh, Tehri Garhwal, Uttarkashi, Chamoli, Dehradun, Nainital districts in Uttaranchal and Tawang, West Kanneng, Lower Subansiri districts in Arunachal Pradesh. Most of the orchards established in these states are on traditional orcharding models. Traditional orcharding system in apple give poor yield due to low density plantations and also the quality of fruit is poor due to improper canopy management. There has been a steady increase in planting density over the last few decades from 625 trees/ha at 4m x 4m spacing to 2222 trees/ha at 3m x 1.5m spacing. The high density plantation system comprising of high yielding apple varieties, dwarfing clonal rootstocks, canopy management systems, micro-irrigation *etc.* for enhancing yield and improving quality of the fruit. Today we have a wide range of rootstock choices that will produce trees of varying sizes, from very large to true dwarfs (less than 10 feet tall at maturity). In addition, the various rootstock choices vary in their winter hardiness, resistance to certain insects and diseases, and performance in the various soil drainage types. Most of these dwarf rootstocks also bear fruit early in the tree's life.

2. Rootstock

(a) Clonal Rootstocks

Seedling (standard) rootstock has been the most commonly used in traditional orchards. This vigorous rootstock often provides a tree that occupies huge space and attains large height. Trees on this rootstock often take 6 to 10 years to begin significant bearing. In contrast, the promising apple rootstocks offer varying degrees of size control and frequently induce early bearing (precocity). Rootstock MM 111 usually produces a tree about 80 percent the size of that on seedling. This stock is quite tolerant to varying soil and climatic conditions, particularly droughty soils, and may be very useful where summer irrigation cannot be provided. M 111 is reported to be resistant to woolly apple aphid. Rootstock MM 106 produces a tree 65 to 75 percent the size of that on seedling and provides more precocity than MM-111 apple rootstocks. It also provides good anchorage, is easily propagated, and is resistant to woolly apple aphid. M 7 produces about a half-size tree and is tolerant to high arsenic levels in the soil but it produces large number of suckers and need high degree of maintenance. M 26, one of the newer clonal apple rootstocks, provides a tree 40 to 50 percent of the size of that on seedling and is quite precocious. Individual trees on this stock may require support under some conditions. M 9 is more dwarfing than M 26, giving a tree about one-third the size of that on seedling. It has a brittle root system with poor anchorage and usually requires support by stakes or trellis. It is very precocious. M 27 and P22 are ultra-dwarfing rootstocks and are generally used for meadow orcharding system which is not been practiced in India at present. Bud-9 is precocious and cold tolerant rootstock and thus suitable to high chill areas. CITH is maintaining and multiplying following apple clonal rootstocks and also evaluating the apple varieties on these rootstocks.

☆ **M111** - Semi-dwarf rootstock. Usually produces a tree 80 per cent the size of the same tree on seedling rootstock. Tolerates many soil conditions. Reported resistant to woolly apple aphid. Imparts earlier bearing fruit than seedling, not as early as more dwarfing stock. Requires irrigation.

☆ **M106** - Semi-dwarf rootstock. Usually produces a tree about 65-75 per cent the size of the same tree on seedling rootstock. Provides good anchorage. Imparts early bearing fruit and is easily propagated. Reported resistant to woolly apple aphid. Requires irrigation. Tree spacing ranges from 10 × 18 ft to 6 × 12 ft.

☆ **M7a** - Semi-dwarf rootstock. Usually produces a tree about 60 per cent the size of the same tree on seedling rootstock. Performs well in irrigated replant situations, but tends to sucker. Spacing is same as M106.

☆ **M9** - Dwarfing rootstock. Usually produces a very small tree less than 30 per cent the size of the same tree on seedling rootstock. Commercially, the most frequently planted rootstock worldwide. However, a poor performer if not adequately managed. Poorly anchored, has brittle root system. Must be trellised.

☆ **Bud 9**: This rootstock is productive, very precocious and when mature, stands seven to eight feet tall. It should be staked to ensure support for heavy crop loads.

(b) High Yielding Varieties

There are more than 10,000 apple varieties which are grown throughout the world both in the northern and southern Hemisphere but four apple groups *i.e.* Fuji, Golden Delicious, Red Delicious and Gala dominates the world apple scenario. Over 700 accessions of apple, introduced from USA, Russia, U.K., Canada, Germany,

Israel, Netherlands, Australia, Switzerland, Italy and Denmark have been tried and tested during the last 50 years. The delicious group of cultivars predominates the apple market. The areas covered under Delicious cultivars are: 83 per cent of the area under apple in H.P., 45 per cent in J&K and 30 per cent in U.P. hills. In more recent times improved spur types and standard color mutants with 20-50 per cent higher yield potential are favoured. The important selections which have been evaluated and characterized by CITH are:

- ☆ **Spur types** – Red spur, Starkrimson, Golden spur, Red Chief, Well Spur, Oregon spur *etc.*
- ☆ **Color mutants** – Vance Delicious, Top Red, Skyline Supreme *etc.*
- ☆ **Low chilling cultivars** – Michal, Schlomit, Mayan, Anna *etc.*
- ☆ **Early cultivars** – Benoni, Irish Peach, Early Shanburry, Fanny *etc.*
- ☆ **Juice making cultivars** – Lord Lambourne, Granny Smith *etc.*
- ☆ **Scab resistant cultivars** – Florina, Liberty, Prima, Firdous, Shireen, Shalimar-1 *etc.*

New Hybrids – Lal Ambri (Red Delicious × Ambri), Sunehari (Ambri × Golden Delicious); Akbar (Ambri × Cox's Orange Pippin),Shireen (Lord Lambourne × Melba × R 12740-7A), Firdous (Golden Delicious × Prima × Rome beauty), Shalimar apple-1 (Sunehari × Prima), Amred (Red Delicious × Ambri), Chaubatia Anupam and Chaubatia Princess (Early Shanburry x Red Delicious) are also being evaluated for different traits at ICAR-CITH, Srinagar. Planting material of these varieties is being supplied to farmers through different schemes. The growth habits and characteristics of various varieties can also influence high density plantings. Spur type apple varieties are strains that have compact growth habits with more spur and less shoot formation than standard strains of the same variety. Varieties like Oregon Spur, Silver Spur, Red Chief, Well Spur, Coe-Red Fuji, Gala Mast, Starkrimson, Super Chief, Red Chief, Red Velox, Spartan, Granny Smith *etc.* have been found to show improvement in yield and quality under HDP system. Following are some standard apple varieties which are showing promise with respect to yield and quality under HDP.

- ☆ **Fuji** - Round to flat apple with a very sweet yellow-orange flesh. Skin color is red if given enough sunlight and cool temperatures. One of the best sweet eating apples. Stores well.
- ☆ **Gala** - Small to medium-sized, conic-shaped red apple with excellent flavor and keeping qualities. The best variety for the early season. Will not cross-pollinate 'Golden Delicious'.
- ☆ **Golden Delicious** - Conic-shaped apple with a long stem, yellow to green skin, yellow flesh, and russet dots. Sweet, juicy, fine-textured. #1 on the North Coast for fresh eating quality and processing. Stores well but susceptible to bitter pit, bruising, russeting. Erratic in self-fruitfulness.

☆ **Granny Smith** - Round, green to yellow-skinned apple that is quite firm. Keeps very well. Crisp flesh. If harvested early, it is green and tart. Late harvested fruit are yellow-colored and sweet.

☆ **Gravenstein** - Medium large fruit with short, fat stem. Skin color is greenish yellow over laid with red stripes. Excellent flavor when fully ripe. Crisp, subacid, and aromatic. A good sauce and pie apple. Stores and ships poorly. High percentage of wind falls. Sterile pollen.

☆ **Jonathan** - Round, red apple with pure white flesh. Crisp, juicy, and slightly subacid. Excellent for eating fresh, sauce, and juice. Highly susceptible to mildew, fire blight, and Jonathan spot.

☆ **Red Delicious** - Conic-shaped apple with tapered base and five distinct lobes. Skin color varies from solid red to a mixture of red and green stripes. Crisp, sweet, mild-flavored yellow flesh. Many strains. Used fresh. Stores well.

☆ **Rome Beauty** - Round fruit with a deep cavity, no lobes, and little russet. Several strains, including the old standard and several new, solid red-skinned strains, such as 'Taylor' and 'Law'. Stores moderately well. Tree leafs out late, flowers late, and produces flowers and fruit on long spur growth that requires modification in pruning. Good for baking.

Other important varieties include Spartan, Jonica, Prima, Mollies Delicious, Silver Spur, Starkrimson, Oregon Spur, Breaburn, Cameo, Pink Lady, York, Jonathan, Ginger Gold, Honey Crisp, Rome, Empire *etc.* which are having excellent fruit quality parameters and are suitable for high density plantation. These varieties need to be evaluated under temperate conditions of J&K to check their performance under different canopy management systems. The demand for quality fruit and size is

| Oregon spur | Miller Sturdy spur | Red Chief | Tydesman's Early Worcester |
| Super Chief spur | Washington | Silver spur | Coe Red Fuji |

high, consumer acceptability depends on juiciness, color and size of fruit, therefore, import of these new generation apple varieties is the need of an hour so that high quality apple can be produced for better revenue and export. These varieties can be utilized in breeding programmes for transfer of traits into indigenous varieties which is more adopted to local conditions.

(c) Early Summer Varieties

These varieties do not have the quality characteristics of standard varieties but ripen early when no other fresh apples are available. They are excellent for eating fresh right off the tree and make a good cooking apple.

- ☆ **Vista Bell** - terminal bearing habit, white-fleshed fruit, stores well
- ☆ **Benoni**- terminal bearing habit, stores well
- ☆ **Mollies Delicious**- terminal bearing habit, high juice content, stores well
- ☆ **Jerseymac** - large, good red color, excellent flavor, firmer than McIntosh, stores 4-8 week
- ☆ **Paulared** - high quality, white flesh, stores fairly well, tree requires thinning
- ☆ **Akane** - similar to Jonathan but earlier, good solid red color, white flesh, good for eating fresh and juice
- ☆ **Jonamac** - similar to McIntosh but has better color, firmness, and storage life

(d) Disease Resistant Varieties

There are several scab resistant varieties developed by different institutions working on apple breeding programmes

- ☆ **Enterprise:** A large fruited, late maturing, dense, crisp variety that has good keeping qualities. The color is dark red over a yellow green background. This is one of the best of the scab resistant varieties.
- ☆ **Florina:** A promising scab resistant selection from France, this variety has large, round-oblong, purple-red colored fruit. It ripens late and has a mixed sweet tart flavor.
- ☆ **Freedom:** Is a late season variety with large fruit and mild flavor; not completely immune to scab.
- ☆ **Goldrush:** A scab immune selection with Golden Delicious parentage, this fruit is late maturing, large, firm textured and tart with an excellent flavor. It stores well.
- ☆ **Pristine:** This moderate to large tart yellow apple is immune to scab and resistant to fire blight and mildew.
- ☆ **Jonafree:** A mid-season apple compares with Jonathan, with soft flesh and uneven coloring.

☆ **Liberty:** One of the best quality apples of the disease resistant varieties, Liberty is very productive and requires heavy early thinning to achieve good size. It ripens in mid-season, has an attractive red color with some striping and a good sweet flavor.

☆ **Prima:** Is an early season, uneven ripening, moderate quality variety.

☆ **Priscilla:** Is a late season variety with small fruit, soft flesh, and mild flavor.

☆ **Red Free:** Is early July maturing, heat sensitive, a small-fruited variety that is susceptible to water core, sunburn and russet.

☆ **Williams Pride:** An early maturing, scab immune variety that is also resistant to fire blight and mildew. The fruit is medium to large with a round-oblique shape. It has an attractive red striped color on a green-yellow background.

☆ **Shireen:** Developed by SKUAST (K), Srinagar, India. This is moderately resistant to scab.

☆ **Firdous**: Developed by SKUAST (K), Srinagar, India. This is moderately resistant to scab.

3. Canopy Management and HDP

Trees in high density apple orchards usually require a different training system than those in standard orchards. Modern canopy management systems like tall spindle, espalier, cordon, single axis, spindle bush, head and spread, two scaffold *etc.* help in accommodating more plants per unit area and enhance yield and quality of fruits by providing more fruiting area and improving the penetration and diffusion of photosynthetically active radiation and PPFD. Some rootstocks like M-9 have brittle root system and thus need support or trellis. The wire trellis aids in tree support, training limbs to the desired angle, and limb support during early bearing. The high cost of materials and labor may preclude many trellis systems. High density orcharding system can be successfully planned by adopting the use of dwarfing rootstock, spur type apple varieties, chemical growth regulators, training system *etc.* Replacement of traditional orcharding system with high density orcharding has been initiated in the apple growing states of India with dwarfing clonal rootstocks like MM-111, MM-106, M-9, M-27 etc and high yielding varieties like Oregon Spur, Starkrimson, Fuji, Gala *etc.* ICAR-Central Institute of Temperate Horticulture, Srinagar has developed the technologies for HDP in apple by identifying suitable rootstocks, varieties, standardized suitable canopy architectural engineering systems, optimizing nutrient and water scheduling *etc.* Institute has developed HDP models on MM-111, MM-106 and M-9 rootstocks at 4m × 4m, 2.5m × 2.5m, 3.0m×1.5m, 3.0m×2.5m, 3.5m × 3.5m *etc.*spacing. Suitable varieties for HDP have been identified like Mollies Delicious, Gala Mast, Coe-Red Fuji, Spartan, Granny Smith, Oregon Spur, Silver Spur, Red Chief, Well Spur, Top Red *etc.* Different canopy management systems suitable for HDP in apple have been evaluated for increasing yield and quality of apple *viz.* spindle bush system, head and spread system, cordon system, vertical axis system and espalier system. Espalier system adopted for varieties like Granny Smith, Spartan, Coe Red Fuji grafted on M-9 rootstock planted at 3.m×1.5m spacing accommodating 2222 plants/ha gave about 60t/ha yield with highest fruit quality index due to high level of PPFD/PAR penetration and diffusion through the canopy. ICAR-CITH, Srinagar is producing and distributing about 20000-30000 grafted apple varieties suitable for HDP from last few decades and therefore, covering an area of about ten hectares every year. HDP of apple has been practiced through demonstrations, network programmes, MGMG, TSP *etc.* in the areas like Gurez, Ganderbal, Srinagar, Shopian, Baramulla, Sopore, Pulwama, Budgam *etc.* in Kashmir valley and also in some parts of Uttarakhand, Arunachal Pradesh, *etc.* Complete HDP package developed by the Institute is being transferred to farmer's field for obtaining higher productivity, quality and returns.

4. Espalier System for HDP in Apple

The High density orcharding system has been standardized for doubling productivity through planting of spur type cultivars *viz.,* Oregon Spur, Red Chief, Red Fuji and Silver Spur in combination with Gold Spur and Golden Delicious as pollinizer (33 per cent) and 5-6 honey bee colonies per hectare. Trees start bearing after 2nd year of planting. In the 8th year, maximum yield ranges from 30-40t/ha. Through this technology yield could be increased to 50-60t/ha by 12th year with benefit cost ratio of 4.10 indicating a highly remunerative enterprise.

5. Development of Micropropagation Protocol

Micro-propagation of woody and semi-woody trees is reported as problematic and to obtain a material without viruses is a long-term, complex process using traditional methods. Many *in vitro* experiments have been conducted for these economically important fruit trees to overcome the problems of virus and large scale planting material production (Sedlak and Paprstein, 2008; Adýyaman *et al.,* 2004; Mir *et al.,* 2010). Apple is an important economic fruit crop widely cultivated in temperate and sub-tropical climate. It belongs to the rose family (Rosaceae) of order

High Density Apple (3m × 3m).

Rosales and class Agnoliopsida. Trees on MM106 are well anchored, do not sucker, are semi dwarfing (60-75 per cent the size of trees on apple seedlings), and very productive. MM106 has many attributes, *i.e.* good induction of cropping, resistance to woolly apple aphid and intermediate vigor. Apple is conventionally propagated by vegetative methods, such as budding or grafting. These traditional propagation methods do not ensure disease-free and healthy plants, they depend on the season; moreover, they typically result in low multiplication rates. Further, if scion wood is taken from healthy disease free mother plants but virus infected rootstocks may lead to disease development. Thus there is an immediate need for development of virus free rootstocks. Micro propagation of apple rootstocks has opened up new areas of research and enabling rapid multiplication of disease-free fruit plants at a commercial scale. Tissue culture has been extensively used for raising multiple clones of apple rootstocks and raising of virus-free planting material in apple (Dobra and Teixeira, 2010). Apple is the major fruit crop of the world and faces many biotic stresses majorly viral diseases. CITH developed a protocol for micropropagation of apple rootstock MM-106 using meristem as an explant. Different phytohormone combinations were tried to optimize the best combination for development of fast and efficient micropropagation protocol. Meristem was cultured on MS medium supplemented with different concentrations of BAP alone and in combination with IBA, phloroglucinol and GA_3 for shoot formation, multiplication and rooting. Best initial establishment was done on MS media supplemented with BAP (2 mg/l) + IBA (0.5 mg/l) + Phloroglucinol (100mg/l). Maximum number of shoots (13.0), length of shoots (9.67 cm), leaf length (3.47cm) and leaf number (15.3) was observed on MS media supplemented with BAP (1mg/l) + GA_3 (0.5 mg/l). Maximum number

(A) *In vitro* Shoot Establishment, (B) Shoot Elongation, (C) Shoot Multiplication, (D) Rooting, (E) Hardening and (F) Transfer into Soil.

(16.3) and length of roots (10.20 cm) was observed on MS media containing IBA (3 mg/l). Acclimatization of rooted plants was observed on moist cotton for 10 -12 days followed by vermiculite for four weeks. About 70 per cent of plants survive after 6 months of transfer to the soil under polyhouse conditions.

6. Rejuvenation of Old and Senile Orchards

CITH developed a technology for rejuvenation of old apple orchards has been recommended for three apple varieties *viz.,* Red Delicious, Early Shanburry and Buckingham involving severe pruning, balanced dose of fertilizers and combination of insecticides and fungicides. The pruning in February and application of Chaubattia paste on the cut pruned surface, Caustic soda (1.0 per cent) spray during dormancy for control of lichens. Sevin (0.1 per cent) spray once in growing season (May) and fungicides (Carbendazim as well as Bayleton) spray twice in the month of May and June. Fertilizers N_2: P_2O_5: K_2O @ 450: 450: 450: g per tree, before fall is recommended. Senile and unproductive orchard could be made productive and more remunerative.

| Old and senile apple tree | Rejuvenated apple tree |

7. Pollination Management

Poor yield in apple is attributed to faulty pollination management with inadequate pollinizer and pollinating insects. They have to be suitably augmented in desired proportion to increase fruit set and yield. Bloom synchronization among the main varieties and pollinizers is essential with fruits of the pollinizers should preferably have commercial value as dessert, processing or ornamental. The pollinizers should be self-fruitful diploid or reciprocally cross compatible, have high bloom density, extended flowering period and should not be susceptible to any diseases or insects.

On the basis of blooming period different pollinizers as indicated below have been identified which can be used for effective pollination.

  a) *Early Bloomer:* McIntosh, Tydeman's Early Worcester, *Malus floribunda.*
  b) *Mid Bloomer:* Yellow Newton, Snow Drift, Red Gold, Gala, Spartan, Yellow Transparent, Jonathan
  c) *Late Bloomer:* Golden Delicious, Golden Spur, Rome Beauty, Granny Smith

a) Crab Apples as Pollinizers

Crab apples are regular in flowering with high bloom index. Most crab apples bear flowers on spurs as well as on one year shoots and have a long flowering duration due to blooming first on spurs followed by flowering on shoots. Trees can be easily trained on pillar shape and tree volume can be regulated by pruning of current year shoots. Crab apples can also be planted as filler tree without interfering main variety spacing. In India Manchurian crab, Snowdrift, Golden Hornet and Japanese crab have been recommended as pollinizers. Blossom colour of crab apples is important as honey bees or bumble bees if become habitual for foraging on pure pink, red or purple blossoms shows a foraging tendency only on trees with such coloured flowers thereby avoid white blooms on main varieties. White or white with slight pinkish ting blooming crabs are preferred as good pollinizers, as bloom colour of main apple cultivars is in this range.

Besides pollinizers, placement of honey bee hives in orchards as pollinators not only increases pollination, fruit set and fruit yield but also provides additional income through honey. For orchards with <15 per cent pollinizers, 8 hives; orchards with >30 per cent pollinizers, 2-3 hives and for high density orchards, because of more plant density per unit and higher bloom density of spur type cultivars 5-8 hives are recommended.

8. Propagation Techniques

Apple do not produce true to type plant if raised by seed hence, vegetative propagation technique has been found only option. Vegetative propagation employs budding and grafting on clonal as well as seedling rootstock. In India, seedling rootstocks are commonly being used, however, with the introduction of clonal rootstocks farmers are establishing the new orchard in High Density Planting system on clonal rootstocks. The apple cultivar Golden Delicious, Granny Smith, Yellow Newton and McIntosh is diploid and can be used for raising the rootstock. Stratification of seeds can be done through refrigerator or at 3-5 °C for 2-3 months in sand helps in good germination or stratification can be done directly sowing in field in November-December which does not require any input. T-budding is most successful technique in apple propagation. This is performed in July-August in Kashmir and June-July in Uttarakhand and Himachal Pradesh for better success. Chip budding is another promising technique of apple propagation. It is performed in dormant season (Oct-March). Wedge or Cleft grafting is very popular technique of apple propagation in India during dormancy (Jan-March). Apple is mainly multiplied on Crab seedling/clonal rootstocks. The clonal rootstocks are uniform in size and trees on these rootstocks are precocious in bearing with intensive planting. The clonal rootstocks recommended in India are M-27 (ultra-dwarf), M-9 (dwarf), M-7 and MM-106 (semi dwarf), MM-104, MM-109, and MM-111 are semi-vigorous in nature. EMLA series of rootstocks are virus resistant and MM series rootstocks are woolly aphids resistant. Pusa Seb Moolvrinth-1 is a dwarf type which has been developed from *Malus baccata* var. Shillong. Propagation of these rootstocks through layering/stooling and cutting has been standardized at CITH, Srinagar for their fast multiplication. Clonal rootstocks are multiplied by layering as well as by cutting. But layering has been found to be most successful method of clonal rootstock. Clonal multiplication through cutting multiplication. The rooted clonal rootstocks are collected from registered nursery as per variety and leveled properly. The well pulverized field should be used for establishing the mother stock. In winter, the clonal rootstocks are planted at 60 x 90cm spacing in the nursery beds and allowed to grow for one season. At the end of next dormancy, the mother stocks are headed back leaving 5 cm collar. In the growing season it will experience profuse shooting which is required to be mounded with soil covering leaving one third of the shoot with only tips open. Frequent irrigation is required for proper rooting. Extra soil is mounded on the layers. In January-February, rooted shoots are separated leaving stub of mother stock inside for sprouting and producing next crop, approximately 15-25 plants can be obtained from one stool.

9. Orchard Irrigation Management

Drip irrigation in apple has been found to be adventitious for efficient water use by 2.6-2.9 times than conventional systems. Under sloppy lands minimum energy is required and for uniform water distribution pressure compensating emitters are required. During the season 3,840 litres/tree through conventional system and 1,695 litres/tree through drip system is required for apple. On an average 90 litres/tree per irrigation by drip is required. Fertigation with water soluble fertilizers like ammonium nitrate, calcium nitrate, urea, potassium chloride, potassium nitrate, and potassium sulphate has become important as it saves lot of nutrients. Urea and ammonium nitrate are highly soluble and clogging is very less. Fertilizers which raise water pH (>7.5) are not desirable as such water reduces micronutrient availability. The work carried out in apple indicated that fertigation through drip irrigation recorded higher fruit set (45 per cent) and fruit yield and the cost of NPK also got substantially reduced to one third. Today, there is a growing awareness for use of drip irrigation for increasing quality apple production. But to achieve wider adoption of this technology, research on the minimal and optimal fraction of the soil to be wetted under varying weather and soil conditions, critical stages of application of nutrients *etc.* needs to be worked out. Water management plays an important role in productivity enhancement of apple. The water must be applied at critical stages of fruit growth and development of apple. Micro-irrigation system is an irrigation system with high frequency application of water in and around the root zone of plant system. It consists of a network of pipes along with a suitable emitting device. It can be installed at surface of the tree row called surface micro-irrigation. The sub surface installations are generally preferred in semi-permanent or permanent installation especially in fruit crops. The laterals are laid 45-60 cm below the soil surface to avoid any damage during the inter-cultural operations. The micro-irrigation saves water up to 50-70 per cent and productivity enhances 40-100 per cent with high quality fruits. The other advantages are saving of labour cost, reduce salt concentration in root zone, reduce incidence of pest and diseases, saves 40-60 per cent fertilizers and increase fertilizer use efficiency. The most commonly applied fertilizer is nitrogen although potassium and chelated forms of micro nutrients are also applied. Drip irrigation is most suitable irrigation system for temperate orchards for enhancing water and fertilizer use efficiency.

10. Rainwater Harvesting and Moisture Conservation Techniques

Rain water harvesting system plays an important role in apple production under rainfed conditions. There is number of rain water harvesting techniques and mulching for conservation of moisture for plain as well as sloppy land conditions. The encouraging findings were obtained at CITH, Srinagar in apple at 4×4 m spacing.

(a) Sloppy Land Condition

(i) Half-moon Water Harvesting System

Half-moon water harvesting system is one of the important water conservation technique for rainfed undulating topographical conditions. In this system, semicircular bunds are created at downstream side of the plant. The shape and design of the structure is semi-circular bunds having 30 cm width and 30 cm high at a radius of 1.7 m away from the tree trunk

Half-moon Water Harvesting System.   **Trench Water Harvesting System.**

and 5 per cent outward slope for rain water collection and storage from micro catchment area of tree. This is a low cost and economically viable technology for conservation of soil moisture during summer period for fruit growth and development.

(ii) Trench Water Harvesting System

Trench water harvesting system is one of the important water conservation technique for sloppy land rainfed topographical conditions. In this system, trench is created upstream side one meter away from the tree trunk. The shape and design of the structure is 30 cm width and 30 cm deep and 1.5 m length of trench created for collection and storage of rain water from the catchment area. This is a low cost and economically viable technology for farmers of hilly area.

(b) Plain Land Condition

(i) Full Moon Water Harvesting System

Full moon water harvesting system is one of the important *insitu* moisture conservation technique for rainfed plain land topographical conditions. In this system, circular bunds are created around the periphery of the plant. The shape and design of the structure is circular bunds having 30 cm width and 30 cm high at the radius of 1.7 m away from the tree trunk and 5 per cent inward slope for

Full Moon Water Harvesting System.   **Cup and Saucer Water Harvesting System.**

rain water collection and storage from micro catchment area of tree. There should be earthing-up around the tree trunk in the radius of 30-40 cm to avoid water stagnation around tree trunk. This is a low cost and economically viable technology for conservation of soil moisture during summer period for fruit growth and development and suitable for resource poor farmers having plain lands.

(II) Cup and Saucer Water Harvesting System

Cup and saucer water harvesting system is one of the important moisture conservation technique for plain land rainfed topographical conditions. In this system, trench is created 1.7 m away from the periphery of the tree having 30 cm width and 30 cm deep for collection and storage of rain water from the catchment area. This is a low cost technology and suitable for farmers having plain lands. The above water harvesting system along with mulching (organic/plastic) conserves more moisture during critical stages of fruit growth and development.

11. Intercropping

During pre-bearing stage of apple orchard, intercropping with several suitable vegetable crops are recommended for realizing some farm income through sustainable utilization of interspaces and available resources. The following crops are recommended as intercrop in apple orchard. Intercropping with legumes increase nitrogen use efficiency of apple crop and thus increases the yield of apple directly. Since intercrop gets ready during the season when apple crop is not ready thus provides the source of income during off-season period. Also crops like lentil, *methi*, red clover *etc.* does not compete for water and nutrients with apple because their critical stages for irrigation and nutrient use are different than for apple. Intercrop increases cropping intensity and benefit: cost ratio.

Apple + Peas   Apple + Methi.

12. Plant Health Management

Pests and diseases in apple are causing heavy losses to an extent of about 30-40 per cent besides impairing their quality and therefore, there effective management is most essential component for increasing production and productivity of quality apples. Most of the varieties in the country have been introduced from developed nations over the past few decades. These varieties have been mainly attacked by some of the important insects like Sanjose scale, aphids, woolly aphid, peach leaf curling aphids, stem and root borer, shot hole borer, tent caterpillar, codling moth and European red mite and diseases like scab, powdery mildew, leaf spot, brown rot, gummosis, canker *etc.* Their management through integrated approach has become priority for producing residue free fruits and their value added products. Involving cultural, biological, and chemical control measures. CITH has developed spray schedules for management of fungal foliar diseases like scab, alternaria, marssonina *etc.* and soil borne diseases for control of root rot in apple and these schedules are being tested at Mukteshwar (Uttarakhand) and Srinagar, (J&K). Canker caused by *Bopyosphaeria* spp. could be successfully controlled through the use of Chaubattia paste (Red Lead: Copper carbonate: Linseed oil, 1:1:1.25) on pruned portions in November-December followed by spray with 0.1 per cent bayleton or 0.1 per cent carbendazim during growing season. To avoid degenerative disorders like apple mosaic, the disease free budwood or grafts be procured only from government nurseries or registered nurseries. Biological agents *Athelia* sp. and *Chaetomium globosum* supplement with urea treatment on pseudothecial density significantly reduced scab pathogen. Pre-mature leaf fall in apple (*Marssonina coronaria*) characterized by formation of brown to dark brown spots of variable size on mature leaves in summer and drop off prematurely was controlled by protective sprays of mancozeb (0.3 per cent), propineb (0.3 per cent), dodine (0,075 per cent), carbendazin (0.05 per cent), dithianon and ziram (0.3 per cent).

13. Pre and Post harvest Management

Despite this fact that India is the 2ⁿᵈ largest producer of fruits and vegetables in the world, but a huge quantity (30-40 per cent) goes waste due to post harvest losses. Inadequate pre and post harvest treatment, short shelf life, lack of maturity indices, inadequate processing infrastructure, grading, packing, packaging, road and transport facilities *etc.* are the key factors causing huge losses. Proper post harvest handling and value addition therefore, are essential for reducing these losses and increasing production, productivity and vis-à-vis the returns. Several modern technologies are invented for pre and post harvest management of temperate fruit crops have greater significance which include preharvest factors such as nutrition, rootstock and environmental factors and post harvest include harvesting, grading, pre-cooling, packaging, CA storage/cool chain, preparation of value added products and their disposal to terminal markets.

14. Crop Regulation and Pre-Harvest Management

(a) Improving Fruit Set

To enhance flowering and fruit set, 3 per cent Dormex (hydrogen cyanamide) 40 days before bud break in Royal Delicious apple was found effective. Application of Boric acid (1 per cent) at the time of bloom can help in better pollen tube growth. Spray of Miraculan (0.75 ml/L water) or Paras (0.6 ml/L water) or Biozyme or Protozyme (2 ml/L water) at bud swell and petal fall stages has been recommended for better fruit set.

(b) Thinning

Hand thinning of flower cluster after every 3-4 cluster or retaining only 2-3 fruitlets per cluster is practiced. At petal fall stage NAA 10 ppm (1ml planofix/4.5L water) 7-15 days after petal fall or at fruit length around 15 mm is most effective. Carbaryl @0.075 per cent (Sevine 50WP/L water) 7-10 days after petal fall and Ethephon (100-200ppm) at full bloom to petal fall is also effective.

(c) Fruit Drop

Most of the fruit varieties have been noticed to have the following three waves of fruit drop namely: Early drop due to inclement weather like hail storm and drought, improper and poor pollination and fertilization, June drop due to moisture stress and competition for growth and food and Pre-harvest drop due to physiological imbalance or any disorder which causes economic loss. The application of 10 ppm NAA (Planofix 1ml/4.5L water) a week before the expected fruit drop or 20-25 days before harvest can check the fruit drop effectively.

(d) Fruit Colour and Maturity

In Delicious varieties and all the red colured improved strains of apple, colour development is generally poor and due to rise in temperature early in the season growth of the fruit takes place faster but conversion of starch into sugars and other necessary physico-chemical changes essential for quality produce does not takes place properly in marginal areas below 1828.8 m elevation above mean sea level which fetch poor market price. On the other hand, fruits at higher altitude areas get sufficient maturity duration accompanied by day time strong solar radiation and cooler night which favours better colour and quality. Application of 250-500 ppm 2-chloroethyl phosphonic acid (Ethrel, CEPA or Ethephon) about 20 days before harvest improves colour of fruit substantially, but impairs shelf-life.

(e) Extension of Shelf-life of Fruits

Apple grown in upper belts can be stored for 90 days under ambient conditions whereas, apples in lower belts can be stored for 60 days. The influence of post harvest infiltration of calcium chloride on keeping quality of Red Delicious apples has also been investigated under various storage conditions. In Himachal Pradesh also pre-harvest three sprays of 0.5 per cent calcium chloride of right from first or second week of July at two weeks interval and addition of carbendazim (0.05 per cent) with calcium chloride solution controls bitter pit, extends shelf life and also reduces blue mould incidence. Post harvest dip (4 per cent) been recommended for improving shelf-life of apples.

(f) Maturity Indices and Harvesting

Maturity indices/harvesting stage determine the quality of fruits and its shelf life. Maturity standards have been calculated and standardized based on days to harvest from full bloom and TSS in apple.

(g) Processing

Apples are processed into various products such as juice, concentrate, vinegar, sauce, butter, preserve, candy, Jam, Jellies and canned products. Apples are also dried as rings, chops, or cubes. They are also used for making fermented beverages such as cider and wine. The waste from the apple processing industry, such as peel, core, or pomace, can be utilized for production of pectin and various edible products. Apple juice is a popular drink and one of the important breakfast items. Apple juice contains a considerable proportion of the soluble components of the original apples such as sugars, acids and various other carbohydrates. Malic acid is the predominant acid in apple juice. Several distinct forms of apple juices available in the market include clarified apple blends with other juices/extracts. In the production of apple jam, good-quality fruits are selected and washed in cold water. The fruits are peeled and the skin and seeds are removed. The peeled fruits are cut into small pieces. The fruit pieces are cooled and crushed with a paddle and made into a fine pulp by sieving. To 1kg of pulp, an equal quantity of sugar and 2.5 g of citric acid are added and the mixture is mixed thoroughly. The mixture is cooked slowly with occasional stirring until it passes a sheeting or drop test. The final weight of jam is in the range of l'5 times the sugar added. The hot jam is filled into clean extract obtained by boiling unpeeled apple pieces in water for 25-30 min. and filtering through muslin cloth.

References

Adýyaman, A.F., Iýkalan, C., Kara, Y. and Baaran, D. 2004. The comparison on the proliferation of lateral buds of *Vitis vinifera* L. cv.Perle de Csaba during different periods of the year *in vitro* conditions. *Int. J. of Agri. and Bio.*2: 328-330.

Dobra nszki, J. and Teixeira da Silva, J.A. 2010. Micropropagation of apple-A review. Biotechnol Adv. doi:10.1016/j.biotechadv.2010.02.008.

FAO,2015.http://www.fao.org/news/archive/news-by-date/2015/en/and http://www.fao.org/3/a-i4646e.pdf

Mir, J.I., Ahmed, N., Verma, M.K., Muneer, A. and Lal, S. 2010. *In-vitro* multiplication of cherry rootstocks. *Indian Journal of Horticulture* 67: 29-33.

National Horticulture Board, 2014. *Indian Horticulture Database, 2014.* Ministry of Agriculture, Government of India.

Sedlak, J. and Paprstein, F. 2008. *In vitro* shoot proliferation of sweet cherry cultivars Karesova and Rivan. *Hort. Sci.*(Prague) 35: 95-98.

Chapter 10

Climate Change Impact on Soils and Apple Based Horticulture of Lesser Himalayas

Mushtaq A. Wani and Shazia Ramzan

Division of Soil Science,
Sher-e-Kashmir University of Agricultural Sciences and
Technology of Kashmir, Srinagar
E-mail: mushtaqb4u@gmail.com

1. Introduction

The naturally occurring thin layer of unconsolidated material on the earth's surface that has been influenced by the parent material, climate, relief, and physical, chemical, and biological agents is known as soil. Soils exhibit differences in their physical and chemical characteristic (soil structure, soil temperature, soil texture, soil humus, soil water), as well as in their capability for growing crops. No wonder, in Kashmir, soil is virtually worshiped as a miracle of divinity as it is a source of wealth.

(a) Soils of Kashmir

(i) Zonal

(ii) Inter-zonal

(iii) Azonal or immature soil

In brief, differences in soils arise from the mineral composition of the parent materials and from differing climatic conditions, which together influence the organic and inorganic processes of soil development. A comprehensive soil study of Kashmir valley and cold arid Ladakh of Jammu and Kashmir State, based on scien-tific data, has been prepared (Wani and Shaista, 2016; Wani and Wani, 2015; Wani *et al.,* 2009, 2010a, 2010b, 2011, 2013a, 2013b, 2014, 2015a, 2015b, 2016a, 2016b and 2016c) (Figure 10.1). Under the existing geo-climatic conditions, a wide range of soils, both of residual and alluvial origin are found in the state. The state of Jammu and Kashmir is essentially hilly and mountainous. The outer plain of Jammu

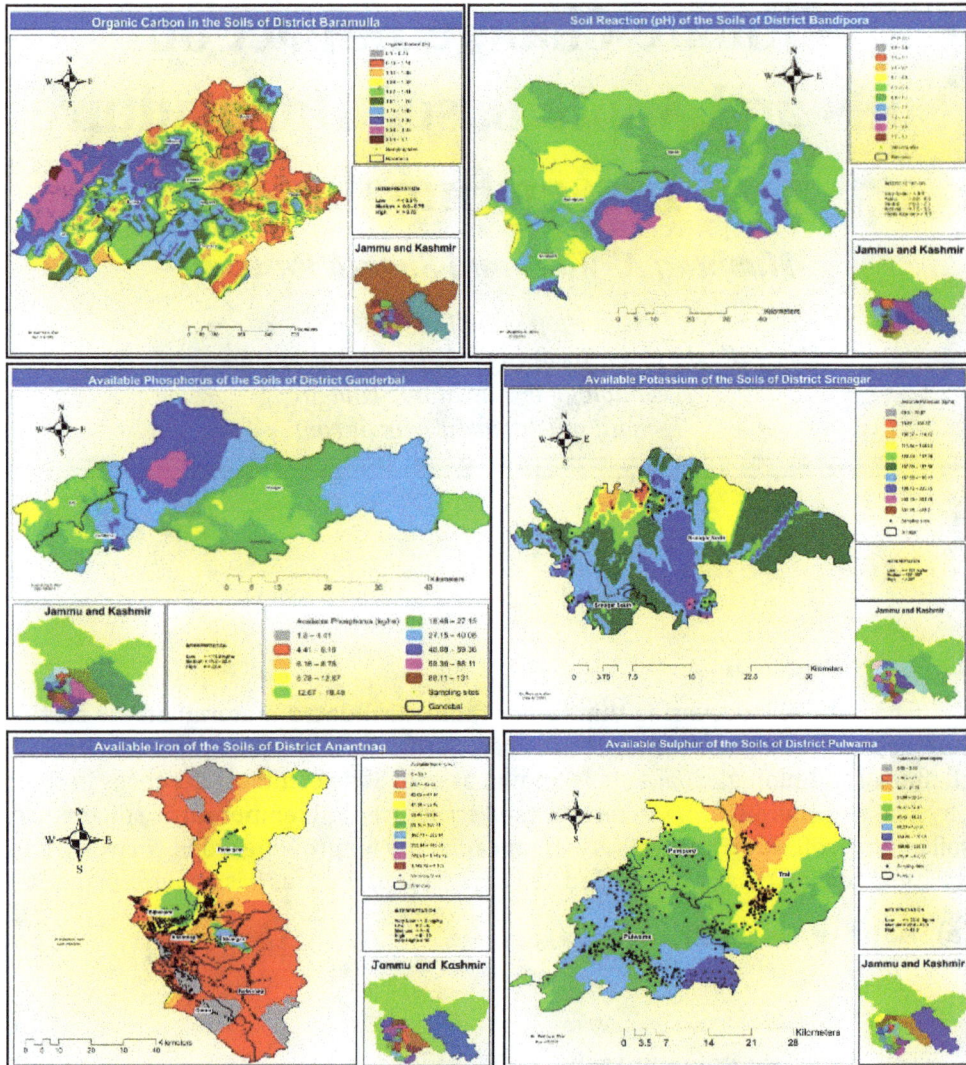

Figure 10.1: Soil Fertility Maps of different Districts of Kashmir.

with alluvial soils was deposited by running water and the fluvio-glacial action. The hilly and mountainous areas are generally covered by the residual soils, while the upper reaches of Chenab and the Jhelum and their tributaries are covered with alluvial and morainic soils. On the basis of rock strata and pedogenic characters, the following major categories of soils may be identified in the state (Figure 10.2).

(i) Hilly and Mountainous Soils

The hilly and mountainous soils are found in the entire state excepting the leveled plain of Jammu adjacent to Punjab and the valley floor of Kashmir. The undulating topography and steep slopes effect the run-off and drainage system. Other things being equal, the run-off is large on steep slopes. As a rule, more the water run-off, lesser is its absorption in the soil on steep slopes. The run-off also washes away more of the weathered rocks on steeper slopes. The depth of soil and soil profiles on steeper slopes are consequently shallower than that on gentle slopes. Many soils in the mountainous areas are shallow, immature and highly susceptible to soil erosion. These soils are generally, acidic in character, deficient in potash, phosphoric and lime and therefore, need regular manuring and fertilization for good yields. The hu-mus content in these soils varies from slope to slope and altitude to altitude. Depending on the availability of sunshine and rains, these soils are generally devoted to the cultivation of maize, pulses, orchards (almonds, apples, peach, and pears), oilseeds, barley, wheat, oats and fodder. The higher altitudes, be-tween 2500 m to 4000 m are reserved as alpine pastures. These pastures develop nutritious grasses during the summer season which are grazed and utilized by the *Gujjars* and *Bakarwals* (nomads) for their flocks of goats, sheep and horses. Scarcity of water, leaching, erosion and avalanches are the main problems of these soils. These problems are getting accentuated owing to the indiscriminate felling of trees and depletion of ecosystems.

(ii) Alluvial Soils

The alluvial soils are deposited by the action of rivers. They are found in the river channels, floodplains, estuaries, lakes and fans at the foot of mountains. The alluvial soil includes all consolidated fragmented material from the coarsest gravels and sands down to the finest clay and silt-sized particles. In other words, sand, silt and mud brought down by rivers in floods and depos-ited on the temporarily submerged land are known as the alluvial soils. These are the most productive soils of the state, found mainly in the Jammu- plain, the valley floor of Kashmir and at narrow river terraces along the tributaries of the Chenab, Jhelum and their tributaries.

The alluvial soils may be classified into:

(a) The old alluviums

(b) The new alluviums

The old alluviums lie above the banks of rivers and are generally free from floods, while the newer alluviums are frequently inundated as they lie in the flood

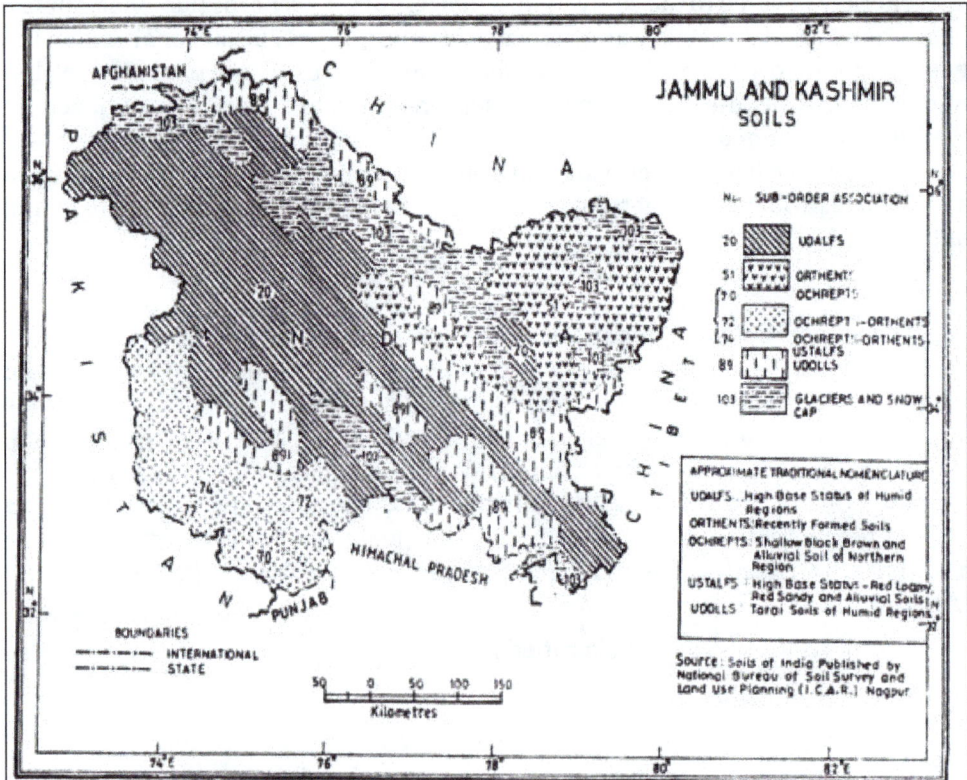

Figure 10.2: Soils of Jammu and Kashmir (NBSS and LUP, Nagpur).

plains of the Jhelum, Chenab and their tributaries. New alluviums (*Khadar*) soils are relatively coarse in texture and contain more sand than that of old alluviums (*Bhangar*) soils. The alluvial soils, wherever, irrigated grow two to three crops in a year. In the Valley of Kashmir, these are largely devoted to paddy, maize and orchards, while in the Jammu plain they are utilized for the cultivation of wheat, gram, pulse, paddy, mustard, barley, sugarcane, oilseeds, potato, bar-seen, vegetation and fodder crops. Being highly productive, these soils are giving good returns of the High Yielding Varieties of wheat and rice in the areas of controlled and assured irrigation.

(iii) Karewa Soil (Wudur)

Karewas are fresh-water (fluviatile and lacustrine) deposits found as low flat mounds or elevated plateaus in the Valley of Kashmir and the Kishtwar and Bhadawah tracts of the Jammu Division. The important Karewas are found in Kulgam, Shopian, Budgam, Qazigund, Tangmarg, Gulmarg, Baramulla, Laithpora, Chandhara, Pampore, Bijbehara, Awantipora, Islamabad (Anantnag), Mattan, Tral and Ganderbal. The Karewa soils are composed of fine, silty clays with sand bouldry gravel, the coarse detritus being as a rule, restricted to the peripheral parts of the valley, while the finer variety prevails towards the central parts. The Karewa soils of Kashmir have enormous agricultural potential. Commercial and cash crops like saffron, almond, apples, walnut, peaches, pears, cherry, plum, *etc.*, with orchards and saffron beds. Moreover, some leguminous and fodder crops are also grown in Karewa. The Pampore Karewa is famous all over the world for saffron cultivation. Soil erosion and depleting soil fertility are the major problems of the karewa soils. It has been reported by the farmers of the Chandhara and Dusu villages (Pampore-Karewa) that with the passage of time the karewa soils are losing their resilience characteristics. As a matter of fact, per unit produc-tion of saffron and almond has gone down substantially during the last three decades. The soil conservation practices need to be adopted to maintain the health of the karewa soils, making them economically more productive and ecologically more sustainable.

(b) Classes of Soil

(i) Gruti (Clayey Soil)

Gruti soils contain a large proportion of clay. Texturally, it resembles to the clayey loam. Its water retaining capacity is high. In years of scanty rainfall, it is considered to be the safest for the cultivation of rice. Contrary to this, if rains are heavy, the *gruti* soil gets compacted and achieves the shape of hard cakes; the ploughing of which becomes difficult and pulverization of soil is an arduous task. In the years of scanty rainfall, these soils give poor yields. The gruti soils are found in the low-lying areas of the Kashmir Valley.

(II) Behil (Loamy Soil)

Behil is a rich loam of great natural fertility. The humus content is high which enriches the soil fertility. Consequently, it does not require heavy manuring.

Figure 10.3: Saffron Cultivation in Pampore Kerewa.

Moreover, there is always a danger that by over-manuring the soil will be too strong, in which the rice crop will show more vegetative growth and will be more susceptible to lodging. It is ideally suited for paddy cultivation.

(iii) Sekil (Sandy loam)

Sekil is a light loam with sandy subsoil. In the *sekil* soil field if artificial irrigation is available, good crops of rice are harvested in the summer season. *Sekil* soil is generally confined to the lower edges of karewas in the Valley of Kashmir.

(iv) Dazanlad (Sandy silt)

Dazanlad soil is chiefly found in the low-lying ground near the swamps, but it sometimes occurs in the higher villages also. The soil has an admixture of sand and clay and becomes warmer in the summer season. A peculiar characteristic of *dazanlad* is that the irrigation water when stands in the fields turns red in colour. If controlled irrigation is provided, high yielding varieties of rice can be grown successfully in dazanlad soils.

(v) Nambal (Peaty soils)

Near the banks of the Jhelum River and in the vicinity of the Wular, Manasbal and Anchar lakes is found the rich peaty soil, locally known as Nambal. In the years of normal rainfall and moderate snowfall, *nambal* soils give good yields of rapeseed, mustard, maize, oats, pulses and fodder.

(vi) Tand (Mountainous soils)

The land on the slopes of mountains, reclaimed from the forests is called *Tand* soil. After reclamation the *tand* gives good yields of maize, pulses and fodder for two or three years, but under the impact of accelerated soil erosion the land loses its natural strength. Consequently, the productivity declines and after a period of about six to ten years the land acquires the shape of a pasture and culturable waste.

(vii) Zabelzamin (Alkaline soils)

Patches of irrigated land if excessively irrigated lose their fertility and develop alkaline formations. Such adversely affected patches of saline and alkaline formations are known as *zabelzamin*. These soils are unproductive from the agricultural point of view unless especially treated with gypsum, water and manures. There are numerous other types of soils recognized by the Kashmiri farmers, such soils are *Kharzamin, Tresh, Limb, Ront, Shath and Tats.*

2. Soil Borne Disease and their Management

Apple is an important fruit crop in the hilly regions of India. Though overall apple production has increased but productivity per unit area is still quite low. Diseases are one of the contributory factors in this regard. Apple plant is prone to diseases right from nursery stage onward and all the plant parts are attacked. Soil borne diseases like white root rot and collar rot cause death of plants from nursery stage onward and even grown up trees are killed. Seedling blight cause death of seedlings at nursery stage, whereas, crown gall and hairy root result in stunted plant growth in nurseries and plantations. Successful management of soil borne disease is a difficult task and involves the integration of all available practices as the pathogen can survive and multiply in the permanent soil medium. In case of apple, the diseases mentioned above have been managed well by following a combination of different methods like cultural practices, use of biocontrol agents and need based use of chemicals.

(a) White Root Rot

It is a disease of roots which remain covered with white fungal mycelial growth in rainy season resulting in death of plants and then attains a serious status (Figure 10.4).

(i) Symptoms

The disease symptoms occur on the underground parts of the tree and the effects are also manifested on the upper ground parts. The earliest above ground manifestation of the disease is bronzing of the leaves, diminution in size and a stunted tree growth resulting in the progressive decline in vigor as a whole or certain branches. These symptoms are usually associated with a heavy blossom and fruiting next year. In succeeding years, few leaves merge and much of fruit fail to reach maturity and dying back of branches is quite evident. Infected trees persist for 2 to 3 years depending upon the infection intensity in the roots but severely infected trees may succumb within a single season.

Figure 10.4: Apple Trees Dying of Root Rot.

The above ground symptoms are not distinctive because of similar symptoms produced by other root maladies. So below ground expression of pathogen is the final diagnostic feature. The lateral roots turn dark brown and are covered with a greenish grey or white mycelial mat having a flocculent web of whitish strands or ribbons during monsoon season. As the disease progresses, all the roots are attacked and fibrous root system almost disappears. The cortical cells are ruptured leading to death of trees. White flocculent web disappears after death of host plant, leaving root surface dotted with small round black sclerotia (Figure 10.5).

(ii) Management

The management of soil borne diseases in general and those in trees specially is difficult because of deep seated infection. Combination of measures help in reducing the disease and recovery of diseased trees. White root rot of apple is best controlled by practicing preventive as well as curative measures consisting of cultural, biological, chemical methods and resistant rootstocks.

i. Cultural Practices

The recommended cultural practices for root rot of apple is hot water treatment of infected seedlings at 45°C for one hour before plantation, digging isolation trenched and removal of rotten roots followed by application of disinfected paste. In order to save trees in early stages, the attacked roots should be trimmed off and destroyed. Replanting should be done in clean soil and infested soil should be

Figure 10.5: White Root Rot of Apple.

followed with frequent cultivation to starve out the infection hyphae. Burning and heat drying is also reported to increase the life of trees. Soil solarization is also effective in reducing the fungal inoculum in soil. Soil moisture plays an important role in affecting the intensity of disease, hence ill drained soils should be improved by following central drainage system. Acidic soils should be amended with lime. Soil amendments like neem cake and deodar needles also reduce incidence of disease.

ii. Chemical Control

The success in controlling soil borne diseases lie in the fact that the required and effective levels of fungicides should reach to the point of infection in soil. Broad spectrum chemicals such as carbon bisulphide, chloropicrin, calcium cynamide, formaldehyde, *etc.* are recommended for checking root rot of apple. With the advent of systemic fungicides old recommendations are replaced by benomyle, aureofungin, carbendazim *etc.* Recently Sharma and Gupta (1996a) reported efficacy of role of phorate and its combination with carbendazim and Dithane M-45 in reducing disease in nursery.

(iii) Resistant Rootstocks

Almost all rootstocks show susceptibility to the pathogen, but some degree of resistance was observed in MM109, M16 and MM104. Modern biotechnological approaches are required to be exploited for combating this pernicious soil borne disease of apple.

(b) Collar Rot

Also known as crown rot is prevalent in apple. The spread of the disease is very fast in orchards planted on genetically uniform rootstocks.

(i) Symptoms

The aboveground symptoms of collar rot are often confused with white root rot, however, examination of underground parts reveal the difference. The infection starts from the collar region and spread mostly to the under ground parts and the above ground stem portion is also infected in highly susceptible scion cultivars. Bark at the soil level becomes slimy and rots resulting in cankered areas. The wounds are irregular in outline but usually roughly oval which extend rapidly, resulting in girdling of the tree. The attacked trees are recognized by chlorotic foliage with red colouration of veins and margins (Figure 10.6).

Figure 10.6: Collar Rot or Crown Rot.

(ii) Management

i. Cultural Practices

The practices which help in avoiding the plant from the pathogen attack are recommended. As the pathogen gets entry through the graft union, hence, the grafting may be practiced about 30 cm from the soil level. General orchard management practices such as improved drainage around the tree base, removal of crop refuse and avoiding injury to the stem especially during bud swell stage are helpful in restricting the disease spread. The disease can also be suppressed by localized superficial heat treatment of infected stem portion with blow lamp flame.

ii. Chemical Control

The fungicides are applied around the affected trees or by painting the wounds with fungicidal paint. Before recommending chemical, proper disease diagnosis is important because of carbendazim or benomyl which are recommended for white root rot, are applied to collar rot affected plants, there may be increase in disease severity. For the application of paints/pastes, the affected portion is first scarified upto healthy tissues and disinfected with methylated spirit or mercuric chloride and after its evaporation, bordeux paint or copper paint or metalaxyl paint or chaubattia paste are applied (Gupta and Mir, 1983). The disease is also controlled by drenching Dithane M-45 (0.3 per cent) or Biltox/Fytolan/Blue copper (0.5-1.0 per cent) in 30 cm radius around the tree trunk. Systemic fungicides are also effective as soil drench and paint (Bleicher, 1994).

(iii) Resistant Rootstocks

Rootstocks such as M2, M4, M110. MM114, MM 115 and crab apple possess high degree of resistance. MMM111 is highly resistant to crown rot.

(c) Seedling Blight

(i) Symptoms

It occurs in nurseries and 2 to 3 year old seedling are attacked.Upon infection of roots, the leaves of affected plants start wilting and show a characteristic reddish or greyish purple discolorations. Ultimately blightening of foliage of infected seedling is the common symptom. The above ground symptoms are generally confused with white root rot but its identity can be confirmed by examination of roots, where mustards seed size sclerotia are seen in the vicinity of the dead seedlings.

Figure 10.7: Seedling Blight.

(ii) Management

Affected seedlings in the nursery beds are treated with thiram, aureofungin, *etc.* to keep this disease under control. As the disease is aggravated by high soil moisture so water logging in nursery beds should be avoided and loam soil should be selected for raising nursery. The nursery site must be rotated every 4 or 5 years. Maize should be planted continuously for 4 or 5 years before site is again selected for nursery raising.

(d) Crown Gall

The disease is caused by bacteria and occurs in nurseries and orchards. Crown gall occurs both in nursery and grown up trees. The disease plant develops tumors at or near the ground line. The disease causes heavy losses in nursery because the diseased seedlings are rendered unfit for transplanting (Figure 10.8).

Figure 10.8: Crown Gall of Apple Trees.

(i) Symptoms

The characteristic symptoms of crown gall are mainly observed on roots or at the soil level. Following infection of tissues, the cells transform into autonomously proliferating tumor cells. The resultant unregulated cell division gives rise to clearly visible galls. The young galls are pale, soft, smooth, composed of light coloured frosty tissue resembling fresh callus growth, but gradually becomes dark, hard and deeply fissured. The galls are globular or elongated or irregular and are generally produced at or near the graft union. The galls may surround the stem or root, or it may be connected to the host stem or root through a narrow neck of tissues. The size of the galls vary from 0.6 to 10 cm in diameter and in extreme case may be up to 0.3 m or more. Few galls may rot during winter and reappear in spring. In addition to galls, affected trees may become stunted, produce small and chlorotic leaves. Such plants are more prone to frost injury. Small galls require careful diagnosis, because they are easily confused with excessive callus growth at wound sites or will galls induced by nematodes and insects.

(e) Hairy Root

Hairy root is generally encountered in nursery in which excessive roots originate from one point, making the root system look like a broom (Figure 10.9).

(i) Symptoms

Hairy root symptoms on young apple trees are characterized by an excessive proliferation of adventitious roots, singly or in clusters. Based on tissue morphology, hairy roots have been divided into three types: namely simple hairy root, woolly knot and broom root. In simple hairy root, large number of small roots appear from stem without any associated callus or tumor tissues. In woolly knot, fibrous roots arise from graft overgrowth or tumors on young tissues. In brown root, fine roots develop from the top of roots, which themselves originate from tumor tissues. In some cases tumors/galls develop first and roots emerge from the tumor tissues.

Figure 10.9: Hairy Root of Apple Tree.

(ii) Management

Nursery sanitation is necessary to avoid the introduction of infected material into nursery stock. The practices like destruction of infected plant material by uprooting and burning and rotation of nursery site are also helpful in preventing the disease to some extent. Most infections result from the grafting union, so incidence can be reduced by budding instead of grafting. To avoid the dissemination of pathogen, the entire root system of apparently healthy grafted plants should be dipped in 1 per cent copper sulphate solution for ½ hour prior to transplanting.

3. Climate

Compared to other parts of the country which are witnessing externalities of climate change, long term annual rainfall in the state showed positive growth momentum with annual growth rate of 1.39 percent during last 28 years. Overall, annual maximum and minimum temperature for the State almost exhibited a uniform trend during last 40 years. The regions of state Jammu, Kashmir and Ladakh have distinct agro climatic characteristics and cultural identity. Jammu region has two different climatic zones depending primarily on altitude. Lower hills and plains bear subtropical climate with hot dry summer lasting from April to July. The summer monsoons coming around middle of July and fading away in early September. This is followed by dry spell from September to November. Winter is mild and temperature seldom touches freezing point. In the high reaches of Chenab valley, the climate is moist temperate, winter are severe and varied quantity of snow is received. The Kashmir valley with Pir Panjal Mountains on its south and Karakoram on its north receives precipitation in the form of snow due to western disturbances. The winter is severely cold and temperature often goes below 0°C. Spring is pleasantly cold. Summers are warm and dry and autumn is again cool and sometimes wet. Ladakh is situated in eastern mountain range of Kashmir. This is one of the highest ranges in the world. It is cold desert receiving very little precipitation. The temperature remains below the freezing point during winter due to its high altitude when people often remain indoors. Drass in Ladakh is the coldest place of the state. It has recorded the temperature of -50°C during winter. During the short period of summer season, the scorching heat of sun often causes sunburns.

(a) Effect of Climate Change on Soils

While the majority of climate change impacts focus on tree health, soil impacts should not be overlooked. Along with changes in temperature, climate change will bring changes in global rainfall amounts and distribution patterns. And since temperature and water are two factors that have a large influence on the processes that take place in soils, climate change will therefore cause changes in the world's soils. The soil responses to climate change variables are multifaceted and rather complicated because of the presence of an intricate network of sequential, simultaneous and/or coupled (often time-dependent) chemical, biological and hydrological reactions and processes. Climate extremes (heat wave and dry spells) induce poorly understood and interconnected long-lasting effects in soils, chemical elements, nutrients and contaminants involved in these reactions and processes are distributed in the soil solid, liquid and gas phases and scale-dependent effects related to the solid phase mineralogical, chemical and physical spatial heterogeneities.

In fact, there are several ways that climate change will affect soil. Soils are also part of the global carbon and nitrogen cycles. The carbon-based gases carbon dioxide (CO_2) and methane (CH_4), and the nitrogen-based gas nitrous oxide (N_2O), are important greenhouse gases. So, as carbon dioxide, methane, and nitrous oxide levels change in the atmosphere, there will be corresponding changes in the soil. In the natural soil formation processes the pedogenic inertia will cause different time-lags and response rates for different soil types developed in various regions of our globe (Scharpenseel *et al.*, 1990; Lal *et al.*, 1994; Rounsevell and Loveland, 1994). The influence of climate change on soil structure (type, spatial arrangement and stability of soil aggregates) is a more complex process. The most important direct impact is the aggregate-destructing role of raindrops, surface runoff and filtrating water, especially during heavy rains, thunder storms and even 'rain bombs', the increasing hazard, frequency and intensity of which are characteristic features of climate change. The indirect influences are caused by changes in the vegetation pattern and land use practices.

Through climate change and anthropogenic activities, many of our world's soils have become or are expected to become more susceptible to erosion by wind and/or water (Zhang *et al.,* 2004; Ravi *et al.,* 2010; Sivakumar, 2011). Increased rainfall due to climate change could lead to significant increases in runoff, with amplification greater in arid areas (up to five times more runoff than the percentage increase in rainfall) than in wet and temperate areas (twice as much runoff as the percentage change in rainfall) (Chiew *et al.,* 1995). Greater runoff would be expected to cause increased erosion. Water erosion models in the United Kingdom predicted that a 10 per cent increase in winter rainfall could increase annual soil erosion by as much as 150 per cent during wet years, but that long-term averages of soil erosion would show a modest increase over current conditions (Favis-Mortlock and Boardman, 1995).

The major challenge for soil scientists today is to figure out exactly how climate change will affect soils, because some of the possible effects counteract each other. For example, organic matter is very important in soils. Many scientists expect increased carbon dioxide levels in the atmosphere to increase plant growth, which would mean more organic matter could potentially be added to the soil.

4. Soil Conservation

The depletion of soil owing to physical and cultural factors is a universal phenomenon. The rate of soil erosion and soil depletion in a given region is largely influenced by following factors

(i) Rainfall erosivity, (ii) Volume of run-off, (iii) Wind- strength, (iv) Relief, (v) Angle of slope, (vi) Slope length, (vii) Length of wind fetch (viii) Shelter belts, (ix) Mode of soil utilization, (x) Intensification of agriculture, and (xi) Management of land for agriculture and allied economic activities.

The rate of soil erosion also depends on the pressure of population, cropping patterns, cropping intensity, crop rotation, irrigation, manuring and tillage practices. Soil erosion not only affects the areas from which soil is removed but also influences the environment where it is deposited. Such deposits if take place in lakes and ponds; destroy the aquatic remedial measures to combat soil erosion and to keep the ecosystems in a healthy condition so that the needs of the present generation be fulfilled without compromising the needs of the future generations.

(i) Steps to Overcome the Problem of Soil Erosion

Soil erosion is more serious in the areas where deforestation has taken place at a large scale. Forest ecosystems have considerable control over pattern of climate, hydrology, circulation of nutrients, erosion and cleansing functions of air and water, as well as over the status of streams, lakes, and underground water supplies. Forests also present a vast reservoir of genetic diversity of plants, animals and micro-organisms, which play significant roles in maintaining the ecosystems.

☆ It is imperative to maintain proper records of land productivity status which should be prepared with the help of soil scientists.

☆ In areas of high irrigation requirement *viz.* paddy growing areas, effective soil and water conservation techniques should be adopted with technical experts and agricultural scientists.

☆ Sustainable cropping patterns and rotation of crops should be followed by the cultivators.

☆ In the areas of water erosion, the agro-forest ecosystem is to be implemented and strengthened.

☆ Industrial activities in the nearby areas which are hazardous to the environment should be regulated through law.

☆ The culturable waste and marginal lands should be amended properly so that further degradation of such lands may be checked.

☆ In the soil erosion areas, adoption of integrated watershed management system need to be implemented.

☆ The auto-regenerating of the soil fertility by adding organic matter through microorganism and using crop residues should be facilitated.

☆ Pressure of population on arable land should be reduced by implementing effective population policy.

5. Impact of Climate Change on Apple

(a) Effects

The IMD monitoring reveals that temperatures are increasing in both Jammu region and Kashmir valley, with significant increase in maximum temperature of 0.05 Celsius per year. The average mean temperature in Kashmir has risen by 1.45° Celsius in last 28 years while in Jammu region, it has increased by the rise is 2.32° Celsius. As a result of rise in temperature and decline in rainfall, the apricot and cherries are fast disappearing from some areas of Kashmir Valley. Due to general rise temperature and less availability of water, the yield and quality of apples in valley and mid temperate region of Jammu are fast deteriorating. Over the last few years, there has been distinct slow growth in production and productivity in rainfed Kashmir's Karewas areas. Due to unusual hailstorms and windstorms in summer fruits like cherry, apple, plum, peach and apricot are getting damaged heavily. In recent years there marked change in the pattern of snowfall in Kashmir which is effecting all the pome and stone fruits. It has been observed that the snowfall and flowering in some years is coinciding leading to great loss in quantity and quality. Due to shortage of water for agronomic crops like rice shift has been recorded from agronomic crops to temperate fruits and nut in J&K as fruit crops are more remunerative as compared to agronomic crops.

(i) Effects on Sprouting

The impact of temperature change is most in apple where trees sprout 2-3 weeks early but normally apples trees sprout in mid-April. As a result last few years about 70 per cent of trees began to open their buds in mid-March. At the end of March it can definitely become very cold again. At this time most trees have their buds open are very susceptible to frost damage.

(ii) Effect on Fruit Color

In Kashmir valley, the failure of apples to change into their specific red shades, or an increase of apples with sunburn. The deep red colour is a result of low temperatures during the night in autumn, just before harvesting. If the temperatures are not low enough, most apples fail to turn into their specific red shades. For many apples their red colour is a trademark of quality but Ladakh province becomes potential area for apple cultivation due to climate change.

(iii) Effect on Chilling Requirements

Most deciduous fruit trees need sufficient accumulated chilling, or vernalization to break winter dormancy Inadequate chilling due to enhanced greenhouse warming may result in prolonged dormancy, leading to reduced fruit quality and yield. The low warming scenario (less than 1°C) is unlikely to affect the vernalization of high-chill fruit (apple, walnut, apricot almond, cherry varieties), and if warming scenario exceeds 1.5 °C and would significantly increase the risk of prolonged dormancy for both stone-fruit and pome-fruit. Periods of mild weather can upset the accumulated chilling requirements (CR) requiring further periods of cold weather to achieve sufficient hours. Mild winters may result in delayed or irregular flowering, reduced fruit set and an extended flowering period. The CR is a major concern in the marginal temperate area of North Western Himalayas where fruit trees with a low CR have to be grown. Therefore, in some areas, it is impossible to grow cherries. If chilling is inadequate, the development and/or the later expansion of leaf and flower buds may be impaired. Problems have already been experienced with poor cropping of blackcurrant after mild winters and the same might happen with raspberry, apple and other fruits as winter temperatures continue to increase.

(iv) Effect on Pollination

More than 70 per cent of orchards have less than 20 per cent pollinizer proportion, whereas, a minimum of 30-33 per cent is required in our agro climatic conditions for good fruit set. Moreover, there is lack of diversity in pollinizing cultivars as mainly Golden Delicious and Red Gold are being predominantly used which have attained biennial bearing tendency and their bloom seldom coincides with the flowering period of Delicious cultivars. The population of natural pollinators has gone down due to indiscriminate use of pesticides and deterioration in ecosystem. Managed bee pollination is very limited and available bee hives during bloom hardly meet 2-3 per cent of the demand. All these factors have led to poor fruit setting of Delicious (Kjohl *et al.*, 2011).

(v) Effect on Fruit Ripening

High temperatures on fruit surface caused by prolonged exposure to sunlight hasten ripening and other associated events. One of the classical examples is that of grapes, where berries exposed to direct sunlight ripened faster than those ripened in shaded areas within the canopy. For fruits exposed to direct sunlight, pulp temperatures reached 35°C and required 1.5 days longer to ripen than those that grew in the shade (Woolf *et al.*, 1999). Cell wall enzyme activity (cellulose and polygalacturonase) was negatively correlated with fruit firmness, indicating that sun exposure, *i.e.*, higher temperatures during growth and development, can delay ripening. However, this delay did not occur via a direct effect on the enzymes associated with cell wall degradation. In apples, treatments of 38 and 40°C for 2-6 days did not have marked effects on respiration, although ethylene production was reduced. High temperatures on fruit surface caused by pronounced exposure to sunlight can hasten ripening and other associated events.

(vi) Physiological Disorders and Tolerance to High Temperatures

Frequent exposure of apple fruit to high temperatures, such as 40°C, can result in sunburn, development of watercore and loss of texture. Moreover, exposure to high temperatures on the tree, notably close to or at harvest, may induce tolerance to low-temperatures in post harvest storage (Buescher, 1979).

(vii) Effects of Higher CO_2 and GHG on Fruit Yield and Quality

Carbon dioxide is important because carbon atoms form the structural skeleton of the plant. A doubling of carbon dioxide levels may increase plant growth by 40-50 per cent though continuous high levels saturate the plant's ability to use carbon dioxide and the benefits decrease with time. If other factors remain favorable, increased carbon dioxide concentrations will lead to greater rates of photosynthesis in plants. Current carbon dioxide concentrations limit plant photosynthesis. Growers of protected horticultural crops have already aware from so many years that artificially raising the concentration of carbon dioxide up to certain stage in greenhouses can substantially increase crop growth and yield. Effect on timing of bud burst, cessation of growth, altered concentrations of carbohydrates and plant hormones in turn altered the dormancy status of trees thereby changing the timing of bud burst and the length of the active growing period. Flowering and fruiting of trees are likely to be hastened under conditions of elevated carbon dioxide. The evidence for an effect of carbon dioxide concentration on leaf senescence and leaf fall is rather contradictory and may be species dependent. Most predictions of the direct effects of carbon dioxide suggested that average yields will increase by about 40-50 per cent with a doubling of carbon dioxide concentrations. Leaves are able to detect and respond rapidly to carbon dioxide concentration. Stomata opening decreases in response to increased carbon dioxide concentration.

(viii) Effect on Incidence of Insect, Pest and Disease

Erratic changes in temperature and precipitation leads to more incidences of insect, pest and disease. In the last few years, the attack of scab, powdery mildew in apple, flee beetle in almost all the temperate fruit crops has been increased.

(ix) Interaction Effect of Temperature and Carbon

The combination of increased temperature and increased carbon dioxide predicted in all climate change scenarios suggests that for some species the growth stimulation may be greater than the 40-50 per cent. The doubling of carbon dioxide concentration combined with a 3°C increase in temperature could lead to 56 per cent stimulation in growth.

(b) Climate Change Impact on Fruit Industry

☆ Changes in the suitability and adaptability of current cultivars as temperatures change, together with changes in the optimum growing periods and locations for fruit crops

☆ Changes in the distribution of existing pests, diseases and weeds, and an increased threat of new incursions.

☆ Increased incidence of physiological disorders such as tip burn and blossom end rot

☆ Greater potential for downgrading product quality *e.g.* because of increased incidence of sunburn

☆ Increases in pollination failures if heat stress days occur during flowering

☆ Increased risk of spread and proliferation of soil borne diseases as a result of more intense rainfall events (coupled with warmer temperatures)

☆ Increased irrigation demand especially during dry periods

☆ Changing reliability of irrigation schemes, through impacts on recharge of surface and groundwater storages

☆ Increased atmospheric CO_2 concentrations will benefit productivity of most fruit crops, although the extent of this benefit is unknown

☆ Increased risk of soil erosion and off-farm effects of nutrients and pesticides, from extreme rainfall events

☆ Increased input costs - especially fuel, fertilizers and pesticides

☆ Additional input cost impacts when agriculture is included in an Emissions Trading Scheme

(c) Priorities for Adapting Climate Change

☆ Identify and build on successful strategies of adaptation by the fruit sector to climate changes already experienced.

☆ Obtain regional climate change scenarios (downscaling) for all fruit growing regions (to 2030) - update as improved scenarios become available.

☆ Develop Impact Assessments for all or major commodities in these regions.

☆ Assess the Vulnerability of all or major regions and/or fruit commodities and Identify current "at risk" production sites (regions) and/or industries.

☆ Identify the long-term (2030 and 2070) opportunities and threats to horticultural regions and cropping systems, as a consequence of climate change - long term adaptation.

☆ Develop (in consultation with growers and their advisors), adaptation strategies which are appropriate, practical, and economically sound.

☆ Review and/or develop where necessary, Best Management Practices (BMP), Good Agriculture Practices (GAP) for fruit cultivation, which include adaptation and mitigation components.

☆ Assess the economic benefits of silvi-horti as well as the benefits it might bring for adaptation and mitigation.

☆ Identify alternative regions that may be suitable for production, to take advantage of these market opportunities.

☆ Investigate the "food miles" concept and the effect decisions on markets and production opportunities for horticulture.

☆ Develop horticulture specific forecasting tools that can be used for climate change and climate variability (especially temperature variability) related decision making at a farm and regional scale

(d) Adaptation Strategies

(i) Breeding Strategies

☆ Pheno-typing of all important fruits genetic wealth to enhancing temperature, moisture stress and genetic enhancement for tolerance to biotic and abiotic stress.

☆ Varieties and rootstocks will be evaluated under controlled temperate moisture stress *etc.* gradient to identify suitable cultivars of all major fruit crops. Experiments on varietal evaluation will also be conducted under natural conditions at different altitudes/conditions.

☆ Marker assisted selection and development of transgenic having resistance to biotic and abiotic resistance.

☆ Development of genotypes having resistance to heat and drought.

☆ Crop diversification.

(ii) Agronomic Strategies

☆ Assessment of the vulnerability and climate risks associated with temperate fruit production system in temperate, tropical and subtropical region.

☆ Development of cropping systems under various agro-climatic conditions.

☆ Improvement in the irrigation and drainage systems.

☆ Development of appropriate tillage and intercultural operations.

☆ Integrated nutrient management.

☆ Integrated pest management.

☆ Integrated weed management.

☆ Development of water harvesting techniques.

(iii) Biotechnological Strategies

☆ Molecular characterization for various traits in relations to biotic and abiotic stress.

☆ Transformation of plants from C3 to C4 plants.

☆ Gene pyramiding against biotic and abiotic stress.

☆ *In vitro* conservation of rare and useful species for future use.

☆ Assessment the carbon sequestration potential of perennial fruit crops production system.

☆ To participate in the international dialogue about greenhouse gas emissions management, global warming and sustainable energy development.

☆ Use of biofuel like diesel from Jetropa and *Pongamia* sp.

☆ The development of nuclear energy.

☆ The use of fuels with lower carbon content, *e.g.*, natural gas, CNG, *Gobber* gas.

☆ Fuel switching, appliance efficiency and use of renewable energy.

☆ Tree planting and forest management.

☆ Waste processing.

(e) Future Research Strategies for Production Optimization

(i) Crop Improvement Strategies

☆ Utilizing the current and future regional climatic scenarios temperate region, a micro-level survey of agro-climatic zones should be conducted to identify sensitive regions with high vulnerability with respect to different fruit crop.

☆ Utilization of wild species and relatives in all the fruit crops.

☆ Introduction of low chilling cultivars of pome, stone and nut fruits.

☆ Diversification with other high value fruit crops like peach, apricot, walnut, kiwi and olive.

☆ Development of new genotypes having resistance to high temperature and CO_2 concentrations.

☆ Development of genotypes having resistance to heat and drought.

☆ Standardization of biotechnological approaches for multiple stress tolerance.

(ii) Development of Agro-techniques

☆ Impact assessment of elevated temperature and CO_2 on growth, development, yield and quality of commercial fruit crop using open top chamber (OTC) and free air CO_2 enrichment system (FACE).

☆ Identification of sensitive stages of crops to weather aberrations.

☆ Monitoring of the phenology of all major fruit crops under changing climate.

☆ Validation of *in-situ* soil moisture conservation practices including indigenous technical know-how to mitigate the impact of drought.

☆ Development of suitable agronomic adaptation measures for reducing the adverse climate related production losses.

☆ Study the impact of climate change and development of technologies on water productivity.

☆ Identify and develop good practices to enhance the adaptation of crop to increased temperature, moisture and nutritional stress.

☆ Identification and mapping of climate resilient as well as climatically vulnerable micro-niches.

☆ Extreme events, such as late spring frost or windstorm, may cause crop failure. Future climate may also increase occurrence of extreme impacts on crops, *e.g.* weather conditions resulting in substantial reduction in yield and quality (for example severe drought or prolonged soil wetness).

☆ To develop a set of high resolution daily based climate change scenarios, suitable for analysis of agricultural extreme events.

☆ To identify climatic thresholds having severe impacts on yield, quality and environment for representative crops and to assess the risks of these thresholds under climate change.

(iii) Plant Protection Strategies

☆ Assessment of the pest and disease dynamics, study of disease triangle and development of prediction models.

☆ Strengthen surveillance of pest and diseases.

☆ Development of ecofriendly pest- ecologies and management strategies and early warning systems.

(iv) Post harvest Management Strategies

☆ Development of cost effective storage techniques.

☆ Development of varieties having longer shelf life.

☆ Studies on mitigation of post harvest spoilage.

(v) HRD and Creating Awareness

☆ Organize seminars/symposia/trainings and conduct field demonstrations, on effective climate resilient technologies.

References

Buescher, R. W. 1979. Influence of high temperature on physiological and compositional characteristics tomato fruits. *Lebensmittel-Wissenen shaft* 26: 237-268.

Bleicher, J.1994. Chemical control of *Phytophtharacactorum*, causal agent of collar rot of apple trees. *Fitopatol Bras.*19:95-98.

Chiew, F.H.S., Whetton, P.H., McMahon, T.A. and Pittock, A.B. 1995. Simulation of the impacts of climate change on runoff and soil moisture in Australian catchments. *Journal of Hydrology* 167: 121–147.

Favis-Mortlock, D. and Boardman, J. 1995. Nonlinear responses of soil erosion to climate change: A modeling study on the UK South Downs. *Catena.*25: 365–387.

Gupta, V. K. and Mir, N. M.1983. Efficacy of fungicides on the survival of *Phytophtharacactorum* propagules in soil and control of disease. *Proc. Int. Cong. Plant. Prot.* Brigton UK. 1:1006.

Kjohl, M., Nielsen, A. and Christian S. N. 2011. Potential effects of climate change on crop pollination. Centre for Ecological and Evolutionary Synthesis (CEES), *Department of Biology, University of Oslo, Norw*ay.

Lal, R.1994. *Soil processes and greenhouse effect.* CRC Lewis Publishers, Boca Raton. 440 pp.

Ravi, S., Breshears, D.D., Huxman, T.E. and D'Odorico, P. 2010. Land degradation in dry lands: Interactions among hydraulic-aeolian erosion and vegetation dynamics. *Geomorphology* 116: 236–245.

Rounsevell, M. D. A. and Loveland, R. J. 1994. Soil responses to climate change. NATO ASI Series I: Global Environmental Change. Vol. 23. *Springer-Verlag.* London. 312 pp.

Scharpenseel, H. W., Schomaker, M. and Ayoub, A. 1990. Soils on a warmer earth. *Elsevier*, Amsterdam, 274 pp.

Sharma, S. K. and Gupta, V. K. 1996a. *Effect of post infection application of chemicals on white root rot of apple in nursery. Natl.Symp.* Plant disease Holistic Approaches Manage. INSOPP, UHF, Nauni-26 (abstract).

Sivakumar, M.V.K. 2011. Climate and land degradation. In Sustaining Soil Productivity in Response to Global Climate Change: *Science, Policy, and Ethics; Sauer, T.J., Norman, J.M., Sivakumar, M.V.K., Eds.;* John Wiley and Sons, Inc.: Oxford, UK. pp. 141–154.

Wani, Mushtaq A. and Shaista Nazir. 2016. Available Nitrogen (N) response through organic Carbon and fertility index of Baramulla district of lesser Himalayas. *Research Journal of Agricultural Sciences* 7(4/5):732-734.

Wani, Mushtaq A. and Wani Zahid M. 2015. Spatial variation of soil available phosphorus in the Dal lake catchment of lesser Himalayas. *International Journal of Environmental Monitoring and Analysis* 3(5):364-372.

Wani, Mushtaq A., Bhat, M. A., Kirmani, N.A. and Shaista Nazir 2013 b. Transformation of zinc and iron in submerged rice soils of Kashmir. *The Indian Journal of Agricultural Sciences* 83(11):1209-16.

Wani, Mushtaq A., J. A. Wani and M. A. Bhat 2009. Potassium forms, their interrelationship and relationships with soil properties in rice soils of Lesser Himalayas. *SKUAST, Journal of Research* 11(1):1-7.

Wani, Mushtaq A., J. A. Wani, M. A. Bhat, N.A. Kirmani, Zahid Mushtaq and Shaista Nazir 2013a. Mapping of soil micronutrients in Kashmir agricultural landscape using ordinary kriging and indicator approach. *Journal of the Indian Society of RemoteSensing* 41(2):319-329.

Wani, Mushtaq A., Shaista Nazir and Shazia Ramzan. 2016b. Evaluation of nutrient index using organic carbon, available P and available K concentrations as a measure of soil fertility in Jhelum River basin, India. *Research Journal of Agricultural Sciences* 7(4/5):709-716.

Wani, Mushtaq A., Shaista Nazir and Shazia Ramzan. 2016c. Potassium supplying power of soils of Lesser Himalayas. *Research Journal of Agricultural Sciences* 7(4/5):720-726.

Wani, Mushtaq A., Shaista Nazir and Zahid M. Wani. 2015a. Spatial variability of available micronutrients in the central parts of lesser Himalayas. *International Journal of Research in Engineering and Applied Sciences* 5(7):201-213.

Wani, Mushtaq A., Shaista Nazir and Zahid M. Wani. 2016a. Spatial variability of some chemical and physical soil properties in Bandipora district of Lesser Himalayas. *Journal of the Indian Society of Remote Sensing.* DOI: 10.1007/s12524-016-0624-z.

Wani, Mushtaq A., Wani Zahid M., Shaista, N., Kirmani, N.A. and Bhat, M.A. 2015b. Spatial variability of DTPA extractable cationic micronutrients in northern part of Lesser Himalayas using GIS approach. *Research Journal of Agricultural Sciences* 6(1):8-14.

Wani, Mushtaq A., Wani, J.A. and Malik, M.A. 2010b. Potassium Sorption Isotherms in relation to its availability in Kashmir rice soils. *SKUAST Journal of Research* 13(1): 93-98.

Wani, Mushtaq A., Wani, J.A., Bhat, M.A., Malik, M.A., Shaista Nazir and Bangroo, S. A. 2011. Macro and Micro-nutrient status of submerged rice soils of Kashmir. *SKUAST Journal of Research* 13(1 and 2):30-37.

Wani, Mushtaq A., Wani, Zahid, M., Bhat, M.A., Kirmani, N.A. and Shaista, N. 2014. Mapping of DTPA extractable cationic micronutrients in soils under rice-and maize ecosystems of Kupwara district in Kashmir-A GIS approach. *Journal of the Indian Society of Soil Science* 62(4):351-359.

Wani, Mushtaq A., Zahid Mushtaq and Shaista Nazir. 2010a. Mapping of micronutrients of the submerged rice soils of Kashmir. *Research Journal of Agricultural Sciences* 1(4): 458-462.

Woolf, A. B., Bowen, J. H., and Ferguson, I. B. 1999. Pre-harvest exposure to the sun influences post harvest responses of 'Hass' avocado fruit. *Post harvest Biology and Technology* 15: 143-153.

Zhang, X.C., Nearing, M. A., Garbrecht, J.D. and Steiner, J.L. 2004. Downscaling monthly forecasts to simulate impacts of climate change on soil erosion and wheat production. *Soil Science Society of America Journal* 68: 1376–1385.

Chapter 11

Macro and Micro Nutrients and their Management in Apple

M. K. Sharma and Rifat Bhat

Division of Fruit Science,
Sher-e-Kashmir University of Agricultural Sciences and
Technology of Kashmir, Srinagar
E-mail: rifat.bhat@rediffmail.com

1. Introduction

Apple plants require all essential nutrients for proper growth and fruiting. Nutrients *viz.* nitrogen, phosphorus, potassium, calcium, magnesium and sulphur are needed in relatively large amounts and are called the major/macro nutrients, while the elements *viz.* boron, iron, zinc, manganese, copper, molybdenum and chlorine are needed in very small amounts and are called minor/trace/micronutrients. Among all these nutrients which are required in higher quantity are nitrogen, phosphorus and potassium (Table 11.1). The main target for the application of nutrients is to achieve good plant growth and fruit quality. A healthy young apple tree must have an annual growth of 40-60 cm and a bearing tree should produce at least 20-30 cm growth. The shoots should be healthy with large number of dark green coloured leaves. Carbon, hydrogen and oxygen are taken up by the plant from soil solution and air. Rest of the elements are either available in the soil or added to the soil through manures and fertilizers. The role of different nutrients, their deficiency symptoms, critical levels, recommendations for nutrient management and correction of nutrient deficiencies through foliar sprays is discussed in this chapter.

Table 11.1: Essential Nutrients for Plant Growth and Fruiting and Forms in which they are taken up by Plant

Nutrient	Symbol	Form taken up by plant
Major/Macro-nutrients		
Primary Nutrients		
Carbon	C	CO_2, HCO_3
Hydrogen	H	H_2O
Oxygen	O	H_2O, O_2
Nitrogen	N	NH_4^+, NO_3^-
Phosphorus	P	$H_2PO_4^-$, HPO_4^{--}
Potassium	K	K^+
Secondary Nutrients		
Calcium	Ca	Ca^{++}
Magnesium	Mg	Mg^{++}
Sulphur	S	SO_2, SO_3, SO_4
Minor/Trace/Micro-nutrients		
Iron	Fe	Fe^{++}, Fe^{+++}, Chelate
Zinc	Zn	Zn^{++}, $Zn(OH)_2$, Chelate
Manganese	Mn	Mn^{++}, Mn^{+++}, Chelate
Copper	Cu	Cu^+, Cu^{++}, Chelate
Boron	B	$B_4O_7^-$, $H_2BO_3^-$, HBO_3^-, BO_3^-
Molybdenum	Mo	MoO_4^-
Chlorine	Cl	Cl^-

2. Major/Macro Nutrients

(a) Primary Nutrients

(i) Carbon, Hydrogen and Oxygen

These three elements constitute about 94 per cent of the dry weight of the plants. They are taken up by the plants from air and soil solution. They are the constituents of proteins, fats and carbohydrates.These non- mineral elements utilize energy from the sun to change carbon dioxide and water into starch and sugars which are foods of the plants, utilized to perform different physiological and biochemical processes.

(ii) Nitrogen

Nitrogen is a constituent of proteins, amino acids and many co-enzymes thus influence the growth, yield and quality of apple. Large single application of nitrogen cause nutrient imbalance in the soil, hence it should be applied with phosphorus and potassium. It is required in large quantity in plants for the formation and enlargement of new cells and tissues as it plays a significant role in the formation of amino acids and proteins.

Leaves of the nitrogen deficient plants turn light green in colour. The oldest leaves are first to lose their green colour and become yellow. The leaves are of small size, shoots are short and thin and the bark of the tree turns light brown to yellow orange in colour. The fruit set is low and fruit size remains small. Fruits become bright red under high light intensity and mature early. Trees attain alternate bearing habit. Fruit bud differentiation and fruit set in the following year may also be reduced. Excessive nitrogen is also harmful to apple plants. The symptoms of excess of nitrogen include excessive shoot growth with abnormally dark green leaves, low fruit set, less flower retention, low fruit quality with less flavour, poor red colour development, less flower bud initiation, less flower bud differentiation and shortened storage life. Most of the calcium related disorders are aggravated by high nitrogen.

Apple trees at their earlier age require a large quantity of nitrogen for producing adequate vegetative growth. Adequate nitrogen help the fruits to retain on the trees. Generally the requirement of nitrogen in apple plant is higher when it becomes active after dormancy and major portion of this requirement is met from nitrogen reserves built up by the plant prior to winter. There should be greater level of nitrogen in apple trees during spring and early summer as at that time the leaves should be of maximum size. Trees grown in sandy soil, receiving less rainfall, does not make adequate growth and those bear heavily require more nitrogen.

(iii) Phosphorus

Phosphorus plays a key role in energy metabolism, enzyme regulation and nucleic acid synthesis. The deficiency of phosphorus causes disruption in metabolism and development of plants. Phosphorus deficient plants remain small with limited root system as phosphorus strengthens the root system and helps in the expansion of roots in the soil. The opening of shoot buds is also affected. The symptoms generally appear on older leaves which turn purple or reddish and fall prematurely. Plants are spindly with small leaves and poor lateral growth. The flowering in phosphorus deficient plants usually occurs early and the fruits remain dull and unattractive with less firmness. Fruit yields are also reduced.

(iv) Potassium

Potassium activates many enzymes in plant system, influences water relations and has indirect involvement in photosynthesis. Thus the deficiency of potassium causes various abnormalities in plant system.In potassium deficient plants, there is reduced growth followed by scorching of leaf margins. Older leaves are affected first and young leaves remain small. Photosynthesis and flowering is reduced. Fruit size and red colour development is poor and plant resistance to drought is also affected. In case of prolonged drought the trees may show potassium deficiency.

Potassium deficiency can also occur in poorly drained soils or when the soil is compact or cool.

(b) Secondary Nutrients

(i) Calcium

Calcium in the form of calcium-pectate has major role in providing mechanical strength to tissue. It maintains cell membrane in functional state. It is also essential for carbohydrate and fat metabolism. Since calcium is immobile in plant so the continuous supply of calcium is must in plants. The deficiency of calcium in apple appear on rapidly growing tissues such as shoot tips, expanding leaves, flowers, fruits and roots. There is necrosis and distortion of young leaves with water soaked areas, drying of marginal tissues and wilting of flower stems. The initial indication of calcium deficiency is upward cupping of margins of youngest leaves. Branches are also malformed and multiple buds appear in plants. Sometimes there is also premature leaf fall. Calcium deficiency is known to cause bitter pit, cork spot, internal break down, cracking, low temperature breakdown, lenticels break down, Jonathan spot, water core, senescence breakdown, superficial scald and softening in storage in apple fruits.

(ii) Magnesium

Magnesium is needed by all the green plants as it is the constituent of chlorophyll and is important for photosynthesis. The deficiency symptoms of magnesium include bright yellow chlorosis in older leaves. The yellowing begins at the tip and margins of the blade and spreads inwards between the veins towards midrib. There is formation of triangular green areas near the leaf base. Chlorosis often appears first in interveinal areas within clearly defined green margins. Leaves become brittle, intercoastal veins are twisted and the leaves fall prematurely and plants become alternate in bearing. In some cases whole leaf may turn yellow before falling.

(iii) Sulphur

Sulphur is a constituent of sulphur containing amino acids like cystine, cysteine and methionine. Other function of sulphur are formation of chlorophyll, oil synthesis and enzyme activation. The deficiency of sulphur occasionally occurs in apple. Younger leaves become pale yellow though neighbouring older leaves have dark green colour. Plants become stunted and exhibit delayed maturity. There is rosette formation of small laterals with small pale leaves develop near terminals. Sometimes interveinal chlorosis also occurs.

(c) Minor/Trace/Micro-nutrients

(i) Iron

Iron is essential for chlorophyll synthesis and enzyme activation. Deficiency of iron shows a distinct pattern of chlorosis of the leaves without affecting the veins.

The veins usually form a green network on the leaf. In case of severe chlorosis, the youngest leaf becomes totally white.The area along the margins of the chlorotic leaves die out, forming brown patches. Growth is reduced and shoots and branches may die back. Tree is less productive and young fruits do not grow. Sometimes, lime induced chlorosis due to iron deficiency also occur. In abnormally cool weather during the spring, leaves may show chlorosis identical to iron deficiency but this temporary deficiency disappears in warm weather as in very cool weather, iron may be temporarily unavailable.

(ii) Zinc

Zinc is the constituent of many enzymes and proteins. Majority of the apple varieties are susceptible to zinc deficiency. Although zinc is available in adequate quantities in most soils, its uptake by the plant may be reduced due to heavy nitrogen and phosphorus application. Cool weather may prevent or reduce zinc deficiency (Kanwar, 2000).In zinc deficient apple plants, buds along the shoots fail to develop, leaves remain small and narrow also known as 'little leaf' and there is rosette formation at tips. Foliage is generally sparse. In severe cases older leaves may fall. Fruits remain small and misshapen (Swietlik, 2002). These symptoms may or may not be accompanied by chlorosis between the veins of the leaves. Zinc deficiency symptoms are produced in spring growth. In the following year the rosetted branch dies back. If rosette symptoms appear early in the growing season, the apple trees may remain unfruitful (Dart, 2007). In zinc deficient plants, the basal leaves on fruit buds also do not attain a good size, which reduces the overall fruit yield. The bark of the tree is generally rough and brittle. Excessive zinc causes iron chlorosis.

(iii) Copper

It is an activator of many enzymes and is required in very low quantities by apple tree. Copper deficiency can occur in sandy, heavy water logged soils as well as in high pH soils. High phosphorus levels in the soil can also increase the deficiency(George and Michael, 2002).In copper deficiency, the terminal leaves are affected and there is cupping of these leaves. Some leaves on tips of shoots turn yellow. There is drying of growing points and twigs. There are little or no fruits.

(iv) Manganese

It plays a key role in many physiological processes particularly in respiration and nitrogen metabolism. The deficiency of manganese causes interveinal chlorosis starting from younger leaves. Yellowing of leaves occur on the tips and the middle of the shoots. Normal green colour bands are generally wider and clearly defined from the chlorate zones.Upper branches remain small and begun to die back. These areas coalesce and bark gets cracked. Severely affected branches may die. In water logged soils manganese toxicity sometimes develop and cause apple measles. These measles occur due to excess of manganese along with low calcium in the bark. There are small eruptions on the bark of the tree similar to measle spots. Low potassium levels may also cause manganese deficiency (Swietlik and Faust, 1984).

(v) Molybdenum

Molybdenum, an important constituent of nitrate reductase assists in the formation of proteins, starch, amino acids and vitamins. Its deficiency is rare in apple trees, as very soils lack molybdenum and also the plant requirement is extremely low. Molybdenum deficiency symptoms show mild chlorosis of the young leaves which progressively affects the leaf margins (Yogaratnam and Greenham, 1982). The leaves remain small and covered with necrotic spots. The lowest leaves showed severe symptoms and drop from the shoots. Flower formation is also reduced.

(vi) Boron

Boron has its role in carbohydrate metabolism, sugar translocation, pollen grain germination and pollen tube growth, nitrogen metabolism and cellular differentiation. Boron is required by the leaves, wood and fruits. The trees cannot absorb boron if the soils are too wet or dry as they have a limited ability to store mobile boron. Under drought or water logged conditions, the trees cannot use any stored boron, which causes boron deficiency. Boron deficiency in apples generally varies with climatic conditions, varieties and the stage of development and the vigour of the tree. The deficiency symptoms usually appear on the fruit before vegetative parts are affected. The deficiency symptom on the fruit is internal cork or brownish dead tissues inside the fruit which may appear any time shortly after bloom until harvest, (Yogaratnam and Johnson, 1982). The internal cork which appears early, is round or irregular in shape and external cork which develops before the fruit is half-grown are irregular lesions rounded in margin, on the surface of the fruit. These are generally found near the calyx end, but may be all over the surface (Kanwar, 2000). Fruits undergo early abscission if affected soon after bloom. There is retarded growth of apical growing parts. Excess of boron causes yellowing of midrib and large lateral veins in leaves, twig dieback, nodes enlargement, enhanced maturity and reduced storage life of fruits.

(vii) Chlorine

The exact role of chlorine in apple trees is not established, however, it is beleived to influence evolution of oxygen in photosynthesis and osmotic and stomal regulation. In chlorine deficient plants, there is chlorosis of younger leaves which progressively leads to overall wilting of plants.

3. Nutrient Management

(a) Leaf Nutrient Ranges in Apple

Leaf nutrient ranges in apple are classified as deficiency range, normal range and toxicity range. Deficiency range is the range which indicates that the supply of the nutrient is so low that there is reduction in plant growth, yield and fruit quality.

Normal range/optimum range is a range in which the plant growth, yield and fruit quality is satisfactory and there is no need to make any change in the schedule of the manures and fertilizers. However, the toxicity range is the range when definite toxicity symptoms occur for some nutrients and the others cause reduction in plant growth and fruit quality.

Table 11.2: Leaf Nutrient Ranges

Element	Time of sampling	Deficiency range	Normal range	Toxicity range
N (per cent)	July	<1.5	1.7-2.5	NE
P (per cent)	July	<0.13	0.15-0.3	NE
K (per cent)	July	<1.00	1.2-1.9	NE
Ca (per cent)	July	<0.70	1.5-2.0	NE
Mg (per cent)	July	<0.25	0.25-0.35	NE
Mn (ppm)	July	<25	25-150	NE
B (ppm)	July	<20	20-60	NE
Cu (ppm)	July	<4	5-12	NE
Zn (ppm)	July	<14	15-200	NE
Mo (ppm)	July	<0.05	0.10-0.20	NE

Source: Shear and Faust (1980), Note: NE is not established.

Nutrient standards are proposed by Childers (1966), Chapman (1975) and Shear and Faust (1980). The leaf nutrient ranges proposed by Shear and Faust (1980) in apple are given in Table 11.2. However under Indian conditions nutrient ranges are established by Karkara (1987) and Upadhayay and Awasthi (1993) in Himachal Pradesh and compiled by Chundawat (1997) are presented in Table 11.3. These ranges vary from place to place keeping soil and climatic conditions into consideration.

Table 11.3: Leaf Nutrient Optimum Range Established by Karkara (1987) and Upadhaya and Awasthi (1993)

Nutrient	Karkara (1987)	Upadhaya and Awasthi (1993)
N (per cent)	2.3-2.6	2.43-2.65
P (per cent)	0.33	0.173-0.203
K (per cent)	2.2-2.3	1.34-1.74
Ca (per cent)	1.7-2.6	1.29-1.47
Mg (per cent)	0.33-0.40	0.41-0.62
Zn (ppm)	39-109	28.5-44.9
Mn (ppm)	82-138	73.1-94.1
Fe(ppm)	120-152	350-482
Cu (ppm)	15	17.2-24.3

Source: Chundawat (1997).

(b) Recommendation of Nutrient Management

In India, most of the recommendations of fertilizer doses are arbitrary and adhoc. However, it has been suggested that the application of fertilizers may be regulated on the basis of soil test and leaf analysis report. In Jammu and Kashmir for a mature apple tree of 15 years age and above, application of 40-60 kg fully decomposed F.Y.M., 832 g N, 340 g P_2O_5 and 1503 g K_2O is recommended. It has also been advocated that the first dose of fertilizers comprising 1/3rd of N along with full of $P_2O_5$and ½ dose of K_2O should be applied about 3 weeks before expected bloom as a basal dose. Second dose comprising 1/3rd of urea and remaining K_2O may be applied about 3 weeks after fruit-set. 3rd dose of N may be applied during June-July (Anonymous, 2011).

In Himachal Pradesh, for an apple tree of 10 years and above of age, 100 kg F.Y.M., 700 g N, 350g P and 700 g K is recommended. In Arunachal Pradesh, 50 kg F.Y.M., 350 g N, 180 g P_2O5 and 180 g K_2O is recommended and in Tamil Nadu, 25 kg F.Y.M., 250 g N, 1000 g P_2O_5 and 1000 g K_2O is recommended (Bal, 1999).

Table 11.4: Concentration and Time of Application of Foliar Nutrient Sprays in Apple

Nutrient	Nutrient form	Spray concentration	Time of Application
N	Urea	0.5-1.0 per cent	After petal fall as per leaf size
		5.0 per cent	Post-harvest
P	KH_2PO_4	0.1 per cent	2-sprays during June-July at fortnightly interval
K	KNO_3	0.7 per cent	2,4,6 weeks after bloom
	K_2SO_4	1.5 per cent	2-sprays during July-August at 3-weeks interval
Ca	$CaCl_2$	0.3 per cent	After petal fall for cork spot/other deficiencies
		0.5 per cent	Post harvest for bitter pit
Mg	$MgSO_4$	0.25 per cent	2-sprays, first during June and second one month later.
Zn	$ZnSO_4$	0.5 per cent	After petal fall
		1.0 per cent	Post-harvest
B	H_3BO_3	0.1 per cent	Before bloom or after petal fall
Mn	$MnSO_4$	0.4 per cent	After petal fall
Cu	$CuSO_4$	0.2 per cent	2-sprays during June-July at fortnightly interval
Fe	Fe-chelates	As recommended by manufacturer	4-weeks after bloom and again 3-weeks later

*Add slaked lime half the quantity of zinc sulphate and manganese sulphate to avoid phyto-toxicity.

Fertilizers should be broadcasted under the canopy of tree and slightly mixed with soil. In high rainfall areas with steep slopes where the size of the basins is small, band application of coated fertilizers should be preferred over broadcasting. Urea and muriate of potash should be spread under the tree canopy area and mixed with the soil. However, phosphorus fertilizer should be applied in bands in root zone.These should be applied 1.5 feet away from the tree trunk, 1 feet in young trees.Fertilizers should not be applied in too wet and too dry soil conditions. In Himachal Pradesh, retention of phosphorus in soil is good, therefore, it should be applied in alternate years.

(c) Correction of Nutrient Deficiencies through Foliar Sprays

In apple, foliar sprays of nutrients like N, K, Ca, B, Zn, Mn to foliage and fruits give quick response. Nitrogen is frequently deficient and most commonly applied fertilizer in orchards, while the soil application of phosphorus and potassium is usually based on plant and soil test results. Calcium sometime is required in large quantity in apple. Magnesium and boron application is sometimes essential while most other micronutrients are rarely applied through soil. Foliar application of urea may be used to supplement soil application. Urea sprays can be made before or after full bloom or after harvest when leaves are in good condition. Fall application of urea should be given only when shoot growth has stopped otherwise the tree do not develop the desired hardiness. Fall sprays or post harvest sprays of urea are given to make spurs high in nitrogen. Fall application extend the growth thus make extra photosynthates for better root and early spring growth. Fruits of apple require more calcium than the plant thus foliar application to fruits is needed. Boron is not available to the plants during early spring due to cool soil, therefore, foliar sprays are given.

References

Anonymous, 2011. *Temperate fruits package of practices* (Manual) Vol. 1. Published by Directorate of Extension, SKUAST-Kashmir, Shalimar, Srinagar (J&K). pp. 123.

Bal, J.S. 1999. *Fruit Growing*. Kalyani Publishers, Ludhiana. pp. 425.

Chapman, H.D. 1975. *Diagnostic criteria for plants and soils*. Eurasia Publishing Houses Pvt. Ltd., New Delhi. pp. 793.

Childers, N.F. 1966. *Fruit nutrition, temperate to tropical*. Horticultural Publication, Rutgers State University, New Brunswick, N.J. pp. 888.

Chundawat, B.S. 1997. *Nutrient management in fruit crops*. Agrotech Publishing Academy, Udaipur, pp. 256.

Dart, J. 2007. Zinc deficiency in apples. *Primefacts*, pp. 395-396.

George, R. and Michael, S. 2002. *Copper for crop production*. In: Nutrient Management, University of Minnesota. pp. 612-624-1222.

Kanwar, S.M. 2000. *Tree Nutrition: Secondary and micro nutrients.*In: Apples, Production Technology and Economics. Tata Mc Graw- Hill Publishing Company Limited, New Delhi. pp. 334-343.

Karkara, B.K. 1987. *Studies on the standardization of nutrient ranges in apple*. Ph.D. Thesis. Dr. Y.S. Parmar University of Horticulture and Forestry, Solan, H.P.

Shear, C.B. and M. Faust. 1980. Nutritional ranges in deciduous tree fruits and nuts. *Horticultural Reviews* 2:142-163.

Swietlik, D. 2002. Zinc nutrition of fruit crops by foliar sprays. *Acta Hort*, 594.

Swietlik, D. and Faust, M. 1984. Foliar nutrition of fruit crops. *Horticultural Reviews* 6:287- 355.

Upadhayay, S.K. and R.P. Awasthi. 1993. Leaf nutrient ranges of plus apple trees in Himachal Pradesh. *Indian J. Hort.*, 50(2): 97-102.

Yogaratnam, N. and Greenham, D.W.P. 1982. The applications of foliar sprays containing nitrogen, magnesium, zinc, and boron to apple trees. I. Effect on fruit set and cropping. *J. Hort. Sci.* 57:151-158.

Yogaratnam, N. and Johnson, D.S. 1982. The applications of foliar sprays containing nitrogen, magnesium, zinc, and boron to apple trees. II. Effects on the mineral composition and quality of fruit. *J. Hort. Sci.* 57:159-164.

Chapter 12

Role of Biotechnology in Apple Crop Improvement

Javid I. Mir[1], Nazeer Ahmed[2], D. B. Singh[1],
O. C. Sharma[1], Anil Sharma[1] and W. H. Raja[1]

[1]ICAR-Central institute of Temperate Horticulture,
Rangreth, Srinagar – 190 007, J&K
[2]Sher-e-Kashmir University of Agricultural Sciences
and Technology, Kashmir, Srinagar
E-mail: javidiqbal1234@gmail.com

1. Introduction

Apple (*Malus × domestica* Borkh.) is one of the most important functional fruit crop of the world having pharmaceutical and nutraceutical values owing to high values of antioxidants and photochemicals that have beneficial effects for human health.Research and development has further improved the qualities of apple through breeding new cultivars, introgression of trait specific genes, mutation for changes trait expression *etc*. Conventional breeding in apple is one of the leading research area playing important role in development of improved varieties in apple but due to lengthy juvenile period of the tree and its large size, requiring a long period of time for evaluation and a large field space, have imposed limitations on apple breeding programs. Due to availability of molecular techniques now it is easy to evaluate the breeding population at very beginning of the life cycle and thus reducing the gap for varietal development by years. Rapid developments in biotechnological breeding have shortened the period of time needed for fruit tree breeding, and such techniques are now being applied to apples. Decoding of the apple genome (Velasco *et al.,* 2010) has provided insight into not only the evolution of this species but also information for clarifying the genetic basis of fruit quality, disease resistance, and growth habit. First, the number of solid markers of disease resistance and fruit character has been increasing, and the development of marker assisted selection (MAS) strategies has accelerated. The development of molecular markers has also facilitated the construction of detailed linkage maps for QTL analysis, revealing chromosome regions associated with various apple traits (Bai *et al.,*2012a; Chagné *et al.,*2012; Devoghalaere *et al.,*2012; Kunihisa *et al.,*2014). Thus due to availability of these molecular markers the breeding programmes have gained the pace and inculcated the accuracy in development of varieties and hybrids in apple. Molecular markers are being used for marker assisted selected, molecular breeding, gene isolation, gene tagging, linkage mapping, map based cloning, association mapping, disease diagnosis *etc.* In addition to molecular markers biotechnological techniques like plant tissue culture and transgenics have played an important role in apple crop improvement (Schaart *et al.,* 2011a; 2011b).

One of the land mark research achievement in apple is the production of fruits showing resistant to browning after cutting an apple and these lines have been allowed to enter the market place in Canada and the USA (Carter, 2012). On the other hand, public concern about GM crops still persists, mainly with regard to the random insertion of a transgene in the genome and the remnant selectable marker gene. Third, in response to public concerns about GM crops, new plant breeding technologies (NPBTs; Lusser *et al.,*2012) have been introduced. NPBTs may allow breakthroughs in crop breeding, and have an enormous impact on apple breeding in the near future. Gene Editing (CRISPER, TALENTS) and other cis genic approaches are getting popular with more acceptability with regard to safety and will play the leading role in future for apple crop improvement with respect to yield and quality. Detailed review about role of biotechnology with special emphasis on progress made in Japan has been discussed by Igarashi *et al.,* 2016. Present review has taken most of the discussion from this review and acknowledging the efforts of these authors made in compilation of the information on role of biotechnology in apple crop improvement. In this chapter role of biotechnology for apple crop improvement will be discussed keeping in view the global prospective.

Plant biotechnology as an interdisciplinary science is able to provide impulses and solutions to important challenges in apple crop improvement programmes by rapid propagation of selected cultivars, conservation of valuable germplasm, genetic improvement, disease management, disease diagnosis, quality improvement *etc.* Since many of these aspects are long term efforts due to perennial nature of apple where breeding progammes take little longer time. The application of biotechnological methods can make considerable contributions, including the application of molecular markers for the identification and conservation of valuable genetic resources, pathogen detection and elimination by *in vitro* methods, new breeding goals in apple *e.g.* resistance breeding against pathogens, molecular determination of internal quality parameters, *etc.*

2. Role of DNA Markers and Genomics

After completion of apple genome sequencing project availability of quality molecular markers is no more a limitation. In apple high density linkage map is available thus making it easy to identify the unlinked markers for mapping and other studies. Due to complete transcriptome analysis of apple genes contributing different traits have been identified and their role has been deciphered to a large extant and thus facilitating the identification of genes responsible for important agronomic traits, and helps the breeding process through MAS. Markers linked to specific traits, such as disease resistance, fruit quality, and growth habit, have been developed (reviewed by Gardiner *et al.,*2007; Keller-Przyby[3]kowicz and Korbin 2013; Mariæ *et al.,* 2010). Therefore with the aid of these molecular markers breeding programmes for much awaited traits like long juvenile period, lateral and spur bearing pattern, tree habit, alternate bearing *etc.* can be addressed.

(a) Disease Resistance

Fungal foliar diseases like apple scab, powdery mildew, marsonina, fire blight *etc.* are the major diseases affecting commercial apple production in many countries. For breeding of resistant apple cultivars, genes and QTLs related to disease resistance, and the linked DNA markers, have been successively identified. Genes for apple scab resistance have been identified, and their global positions have been located on the apple genetic map (Bus *et al.,*2011). Among them, the most intensively studied has been the *Rvi6* (*Vf*) gene from *M. floribunda* 821. The *Rvi15* (*Vr2*) locus was found to contain three candidate genes (of the Toll and mammalian interleukin-1 receptor protein nucleotide-binding site leucine-rich repeat structure resistance gene family) (Galli *et al.,* 2010a, 2010b), and the *Rvi1* (*Vg*) locus was shown to contain 6 ORFs of four putative TIR-NBS-LRR (TNL) genes, a TNL pseudogene, and a serine/threonine protein phosphatase 2A gene (Cova *et al.,* 2015). Clark *et al.,*(2014) have identified two novel scab resistance loci in 'Honeycrisp', and mapped the loci as *Rvi19* and *Rvi20* on LG1 and LG15, respectively. Among the identified scab-resistance genes, *Rvi15*(*Vr2*) and *Rvi6* (*Vf*) have been proven to be practical for transformation of common susceptible cultivars (*Vf*: Belfanti *et al.,*2004; Joshi *et al.*2011).Scab resistant varieties of apple like Shireen, Firdous, Florina, Prima, Liberty, Red Free *etc.* are the source of some scab resistant genes are now widely being used for apple breeding programme.

(b) Fruit Quality

Fruit quality traits like skin color, fruit size, juiciness, acidity, sugar content, luster, fruit shape, texture *etc.* have been studied at biochemical and molecular level. Since these traits are controlled by multiple genes unlike disease resistance which is monogenic in nature. Therefore, association mapping or QTL mapping for tagging these genes is required. Due to long juvenile phase development of mapping population takes a long time which becomes impossible field of study in perennial crops like apple. Thus association mapping is one of the important area of research for gene association for fruit quality traits. Genes responsible for these traits improvement of fruit quality and growth habit is also amajor goal of apple breeding programs worldwide. Some studies have been done where molecular markers have been associated with different fruit quality traits in apple (Gadiner *etal.,*2007; Keller-Przyby[3]kowicz and Korbin 2013; Mariæ *etal.,*2010). In addition traits like bitter pit, skin russeting, fruit size, fruit texture, ingredients and tree habit have been associated with some QTLs (Buti *et al.,*2015; Chang *et al.,*2014; Devoghalaere *et al.,*2012; Falginella *etal.,*2015; Kunihisa *et al.,*2014; Sun *et al.,* 2015;Chagné *et al.,*2014; Kunihisa *et al.,*2014; Longhi *et al.,*2012; Sun *et al.,*2015; Bai *et al.,*2012b; Guan *et al.,*2015; Kunihisa *et al.,* 2014; Morimoto *et al.,*2014; Sun *et al.,* 2015; Bai *et al.,*2012a; Celton *et al.,*2014; Guitton *et al.,*2012; Morimoto and Banno, 2015). These QTLs need to tested for different varieties and gene expression studies for candidate genes is to be done in order to obtain a clear picture regarding trait.

Expression and Follow-up

Apple fruit crop has great level of diversity with respect to fruit quality traits and the biochemical pathways. Contributing for these traits are cross linked and most of these genes cross talk in between the pathways. Therefore, high throughput sequencing followed by transcriptome profiling of different varieties with contrasting traits will further fine tune the role of genes involved for fruit quality traits. Advanced technologies like genomic selection (GS) can be used to obtain genomic breeding values for selection of next-generation parents or potential cultivars for further testing at a very early stage (Desta and Ortiz, 2014). By using an 8K SNP array and a population of 1,200 seedlings, Kumar *et al.,*(2012) evaluated the accuracy of GS, and demonstrated its suitability as analternative approach for fruit trait selection. Furthermore, Bianco *et al.,*(2014) have developed a more high-through putwhole-genome genotyping array (20K) for apple.Due to availability of apple genome after completion of apple genome sequence project achievements in identification of QTLs and genes associated with apple fruit character and growth are significantly higher. Fruit quality traits like skin color, cortical intensity, fruit dimensions, texture, sugars, acidity, esters and other biochemical compounds have been studied recently and genes have been identified for these traits (Kumar *et al.,* 2013; Chang *et al.,* 2014; Kunihisa *et al.,* 2014; Guan *et al.,* 2015; Sun *et al.,* 2015).

(c) Plant Architecture and Growth Habit

Due to modernization of horticultural industry traditional orcharding system is being replaced with modern high density orcharding system wherein plant architecture is being maintained scientifically with minimum vegetative spread and maximum fruit bearing canopy coverage. Thus traits like regular bearing, early maturing, precocity, less pre-harvest drop, wide crotch angles, higher spur development potential, higher rooting capability, columnar nature *etc.* are being targeted and genes are being identified for these traits to develop a designer plant architecture system with natural tendency for bearing higher quality fruits. Studies on biennial bearing (Guitton *et al.,* 2012) have revealed the potential of some genes which can be utilized for induction of regular bearing potential in varieties shows alternate bearing habit. Precocity is one of the important traits which decide when plant will come into after and minimum the age of the bearing plant shows maximum precocity. Flowering genes like FLT, AP-3, AP-2, AG, SEP, FT *etc.* are important in deciding when plant becomes ready for flowering and fruiting. Studies of Kunihisa have shown genes responsible for flowering date and thus can play an important role in utilizing these genes for breeding of varieties with higher precocity. Apple crop shows gametophytic self-incompatibility which is contributed by a locus called as "S allele". If two genotypes share common S- allele pairs then they are incompatible and if they share one allele common then they are partially compatible and if two genotypes have different S-allele pairs then they are fully compatible. Thus S- allele typing is also important factor which has been studied in most of the apple varieties and need to be done for all the commercially grown cultivars along with studies for metaxenia. Fruit size in apple depends on crop density (number of fruits per unit TCSA) and therefore, thinning of fruits sometimes becomes important to obtain desired size of fruits. Some cultivars possess genes responsible for self-thinning and therefore, provide the potential source for this trait (Celton *et al.,* 2014). Genes associated with pre-harvest fruit drop, harvest date and earliness of fruit maturity have been identified in apple (Kunihisa, 2014; Morimoto *et al.,* 2013). Plant architecture is a complex trait which is controlled by multiple genes but single gene mutation has induced columnar nature of growth habit in plants like Moonlight, Sunlight, Redlane, Goldlane *etc.* and all these varieties are compact with columnar growth habit contributed by single "Co" gene. Role of "Co" gene locus in contributing columnar trait in apple has been discussed in detail (Morimoto and Banno, 2015). Development of designer plant canopy does not only involve upper ground canopy management but also root system should be taken care of and there should be balanced growth of shoot and root. Genes involved in rooting capability in apple have been identified (Moriya *et al.,* 2015).

(d) Higher Shelf Life

Apple being climacteric fruit and thus its shelf life depends upon the levels of ethylene and its maturity continues even after harvest and during that period ethylene biosynthesis is also there unlike non-climacteric fruits like kiwi fruit, cherry, strawberry, olive *etc.* Apple shows cultivar wise variation with respect to ethylene content/emission. It is known that apples of high storability like Fuji emit low levels of ethylene. Indeed, it has also been shown that high release levels of endogenous ethylene are associated with a more pronounced flesh softening over ripening. Knowledge of the markers for the genes determining the biosynthetic pathways of ethylene is thus especially important for MAS. The enzymes ACS (1amino cyclopropane 1carbioxylate synthase) and ACO (1aminocyclopropane 1carbioxylate oxidase) or EFE (ethylene forming enzyme), preside over the last two stages of ethylene synthesis and are encoded by two gene families that have been widely studied. Several homologous and heterologous sequences of *ACO* found in the data banks have led to the identification of a functional marker, *MdACO1* specifically involved in ethylene biosynthesis. A specific allele of *ACC synthase* (*MdACS1*) involved in apple ripening has also been reported (Harada *et al.*, 2000). Thus molecular markers can play an important role in tagging these genes for development of low ethylene varieties in apple. Antisense construct development against the genes responsible for ethylene biosynthesis and transformation of those constructs into the apple plants will be the strategy for transgenic low ethylene apple variety development. Complete biosynthetic pathway of ethylene has been deciphered and the gene sequences of key regulatory genes of the pathway are available and can be used for identifying cultivars with minimum potential of ethylene biosynthesis and also cis genetic approaches like CRISPR/Cas9/TALENES *etc.* can be used to down regulate gene expression of those genes for extended shelf life in apple.

3. Plant Tissue Culture Advances

Plant tissue culture refers to the *in vitro* culture of plants from plant parts (tissues, organs, embryos, single cells, protoplasts, *etc.*) on nutrient media under aseptic conditions (Altman, 2000). *In vitro* cultures are now being used as tools for the study of various basic problems in plant sciences. It is now possible to propagate all plants of economic importance in large numbers by tissue culture.

(a) Micropropagation

In apple plant tissue culture techniques have made significant contributions in micro propagation of clonal rootstocks like MM-106, M-9, MM-111, Bud-9 *etc.* and through this technology large scale multiplication and production of disease free clonal rootstocks are now available. Large scale multiplication of elite genotypes through commercial micropropagation holds the promise for regular and ample supply of disease free quality planting materials of scion and rootstocks. Micropropagation on one side provide the supply of disease free clonal rootstocks in large number and on other side the season independent multiplication can be done. Traditional multiplication techniques need space and are time dependent and provide very small number of plants in this crop. Under *in-vitro* conditions multiplication rate can be achieved as high as lakhs of plants can be produced in a single season in the laboratory having very good facilities for multiplication and hardening. The initial cost for establishment of these facilities is high but final out put in the form of large number of quality planting material is very high. Different explants have been used for tissue culture work in apple and good response has been achieved (Table 12.1).

(b) Somaclonal Variation, Somatic Embryogenesis, Double Haploidy, Embryo Rescue and Cryopreservation

In apple induction of mutation under tissue conditions leading to expression of some phenotype or character will provide somaclonal variant. Traits for biotic and a-biotic stress tolerance and quality improvement through somaclonal variation have been obtained. This type character needs to be screened for stability testing either by field testing or by the aid of molecular markers.

Table 12.1: Summary of Achievements in Apple Tissue Culture using different Explants with Various Media

Explant	Media	Response	Reference
Auxiliary bud	MS	Genetic diversity of micro propagated apple plants was checked using RAPD markers	Modgil *et al.*, 2005
Shoot tip	MS	Successful shoot and root proliferation was observed in apple on MS media supplemented with TDZ and IBA	Zimmermen, 1984
Leaf	MS	Leaf explants cultivated on MS medium supplemented with TDZ regenerated better over intenodal shoot segments cultured on medium supplemented with BAP in terms of organogenesis	Erig *et al.*, 2004
Leaf	MS	The best results of shoot regeneration was achieved when using explants of plants grown in plastic vessels with a gas permeating cover in modified MD nutrient medium	Gercheva *et al.*, 2009
Meristem	MS and QL	Meristems grown on MS and QL media supplemented with BAP (0.9-1. 5 mg l⁻¹) formed leaf rosettes after 6 weeks and shoot multiplication.	Golosin and Rodojevic, 1987
Meristem	MS	Shoot proliferation was observed on ½ MS medium supplemented with BAP (0.5-5 mg l⁻¹) in aplle scion cultivar "ougnoe" and "chernomorskoe letneey"	Kataeva and Butenko, 1987
Meristem	MS	Successful *in-vitro* multiplication of clonal rootstocks of apple (M-9, M-26 and MM-106) was achieved on ½ MS supplemented with BA, IBA and GA3	Alkan *et al.*, 1997
Shoots	N6 and MS	MS medium was better than N6 in proliferation of apple dwarf rootstock "Gami Almosi". Active charcoal did not effect *in vitro* proliferation, Phloroglucinol induced transient proliferation and GA induced stable proliferation of shoots.	Rustaei *et al.*, 2009
Auxillary	MS	*In vitro* clonal multiplication of apple rootstock buds and MM-111 was successfully achieved using auxiliary shoot apices buds and shoot apices as explants and MS media supplemented with BA (0.5-1 mg l⁻¹), GA3 (0.5mg l⁻¹) and IBA (0.05-0.1 mg l⁻¹)	Kaushal *et al.*, 2005
Shoot tips	MS	Partial success was achieved in the whole system of virus elimination by a combination of *in vitro* cultures, heat treatment at 39ºC and subsequent removal of apical meristematic region.	Paprstein *et al.*, 2008
Leaf	Media containing BAP and TDZ	The wounded leaf and leaf segments showed best segments response to regeneration. Pre-conditioning the and internodes explants source during a week in darkness conditions	Wulff and Antonio 2002
Leaf	Regeneration medium "containing TDZ and BA or Zeatin	The best results of regeneration were achieved in., Royal Gala" by using 0.5 mg l-1 TDZ.	Dobranszki *et al.*, 2006

Explant	Media	Response	Reference
Auxiliary	MS	Auxiliary buds of apple cv. Tydeman's Early buds collected during the spring season produced significantly higher percentage of explant establishment (75 per cent) as compared to buds collected during other seasons.	Modgil *et al.*, 1999
Anther	Anther media	Homozygous genotypes of apple were induced by anther culture but the rate of embryo production was very low	De Wittee *et al.*, 1998
Anther	Regeneration medium	A successful regeneration has been obtained from anthers by formation of adventitious shoots	Hofer, 1997
Shoots	MS	Survival rates on apple cryopreservation decreased when shoot were cultured on MS medium containing higher concentration of DMSO.	Wenhuang and Sanada, 1995
Cotyledon	MS	Optimum secondary somatic embryogemnesis was obtained by culture of large size somatic embryos or cotyledon like structures on medium containing a combination of NAA/BAO/KIN or TDZ (10µM)	Daigny *et al.*, 1996
Protoplast	MS	TDZ increased the frequency of proto-calluses that develop shoots.	Perales and Schleder, 1993

Source: Nazeer *et al.*, 2011.

Another important area in tissue culture research is induction of somatic embryos (synthetic seeds) which can play an important role in crops like apple which is vegetatively propagated. SEs can maintain the fidelity of the apple cultivars and provide an alternate route of large scale multiplication of true to type varieties/cultivars. Synthetic seed technology is an alternative to traditional micropropagation for production and delivery of cloned plantlets.Several aspects of the technique are still underdeveloped and hinder its commercial application. One of these aspects is the high hand labor requirements and costly procedures for the production of encapsulated explants. Direct organogenesis is an efficient shoot regeneration method and uninodal explants can be used for encapsulated synseed production. Comparison of performance of encapsulated organogenetic explants of M. 26 apple rootstock that were prepared by handand mechanical manipulation revealed that synthetic seeds of M. 26 apple rootstock can be produced through organogenesis from machine processed explants followed by root induction and encapsulation of differentiating propagules (Brischia *et al.*, 2002). The main factors affecting morphogenesis of apple explants are BA concentration, basal medium, leaf explants origin and maturity, explant orientation, and photosynthetic photon flux (Yepes and Aldwinekle, 1994).

Double haploidy is the technique used for production of homozygous lines in short duration and thus has wide application in breeding programmes of apple. Homozygous plants are interesting for genetic studies and to improve breeding efficiency. Production of homozygous apple plants via antherculture was mainly assessed by regeneration of androgenic embryos. Calli derived from antherexplants can be used as a starting material for regeneration and higher percentage of homozygous shoots can be obtained (Norelli *et al.*, 1996).

Breeding programmes involving distant hybridization encounter the problems of incompatibility and embryo degeneration which can be saved or rescued through embryo rescue technology. The plant breeders usually rescue inherently weak, immature or hybrid embryos to prevent degeneration. The successful production of plants from the cultured embryos largely depends upon the maturation stage and the composition of the medium. Abortion of embryos at one or the other stage of development is a characteristic feature of distant hybridization. The continuous demand for new plant materials requires the upgrading of traditional breeding methods to accelerate the production of new and improved genotypes. Biotechnological tools and traditional breeding techniques may successfully be coupled. These new strategies allow the early selection of useful traits in apple trees, while avoiding the problems of the long juvenile period and the long generation times (Roen, 1994). The rescue of hybrid embryos resulted from intra and inter-specific crosses is commonly applied in apple breeding programs aimed at increasing the efficiency of the seed germination and the number of individuals obtained through sexual hybridization (Rubio Cabetas *et al.*, 1997). Successful embryo rescue methodologies have been developed for making distant hybridization a success in apple (Lu and Bridgen, 1996; Palmer *et al.*, 2002; Dantas *et al.*, 2006).

Apple crop is vegetatively propagated and thus lone term storage of seeds will not maintain the fidelity of this crop. Therefore, propagating material in the form of dormant buds is stored for long term. *In vitro* tissue culture techniques offer an alternate way of preserving vegetatively propagated apple cultivars through slow growth but this preservation is for short to medium term. For long term storage dormant buds of apple are stored at -194°Cin liquid nitrogen. The quality and viability of dormant buds under storage is to be checked at intervals so as to confirm their status of viability. Although viability under these conditions is retained but the regeneration power/potential of these buds need to be ascertained at times. *In-vitro* grown apple shoot tips were successfully cryopreserved by vitrification, with an average survival and shoot formation of 80 per cent and 76 per cent respectively. Surviving shoots showed the same rate and regrowth patterns as those of non-treated controls. Shoot tips excised from apple (*Malus domestica* Borkh. cv. Golden Delicious) dormant buds immediately before freezing were successfully cryopreserved using three different methods (vitrification, encapsulation-dehydration and two-step freezing), (Wu *et al.*, 2001).

4. Gene Cloning, Genetic Engineering/Transgenic Technology

Gene cloning for expression of specific genes from apple provided the base for further transgenic studies. Polygalactouronase (PG) was over expressed in transgenic apples, novel phenotypes were observed, reflecting changes in cell adhesion. Apple was found to have at least two homologues of DAD1, an inhibitor of programmed cell death. The cDNA expression of DAD1 was studied by Dong *et al.* (1998). In apple three full length genes *Jt*1, *Jt*2 and *Jt*3 were cloned in pollen cDNA of apple cultivar 'Jonathan'. These three genes are the members of the F-box protein family through the alignment of the sequence which showed that they all contain conserved F-box motifs in N-terminus. Using RT- PCR tissue-specific, haplotype-specific, linkage inheritance indicated that three genes are the novel gene expressed specifically in pollen of Jonathan, *Jt*1 possesses the haplotype diversity, which genetic linkage with *S*9-RNase in hybrid progeny of 'Granny Smith' × 'Jonathan', it was suggested that *Jt*1 is the good candidate for pollen *S*-genes in apple, and was named *MdSLFB9*. *Jt*2 and *Jt*3 has high genetic sequence similarity with the *S*-genes sequences, but without any specific expression. These are apple pollen *SLFB-like* genes and were named *MdSLFR-like* 1 and *MdSLFB-like* 2 respectively (Zhu *et al.*, 2009). In another recent study two *Flowering Locus T* (*FT*)-like genes of apple (*Malus×domestica* Borkh.), namely *MdFT1* and *MdFT2*, have been isolated and characterized (Kodota *et al.*, 2010). *MdFT1* and *MdFT2* were mapped, on distinct linkage groups (LGs) LG 12 and LG 4 with partial homoeology. Furthermore, over expression of *MdFT1* conferred precocious flowering in apple, with altered expression of other endogenous genes such as *MdMADS12*. Wei Yang *et al.* (2011) reported cloning and characterization of a novel gene encoding a DREB1 transcription factor from dwarf apple, *Malus baccata* (Gen Bank accession number: EF582842). Expression of MbDREB1 was induced by cold, drought, and salt stress, and also in response to exogenous ABA.Genetic engineering of plants has become a reality and plant gene transfer is now a fertile field of crop improvement.

Genetic engineering involves manipulation of the genetic material towards a desired end in a directed and pre decided mode. Alternately, this is termed as recombinant DNA technology or gene cloning. Cloning genes from apple has accelerated greatly at present. Current interest in gene cloning in apple centers on its various applications such as (i) development of varieties with columnar growth habit; (ii) development of varieties with stress tolerance; (iii) development of varieties with higher shelf life/delayed ripening features; (iv) longevity of trees; (v) study on self-incompatibility system; (vi) flowering and fruiting regulation. Genetic transformation holds the enormous potential for crop improvement. Key characteristics such as disease resistance can be changed, yet the phenotype remains the same as name recognition is important in apple for marketing. The self-incompatibility and inbreeding depression that exists in Apple (Malus domestica) do prevents breeders from using a back cross technique to introgress genes. The gene(s) of interest can be introgressed but, because breeders must cross back to a different cultivar each time, cultivar identity is lost and a new hybrid is produced. First transgenic crop among fruits/vegetables dates back to 1994 when gene modified (GM) crop (Flavor Savor tomato) was commercialized (Kramer and Redenbaugh, 1994). Cultivation of GM crops has continued to increase steadily and globally over the past few years, and at present, over 15 GM crops are being cultivated in about 25 countries. First genetically modified apple which does not turn brown when cut or bruised, was approved for consumption by the United States Department of Agriculture (Xu, 2013). Transgenic technology can introduce any gene for any character in the crop and thus can solve number of problems we are facing at present. Genes identified for shelf life, texture, skin color, columnar nature, rooting, acidity, ascorbic acid, regular bearing, flowering time, dwarfing etc are potential candidates which can be introduced into apple cultivar through gene stacking/pyramiding for development of single cultivars will all desired traits along with genes for diseases resistance like Vf gene for scab resistance. Transgenic studies in apple for characters like storage quality and shelf-life through incorporation of genes like endo-polygalacturonase (MdPG1) and RNAi technology for inhibition of expression of polyphenol oxidases (PPOs) leading to development of apple cultivars resistant to browning are land mark achievements in apple under transgenic technology. In addition transgenic for development of scab resistant cultivars through introduction of genes like Rvi6/Vf, Lc gene, a bHLH transcription factor hordothionin gene (hth) *etc.* are the significant achievement in this area of research. In addition to transgenic approach cis-genic approach for development of desired plant with specific traits has been utilized in apple (Krens *et al.,*2015). Approaches like CRISPR/Cas9, ZFNallow direct manipulation of target genes for expressing suitable phenotype and involves DNA insertion, replacement, or removal from a genome using artificially engineered nucleases. Trans-grafting approaches in which genetically engineered rootstock for improvement of root specific traits can be grafted with genetically unmodified scion to avoid gene flow and other constraints of transgenics in apple. Genes like rolC, gai, rhi, rin *etc.* can improve the rooting efficiency in rootstocks and thus providing better lower ground status to the plant for anchorage, water utilizing potential, nutrient utilizing potential, disease resistance, drought tolerance *etc.* Thus this technology plays an important role in designing rootstocks with desirable traits for development of transgrafts without the threat of gene transfer/gene flow.

The aim of genetic transformation in apple varieties/cultivars focus on imparting disease/pest resistance, delay in fruit ripening using antisense technology to silence the target gene, impart/improve fruit color, increase sugar level, maintain/balance acid/sugar ratio, improve texture, increase fruit size, induce dwarfing, enhance precocity, improve plant canopy architecture, in duce earliness, prevent physiological disorders *etc.* and the objectives for rootstocks are similar to that of scion cultivars but also include gene for dwarfing, resistance to viral pathogens and genes influencing rooting and propagation. Genetic manipulation in fruit trees, especially the apple, has been the subject of intense research. Several transgenic plants have been produced by *Agrobacterium* mediated gene transfer. Most studies on genetic transformation in apples have focused on improving agronomically important traits in commercial cultivars. A high priority has been given to develop resistance to bacterial and fungal pathogens, such as *Erwinia amylovora, Venturia inaequalis,* and *Podosphaera leucotricha.* In apple, several genes have been targeted for transformation using Agrobacterium-mediated gene transfer, including those of insecticidal crystal protein (ICP) from Bacillus thuringensis and the cowpea trypsin inhibitor (cpTi) (James *et al.,* 1993), lytic peptides for bacterial disease resistance, Vf gene from M. floribunda against scab (Barbieri *et al.,* 2003), npt II and ipt for cytokinins biosynthesis, acetolactate synthase gene against herbicides, endochitinase from Trichoderma against scab and for reduced vigour and rol genes from Agrobacterium rhizogenes. Transformation of 'Jonagold' with antimicrobial peptide genes (A1 AMP) resulted in 28 independent transgenic lines, which are being tested for resistance to apple scab using artificial inoculation assays (Broothaerts *et al.,* 2003). Likewise, the cv. 'Orin' and rootstock 'JM-7' have been transformed with genes encoding the sorbitol metabolizing enzyme, sorbitol-6 phosphate dehydrogenase from apple, chitinase from rice, glucanase from soybean and sacrotoxin from the flesh fly. Szankowski *et al.* (2003) transformed the apple cultivars with stilbene synthase gene from *Vitis vinifera* L. and with a polygalacturonase-inhibiting protein (PGIP) gene from kiwi fruit to impart fungal resistance. A total of 13 transgenic lines were obtained, with some lines showing the evidence of gene silencing. Broothaerts *et al.* (2003) used S-RNAase gene silencing to obtain self-fertile transgenic lines. PG enzyme in ripe apple fruit has been isolated and biochemically characterized as an endo-PG (Wu *et al.,* 1993). The corresponding DNA (MdPG1, formerly GDPG1; Atkinson, 1994) was isolated from apple cv Golden Delicious and shown to hybridize to an mRNA presentin ripe fruit but not in developing fruit or flowers (Atkinson *et al.,* 1998). MdPG1 encoded a protein with 52 per cent amino acid identity to the tomato fruit-specific clone pTOM6. Analysis of the promoter of MdPG1 showed that 532- and 1,460-bp fragments conferred -glucuronidase expression in ripe tomato fruit, but not in flowers, leaves, or developing fruit (Atkinson *et al.,* 1998). Transgenic apple trees were produced containing additional copies of a fruit-specific apple PG gene under a constitutive promoter.

Apple genome project has deciphered most the candidate genes responsible for different traits in apple. In this project putative association of these genes has been given which need to be validated in other apple varieties through gene expression studies. Apple GFDB (apple gene function and gene family) database v1.0 is available at www.applegene.org. This site provides genome sequence of apple and annotation of the 17 apple chromosomes. The genome sequence and the peptide sequence can be downloaded from GDR database. These domains will help in studying function of apple genes (GO analysis), protein domain and functional motif studies (conserved domain analysis), gene family classification, interpro, gene evolution analysis, miRNA, BLAST sequence serach *etc.*

5. Preharvest Value Addition

Since present focus on exploitation of nutraceutical potential of crops and development of nutrient rich crop varieties. Although apple crop is rich in bioactive compounds having medicinal and nutritional values. Apple is having high antioxidative and free radical scavenging potential due to presence of phenolics, flavanoids and other biochemically active molecules. But still this crop can be fortified with genes which can improve this fruit crop for traits like high vitamin content, high mineral content, plantibodies *etc*. Genes can be introduced for expression of antibodies or antigens so that apple can be used as an edible vaccine. Value addition can be done through gene stacking or gene pyramiding experiments for fruit quality traits like disease resistance, uniform fruit size, improved fruit texture, enhanced aroma, resistant to browning, high juice content, better color values, fruit shape, acid/sugar balance *etc*.

6. Post Harvest Management

Biotechnology has played an important role for development of designer fruits with better post harvest characteristics. Post harvest management needs grading of fruits where the need for uniformity is mandatory and this character can be imparted by introducing genes for development of uniform size, shape and color parameters in apple. Uniformity of planting material is ensured by large scale multiplication of apple plants through micropropagation sand uniformity of the produce which is more important can be achieved by regulating the gene expression of quality traits to uniform level. Second most important factor in post harvest management in apple is shelf life which can be extended in climacteric fruits like apple by manipulating the expression of genes involved in the biosynthesis of ethylene. Ethylene is the major contributing factor for ripening and thereby decreasing shelf life and thus inhibiting genes responsible for ethylene biosynthesis (ACC oxidase, ACC synthase, EFP, PG-aseetc.). PG-ase is one of the key enzymes involved in the ripening associated softening of fruits. Apple plant engineered to have the PG-ase antisense RNA did not express PG-ase activity, as result fruits did not soften. But due to this single gene modification ethylene biosynthesis cannot be reduced to high level. But anti sense technology for silencing genes like ACC synthase along with PG-ase can tremendously reduce ethylene metabolism and hence increase shelf life of apples. Due to improved shelf life of apples transportation and storage becomes easy and total cost incurred on storage, transport and development of cold chain facilities for apple can be reduced to high level. Another problem during post harvest storage and transport is occurrence of diseases. Since the use of fungicides under storage conditions is not feasible and therefore development of disease resistant varieties using molecular breeding approaches is required. In apple less emphasis on post harvest disease and pest control as been given but the genes which can contribute resistance to these pests and diseases need to be identified and transferred to commercial varieties for control. The role of genes like *Md-ACO1, Md-ACS1, Md-EXP71* and*Md-PG1* which are texture-related genes has been deciphered and their role in explaining fruit texture characteristics involved for extended shelf life and disease and pest tolerance in fruit crops has been studies. In addition genes like *FLS, LDOX, CHS etc*. which are involved in biosynthesis of flavanoids PGIP, TT10, WAK and CTL1 genes related to cell wall structure indicate the importance of fruit characteristics and biochemical compounds in resistance mechanism and extension of fruit shelf life. One of the landmark achievements in post harvest management of apple fruit was achieved when first genetically modified apple which does not turn brown when cut or bruised, was approved for consumption by the United States Department of Agriculture (Xu, 2013). Since browning in apple occur due to polyphenol oxidation by an enzyme polyphenol oxidase and the product of that reaction gives brown color to the apples when exposed to oxygen. RNA interference mediated gene silencing was done to PPO gene which lead to silencing of expression of this genes and this product is now commercially available in the most GM grown countries.

7. Future Perspectives

Biotechnology has significantly contributed in the areas of molecular marker development, marker assisted selection, gene identification, gene tagging, gene pyramiding, regeneration, micropropagation, somaclonal variation, somatic embryo development, development of transgenics *etc.* in apple. In micropropagation in addition to quality planting material production the conservation of wild species of apples, which hither to are unexploited for their possible use in breeding and face the threat of extinction can be conserved through tissue culture to conserve the bio-diversity. Since now whole apple genome is available and this will provide the base for designing new experimentations for deciphering the role of each and every gene in controlling traits like growth habit, alternate bearing, spur type bearing *etc.* which are controlled by multiple genes. Biotechnology will focus on identifying the pathways lying behind the fruit quality traits like color, aroma, shelf life, stress tolerance, precocity *etc.* In addition to research areas like tissue culture, use molecular markers and transgenics focus need to be imposed upon cis genic approaches like CRISPR/Cas9, TALENS *etc.* which have less regulatory check points and therefore, the scope for further application in the improvement of apple.

References

Aklan, K., Centiner, S., Aka-Kacer, Y., and Yalcin-Mendi, Y., 1997. *In vitro* multiplication of clonal apple rootstocks, M9, M-26 and MM-106 by meristem culture. *Acta Hort.* 441:325-328.

Altman, A., 2000. *Micropropagation of plants, principles and practice.* In: SPIER, R. E. Encyclopedia of Cell Technology. New York: John Wiley and Sons, pp. 916-929.

Atkinson, R.G., Bolitho, K.M., Wright, M.A., Iturriagagoitia -Bueno, T., Reid, S.J, and Ross G.S. 1998. Apple ACCoxidase and polygalacturonase: ripening specific gene expression and promoter analysis in transgenic tomato. *Plant Mol. Biol.* 38:449-460.

Bai, T., Y. Zhu., F. Fernández-Fernández., J. Keulemans., S. Brown and K. Xu 2012a. Fine genetic mapping of the *Co* locus controlling columnar growth habit in apple. *Mol. Genet. Genomics* 287: 437– 450.

Bai, Y., L. Dougherty, M. Li., G. Fazio., L. Cheng and K. Xu 2012b. A natural mutation-led truncation in one of the two aluminum activated malate transporter-like genes at the *Ma* locus is associated with low fruit acidity in apple. *Mol. Genet. Genomics* 287: 663–678.

Barbieri M., Belfanti E., Tartarini S., Vinatzer B., Sansavini S.,Silfverberg-Dilworth E., Gianfranceschi L., Hermann D., Patocchi A and Gessler C 2003. Progress of map-based cloning of the *Vf* resistance gene and functional verification: Preliminary results from expression studies in transformed apple. *Hort. science* 38:329-331.

Belfanti, E., E. Sifverberg-Dilworth., S. Tartarini., A. Patocchi., M. Barbieri., J. Zhu., B.A. Vinatzer., L. Gianfranceschi., C. Gessler and S. Sansavini 2004. The *HcrVf2* gene from a wild apple confers scab resistance to a transgenic cultivated variety. *Proc. Natl. Acad. Sci. USA* 101: 886–890.

Bianco, L., A. Cestaro., D.J. Sargent., E. Banchi., S. Derdak., M. Di Guardo., S. Salvi., J. Jansen., R. Viola., I. Gut *et al.*, 2014.Development and validation of a 20K single nucleotide polymorphism (SNP) whole genome genotyping array for apple (*Malus × domestica* Borkh). *PLoS ONE 9: e110377.*

Brischia, R., Piccioni, E. and Standardi, A. 2002. Micropropagation and synthetic seed InM. 26 apple rootstock (II): A new protocol for production of encapsulated differentiating propagules. *Plant Cell, Tissue and Organ Culture* 68:137-141.

Bus, V., E. Rikkerink., V. Caffier., C.E. Durel and K.M. Plummer 2011. Revision of the nomenclature of the differential host-pathogen interactions of *Venturia inaequalis* and *Malus. Annu. Rev.Phytopathol* 49: 391–413.

Buti, M., L. Poles., D. Caset., P. Magnago., F. Fernández-Fernández., R.J. Colgan., R. Velasco and D.J. Sargent 2015. Identification and validation of a QTL influencing bitter pit symptoms in apple (*Malus × domestica*). *Mol. Breed* 35: 29.

Carter, N. 2012.Petition for determination of nonregulate status: Arctic apples (*Malus × domestica*) event GD743 (Arctic Golden Delicious) and Gs784 (Arctic Granny Smith). Okanagao specialty fruits iNC. Summerland, BC VoH1ZO, Canada.

Celton, J.M., J.J. Kelner., S. Martinez., A. Bechti., A. Touhami, M. James., C.E. Durel., F. Laurens and E. Costes 2014. Fruit self-thinning: A trait to consider for genetic improvement of apple tree. *PLoS ONE 9: e91016.*

Chagné, D., C. Krieger., M. Rassam., M. Sullivan., J. Fraser., C. André.,M. Pindo., M. Troggio., S.E. Gardiner, R.A. Henry *et al.,* 2012. QTL and candidate gene mapping for polyphenolic composition in apple fruit. *BMC Plant Biol.* 12: 12.

Chagné, D., D. Dayatilake., R. Diack., M. Oliver., H. Ireland., A. Watson., S.E. Gardiner., J.W. Johnston., R.J. Schaffer and S. Tustin 2014. Genetic and environmental control of fruit maturation, dry matter and firmness in apple (*Malus × domestica* Borkh.). *Hortic. Res.* 1:14046.

Chang, Y., R. Sun., H. Sun., Y. Zhao., Y. Han., D. Chen., Y. Wang., X. Zhang and Z. Han.2014. Mapping of quantitative trait loci corroborates independent genetic control of apple size and shape. *Sci. Hortic.* 174: 126–132.

Clark, M.D., J.J. Luby., J.M. Bradeen and V.G.M. Bus 2014.*Identification of candidate genes at Rvi19 and Rvi20, two apple scab resistance loci in the 'Honeycrisp' apple*

(*Malus × domestica*). In Plant and Animal Genome XXII Conference. Plant and Animal Genome

Cova, V., P. Lasserre-Zuber., S. Piazza., A. Cestaro., R. Velasco., C.E. Durel and M. Malnoy 2015. High-resolution genetic and physical map of the *Rvi1* (*Vg*) apple scab resistance locus. *Mol. Breed* 35: 16.

Daigny, G., Paul, H., Sangwan, R.S. and Sangwan-Norreal, B.S. 1996. Factors influencing secondary somatic embryogenesis in *Malus × domestica* Borkh (cv Gloster 69). *Plant Cell Reports* 16:153-157.

Dantas, A.C.M., Boneti, J.I., Nodari, R.O. and Guerra M.P., 2006. Embryo rescue from interspecific crosses in apple rootstocks. *Pesq. Agropec. Bras.Brasília* 41(6):969-973

De Wittee, K., Schroeven S., Broothaerts, W. and Keulemans, J. 1998. Anther culture in apple. *Acta Hort.* 484:527-530.

Desta, Z.A. and R. Ortiz 2014. Genomic selection: Genome-wide prediction in plant improvement. *Trends Plant Sci.* 19: 592–601.

Devoghalaere, F., T. Doucen., B. Guitton., J. Keeling., W. Payne, T. Ling., J.J. Ross, I.C. Hallett., K. Gunaseelan., G.A. Dayatilake *et al.*, 2012. A genomics approach to understanding the role of auxin in apple (*Malus × domestica*) fruit size control. *BMC Plant Biol.* 12: 7.

Dobranszki, J., Hudak I., Magyan T.K., Jambor B.E., Galli, Z., and Kiss, E. 2006. How can different cytokinin influence the process of shoot regeneration from apple leaves in "RoyalGala" and "M26". *Acta Hort* 726:191-196.

Dong, Y.-H., Zhan, X.C., Kvarna, A., Atkinson, R.G., Morris, B.A. and Gardner, R.C. 1998. Expression of cDNA from apple encoding a homologue of DAD1, an inhibitor of Programmed cell death. *Plant Science* 139:165-174.

Erig, A.C., Schuch, M.W. and Silva, L.C. da, 2004. Effect of cytokinins and of the origin of the explant in the *in vitro* organogenesis of shoots of apple tree. *Revista Científica Rural 9:113*-120

Falginella, L., G. Cipriani., C. Monte., R. Gregori., R. Testolin., R. Velasco., M. Troggio and S.Tartarini 2015.A major QTL controlling apple skin russeting maps on the linkage group 12 of 'Renetta Grigia di Torriana'. *BMC Plant Biol.* 15: 150.

Galli, P., A. Patocchi., G.A.L. Broggini and C. Gessler 2010b. The *Rvi15* (*Vr2*) apple Scab resistance locus contains three TIR-NBSLRR genes. *Mol. Plant Microbe Interact* 23: 608–617.

Galli, P., G.A.L. Broggini., M. Kellerhals., C. Gessler and A. Patocchi 2010 a. High-resolution genetic map of the *Rvi15* (*Vr2*) apple scab resistance locus. *Mol. Breed.* 26: 561–572.

Gardiner, S.E., V.G.M. Bus., R.L. Rusholme., D. Chagné and E.H.A Rikkerink 2007. Chapter 1 Apple. *In*: Kole, C. (ed.) *Genome mapping and molecular breeding in plants* vol. 4. Fruits and Nuts, Springer.

Gercheva, P., Nacheva L., and Dineva, V., 2009. The rate of shoot regeneration from apple (*Malus domestica*) leaves depending on the in-vitro culture conditions of the source plants. *Acta Hort.*825:71-76

Golosin, B. and Radojevic, L., 1987. Micropropagation of apple rootstocks. *Acta Hort.* 212:589-594

Guan, Y., C. Peace., D. Rudell., S. Verma and K. Evans 2015. QTLs detected for individual sugars and soluble solids content in apple. *Mol. Breed.* 35: 135.

Guitton, B., J.J. Kelner, R. Velasco, S. Gardiner, D. Chagné and E. Costes 2012. Genetic control of biennial bearing in apple. *J. Exp. Bot.* 63: 131–149.

Harada, T., Sunako, T., Wakasa, Y. and Niizeki, M. 2000. An allele of the 1-ACC synthase gene (Md-ACS1) acconts for the low level of ethylene production in climacteric fruits of some apple cultivars. *Theoretical and Applied Genetics* 101:742-746.

Hofer, M. 1997. In-vitro anrogenesis on apple: Optomization of the anther culture. *Acta Hort.* 447:341-344.

Igarashi M., Hatsuyama Y, Harada T and Fukasawa-Akada T 2016. Biotechnology and apple breeding in Japan. *Breeding Science* 66: 18–33

James, R.R., Miller, J.C., and Lighthart, B. 1993. *Bacillus thuringiensis* var. *kurstaki* affects a beneficial insect, the Cinnabar moth (Lepidoptera:Arctiiae). *J. Econ. Entomol.* 86:334–339.

Joshi, S.G., J.G. Schaart., R. Groenwold., E. Jacobsen., H.J. Schouten and F.A. Krens 2011. Functional analysis and expression profiling of *HcrVf1* and *HcrVf2* for development of scab resistant cisgenic and intragenic apples. *Plant Mol. Biol.* 75: 579–591.

Kataeva, N.V. and Butenko R.G., 1987. Clonal micropropagation of apple trees. *Acta Hort.* 212:585-588

Kaushal, N., Modgil M., Thakur, M. and Sharma, D.D. 2005. In-vitro clonal multiplication of apple rootstock by culture of shoot apices and auxillary buds. *Indian Journal of Experimental Biology* 43(6):561-565.

Keller-Przyby[3]kowicz, S. and M. Korbin 2013. The history of mapping the apple genome. *Folia Hort.* 25: 161–168.

Kotoda, N., H. Hayashi., M. Suzuki., M. Igarashi., Y. Hatsuyama., S. Kidou., T. Igasaki., M. Nishiguchi., K. Yano., T. Shimizu *et al.,* 2010. Molecular characterization of *FLOWERING LOCUS T*-like genes of apple (*Malus × domestica* Borkh.). *Plant Cell Physiol.* 51: 561–575.

Kramer, M. and K. Redenbaugh 1994. Commercialization of a tomato with an antisense polygalacturonase gene: The FLAVR SAVRTM tomato story. *Euphytica* 79: 293–297.

Krens, F.A., J.G. Schaart., A.M. van der Burgh., I.E.M. Tinnenbroek-Capel., R. Groenwold., L.P.Kodde., G.A.L. Broggini., C. Gessler and H.J. Schouten 2015.

Cisgenic apple trees; development, characterization, and performance. *Front Plant Sci.* 6: 286.

Kumar, S., D. Chagné, M.C.A.M. Bink., R.K. Volz, C. Whitworth and C. Carlisle 2012. Genomic selection for fruit quality traits in apple (*Malus* × *domestica* Borkh.). *PLoS ONE 7: e36674.*

Kumar, S., D.J. Garrick., M.C. Bink., C. Whitworth., D. Chagné and R.K. Volz 2013. Novel genomic approaches unravel genetic architecture of complex traits in apple. *BMC Genomics* 14: 393.

Kunihisa, M., S. Moriya., K. Abe., K. Okada., T. Haji., T. Hayashi., H. Kim., C. Nishitani., S. Terakami and T. Yamamoto 2014. Identification of QTLs for fruit quality traits in Japanese apples: QTLs for early ripening are tightly related to preharvest fruit drop. *Breed. Sci.* 64: 240–251.

Longhi, S., M. Moretto., R. Viola., R. Velasco and F. Costa 2012. Comprehensive QTL mapping survey dissects the complex fruit texture physiology in apple (*Malus* × *domestica* Borkh.). *J. Exp. Bot.* 63: 1107–1121.

Lu, C. and Bridgen, M.P.1996. Effects of genotype, culture medium and embryo developmental stage of the *in vitro* responses from ovule culture of interspecific hybrids of *Alstroemeria. Plant Science* 116:205-212.

Lusser, M., C. Parisi., D. Plan and E. Rodriguez-Cerezo 2012. Deployment of new biotechnologies in plant breeding. *Nat. Biotechnol.* 30: 231–239.

Mariæ, S., M. Lukiæ., R. Cerovíæ., M. Mitrovíæ and R. Boškovíæ 2010. Application of molecular markers in apple breeding. *Genetika* 42: 359–375.

Modgil M., Mahajan, K., Chakrabarti, S.K.,Sharma, D.R. and Sobti, R.C. 2005. Molecular analysis of genetic stability in micro propagated apple rootstock MM106. *Scientia Horticulturae*104:151-160.

Modgil, M., Sharma, D.R. and Bhardwaj, S.V., 1999. Micropropagation of apple cv. Tydeman's Early Worcester. *Scientia Horticulture* 81(2):179-188

Morimoto, T. and K. Banno 2015. Genetic and physical mapping of *Co*, a gene controlling the columnar trait of apple. *Tree Genet. Genomes* 11: 807.

Morimoto, T., K. Yonemushi., H. Ohnishi and K. Banno 2014. Genetic and physical mapping of QTLs for fruit juice browning and fruit acidity on linkage group 16 in apple. *Tree Genet. Mol. Breed.* 4: 1–10.

Morimoto, T., Y. Hiramatsu and K. Banno 2013. A major QTL controlling earliness of fruit maturity linked to the red leaf/red flesh trait in apple cv. 'Maypole'. *J. Japan Soc. Hort. Sci.* 82: 97–105.

Moriya, S., H. Iwanami, T. Haji, K. Okada, M. Yamada, T. Yamamoto and K. Abe 2015. Identification and genetic characterization of a quantitative trait locus for adventitious rooting from apple hardwood cuttings. *Tree Genet. Genomes* 11: 59.

Nazeer Ahmed, J. I. Mir, M. K. Verma and Hare Krishna 2011. Apple. Ed H. P. Singh, V. A. Parthasarathy and K. Nirmal Babu. *Advances in Horticulture Biotechnology — Regeneration Systems — Fruit Crops, Plantation Crops and Spices* (Volume I), pp. 484, Westville Publishing House, New Delhi. Pp. 191-199.

Norelli, J., Mills J., and Aldwinckle, H. 1996. Leaf wounding increases efficiency of *Agrobacterium*-mediated transformation of apple. *Hortscience* 31:1026-1027.

Palmer, J.L., Lawn, R.L. and Adkins, S.W., 2002. An embryo-rescue protocol for *Vigna* interspecific hybrids. *Australian Journal of Botany* 50: 331-338.

Paprstein, F., Sedlak, J., Polak, J., Svobodova, L., Hassan, M. and Bryxiova, M. 2008. Results of *in vitro* thermotherapy of apple cultivars. *Plant Cell Tissue Organ Cult.* 94:347-352

Perales, E.H.and Schieder, O., 1993. Plant regeneration from leaf protoplasts of apple. *Plant Cell, Tissue and Organ Culture* 34(1):71-76.

Roen, D., 1994. Prospects for shortening the breeding cycle of apple (*Malus x domestica* Borkh.) using embryo culture. I. Reducing the period of cold treatment by hormone application. *Gartenbauwissenschaft* 59: 49-53.

Rubio Cabetas, M.J., Carrera Morales, M. and Socias Company, R.1997. Cultivos de óvulos y embriones en programas de mejora genética de frutales. *Fruticultura Profesional* 84: 58-68.

Rustaei, M., Nazeeri S., Ghadimzadeh M and Hemmaty S., 2009. Effect of phloroglucinol, medium type and some component on in-vitro proliferation of dwarf rootstock of apple (*Malus domestica*). *International Journal of Agriculture and Biology* 11(2):193-195.

Schaart, J.G., F.A. Krens., A.-M.A. Wolters and R.G.F. Visser 2011b.Transformation methods for obtaining marker-free genetically modified plants. *Plant Transformation Technologies* 15: 229–242.

Schaart, J.G., *I.E.*M. Tinnenbroek-Capel and F.A. Krens 2011a. Isolation and characterization of strong gene regulatory sequences from apple, *Malus × domestica. Tree Genet. Genomes* 7: 135–142.

Sun, R., Y. Chang., F. Yang., Y. Wang., H. Li, Y. Zhao., D. Chen, T. Wu., X. Zhang and Z. Han 2015. A dense SNP genetic map constructed using restriction site-associated DNA sequencing enables detection of QTLs controlling apple fruit quality. *BMC Genomics* 16: 747.

Szankowski, I., Briviba, K., Fleschhut, J., Schönherr, J., Jacobsen, H. J. and Kiesecker, H. 2003. Transformation of apple (*Malus domestica* Borkh.) with the stilbene synthase gene from grapevine (*Vitis vinifera* L.) and a PGIP gene from kiwi (*Actinidia deliciosa*). *Plant Cell Rep.* 22:141–149.

Velasco, R., A. Zharkikh., J. Affourtit., A. Dhingra., A. Kalyanaraman., P. Fontana., S.K. Bhatnagar., M. Troggio., D. Pruss., S. Salvi *et al.*, 2010. The genome of the domesticated apple (*Malus × domestica* Borkh.). *Nat. Genet.* 42: 833–839.

Velasco, R., Zharkikh A., Affourtit J *et al.,* 2010. The genome of domesticated apple (*Malus x domestica* Borkh). *Nature Genetics* 42(10):833-843

Wei, Yang., Xiao-Dan, Liu., Xiao-Juan, Chi., Chang-Ai, Wu., Yan-Ze, Li., Li-Li, Song., Xiu-Ming, Liu., Yan-Fang, Wang., Fa-Wei, Wang., Chuang, Zhang., Yang, Liu., Jun-Mei, Zong and Hai-Yan, Li 2011. Dwarf apple MbDREB1 enhances plant tolerance to low temperature, drought, and salt stress via both ABA-dependent and ABA-independent pathways. *Planta* 233:219-229.

Wenhuang, M. and Sanada T.1995. Experiment on the cryopreservation of apple shootys in-vitro. *Acta Hort* 403:59-62.

Wu Y., Zhao, Y., Florent, E., Zhou, M., Zhang, D. and Chen, S. 2001. Cryopreservation of apple dormant buds and shoot tips. *Cryo-letters* 22:375-380.

Wu, Q., Szakacs-Dobozi, M., Hemmat, M., and Hrazdina, G. 1993. Endopolygalacturonase in apples (*Malus domestica*) and its expression during fruit ripening. *Plant Physiol.*102:219-225.

Wullf, S.M. and Antonio P.J., 2002. Regeneration of apple shoots (*Malus domestica* Borkh) cv. Gala. *Rev Bras. Frutic.*24 (2):301-305.

Xu, K. 2013. An overview of Arctic apples: Basic facts and characteristics. *New York Fruit Quarterly* 21: 8–10.

Yepes L.M. and Aldwinekle, H.S., 1994. Factors that effect leaf regeneration efficiency in apple, and effect of antibiotics in morphogenesis. *Plant Cell, Tissue and Organ Culture* 37(3).

Zimmerman, R.H., 1984. Apple. p. 369-395. (Eds. W. R. Sharp., D. A. Evans., P. V. Ammirato and Y. Yamadav.) *Hand book of plant cell culture,* Vol 2. Crop Species. Macmillan, New York.

Chapter 13

Role of Arbuscular Mycorrhizal Fungi in Growth and Mineral Nutrition of Apple

Baby Summuna¹, P. A. Sheikh¹ and Sachin Gupta²

¹Division of Plant Pathology,
Sher-e-Kashmir University of Agricultural Sciences and
Technology of Kashmir, Shalimar, Srinagar – 190 025 J&K
²Division of Plant Pathology,
Sher-e-Kashmir University of Agricultural Sciences and
Technology of Jammu, Chatha – 180 009, J&K
E-mail: sheikh.spa786@gmail.com

1. Introduction

Mycorrhizal fungi are species of fungi that intimately associate with plant roots forming a symbiotic relationship, with the plant providing sugars for the fungi and the fungi providing nutrients such as phosphorus, to the plants. Mycorrhizae are said to be the most important form of symbiosis on earth (Bucking *et al.,* 2012). Mycorrhizal fungi can absorb, accumulate and transport large quantities of phosphate within their hyphae and release to plant cells in root tissue. In natural ecosystems AMF are regular component of rhizosphere microflora and are desirable for sustainable plant-soil system by establishing symbiotic associations with most terrestrial plants (Sharma *et al.,* 2009). AMF inhabit a variety of ecosystems including agriculture lands, forest, grasslands and many stressed environments. They colonize roots of most plants including bryophyte, pteridophyte, gymnosperms and angiosperms (Wang and Zao, 2008).

While various epiphytes and endophytes may contribute to biological control, the ubiquity of mycorrhizae deserves special consideration. Mycorrhizae are formed as the result of mutualist symbiosis between fungi and plants and occur on most plant species. Because they are formed early in the development of the plants, they represent nearly ubiquitous root colonists that assist plants with the uptake of nutrients (especially phosphorus and micronutrients). The vesicular arbuscular mycorrhizal fungi (AMF, also known as arbuscular mycorrhizal or endomycorrhizal fungi) are all members of the zygomycota and the current classification contains one order, the Glomales, encompassing six genera into which 149 species have been classified (Morton and Benny, 1990). Arbuscular mycorrhizae involve aseptate fungi and are named for characteristic structures like arbuscules and vesicles found in the root cortex. Arbuscules start to form by repeated dichotomous branching of fungal hyphae approximately two days after root penetration inside the root cortical cell. Arbuscules are believed to be the site of communication between the host and the fungus. Vesicles are basically hyphal swellings in the root cortex that contain lipids and cytoplasm and act as storage organ of AMF. These structures may present intra- and inter- cellular and can often develop thick walls in older roots. These thick walled structures may function as propagules. During colonization, AMF fungi can prevent root infections by reducing the access sites and stimulating host defense. AMF fungi have been found to reduce the incidence of root-knot nematode (Linderman, 1994). Various mechanisms also allow AMF fungi to increase a plant's stress tolerance. This includes the intricate network of fungal hyphae around the roots which block pathogen infections. Inoculation of apple-tree seedlings with the AMF fungi *Glomus fasciculatum* and *G. macrocarpum* suppressed apple replant disease caused by phytotoxic myxomycetes. AMF fungi protect the host plant against root-infecting pathogenic bacteria. The mechanisms involved in these interactions include physical protection, chemical interactions and indirect effects. The other mechanisms employed by AMF fungi to indirectly suppress plant pathogens include enhanced nutrition to plants; morphological changes in the root by increased lignification; changes in the chemical composition of the plant tissues like antifungal chitinase, isoflavonoids, *etc.*, alleviation of abiotic stress and changes in the microbial composition in the mycorrhizosphere (Linderman, 1994).

2. Benefits of Mycorrhizal Biofertilizer

Mycorrhiza plays a very important role in enhancing the plant growth and yield due to an increased supply of phosphorus to the host plant. Mycorrhizal plants can absorb and accumulate several times more phosphate from the soil or solution than non–mycorrhizal plants. Plants inoculated with endomycorrhiza have been shown to be more resistant to some root diseases. Mycorrhiza increase root surface area for water and nutrients uptake. The use of mycorrhizal biofertilizer helps to improve higher branching of plant roots, and the mycorrhizal hyphae grow from the root to soil enabling the plant roots to contact with wider area of soil surface, hence, increasing the absorbing area for water and nutrient absorption of the plant root system. Therefore, plants with mycorrhizal association will have higher efficiency for nutrients absorption, such as nitrogen, phosphorus, potassium, calcium, magnesium, zinc, and copper; and also increase plant resistance to drought. Some important benefits of mycorrhizal biofertilizer are as follows:

1. Allow plants to take up nutrients in unavailable forms or nutrients that are fixed to the soil. Some plant nutrients, especially phosphorus, dissolve in water in neutral soil. In the extreme acidic or basic soil, phosphorus is usually bound to iron, aluminum, calcium, or magnesium, leading to water insolubility, which is not useful for plants. Mycorrhiza plays an important role in phosphorus absorption for plant. In addition, mycorrhizae help to absorb other organic substances that are not fully soluble for plants to use, and also help to absorb and dissolve other nutrients for plants by storage in the root it is associated with.

2. Enhance plant growth, improve crop yield, and increase income for the farmers. Arising from improved water and essential nutrients absorption for plant growth by mycorrhiza, it leads to improvement in plant photosynthesis, nutrient translocation, and plant metabolic processes. Therefore, the plant has better growth and yield, reduce the use of chemical fertilizer, sometimes up to half of the suggested amount, which in turn increases income for the farmers. As in the trial involving mycorrhizal biofertilizer on asparagus it was observed that, when the farmers used suggested amount of chemical fertilizer together with mycorrhizal biofertilizer, it was found that the crop yield improved by more than 50 per cent, and the farmers' income increased 61 per cent higher than when chemical fertilizer alone was used.

3. Improve plant resistance to root rot and collar rot diseases. Mycorrhizal association in plant roots help plant to resist root rot and collar rot diseases caused by other fungi.

4. It can be used together with other agricultural chemicals. Mycorrhiza are endurable to several chemical substances; for example; pesticide such as endrin, chlordane, methyl parathion, methomyl carbofuran; herbicide such as glyphosate, fuazifopbutyl; chemical agents for plant disease elimination such as captan, benomyl, maneb triforine, mancozed and zineb.

(a) Mutualist Dynamics

Mycorrhizal fungi form a mutualistic relationship with the roots of most plant species. In such a relationship, both the plants themselves and those parts of the roots that host the fungi, are said to be mycorrhizal. Relatively few of the mycorrhizal relationships between plant species and fungi have been examined to date, but 95 per cent of the plant families investigated are predominantly mycorrhizal either in the sense that most of their species associate beneficially with mycorrhizae, or are absolutely dependent on mycorrhizae. The Orchidaceae are notorious as a family in which the absence of the correct mycorrhizae is fatal even to germinating seeds.

(b) Sugar-water/Mineral Exchange

The mycorrhizal mutualistic association provides the fungus with relatively constant and direct access to carbohydrates, such as glucose and sucrose. The carbohydrates are translocated from their source (usually leaves) to root tissue and on to the plant's fungal partners. In return, the plant gains the benefits of the mycelium's higher absorptive capacity for water and mineral nutrients, partly because of the large surface area of fungal hyphae, which are much longer and finer than plant root hairs, and partly because some such fungi can mobilize soil minerals unavailable to the plants' roots. The effect is thus to improve the plant's mineral absorption capabilities (Sellosse *et al.*, 2006).

Unaided plant roots may be unable to take up macronutrients that are chemically or physically immobilized; examples include phosphateions and micronutrients such as iron. One form of such immobilization occurs in many types of clayey soils, or soils with a strongly basic pH. The mycelium of the mycorrhizal fungus can, however, access many such nutrient sources, and make them available to the plants they colonize (Li *et al.*, 2006). Thus many plants are able to obtain phosphate, without using soil as a source. Another form of immobilization is when nutrients are locked up in organic matter that is slow to decay, such as wood, and some mycorrhizal fungi act directly as decay organisms, mobilizing the nutrients and passing some onto the host plants; for example, in some dystrophic forests, large amounts of phosphate and other nutrients are taken up by mycorrhizal hyphae acting directly on leaf litter, bypassing the need for soil uptake. Inga alley cropping, proposed as an alternative to slash and burn rainforest destruction, relies upon mycorrhiza within the Inga Tree root system to prevent the rain from washing phosphorus out of the soil. In some more complex relationships mycorrhizal fungi do not just collect immobilized soil nutrients, but connect individual plants together by mycorrhizal networks that transport water, carbon, and other nutrients directly from plant to plant through underground hyphal networks (Simard *et al.*, 2012).

(c) Disease, Drought and Salinity Resistance and its Correlation to Mycorrhizae

Mycorrhizal plants are often more resistant to diseases, such as those caused by microbial soil-borne pathogens. Furthermore, AMF was significantly correlated with soil physical variable, but only with water level and not with aggregate stability and are also more resistant to the effects of drought. The significance of arbuscular mycorrhizal fungi in alleviation of salt stress is well known and also their beneficial effects on plant growth and productivity. Although salinity can affect negatively arbuscular mycorrhizal fungi, many reports show improved growth and performance of mycorrhizal plants under salt stress conditions (Nikolaou *et al.,* 2003).

(d) Resistance to Insects

Recent research has shown that plants connected by mycorrihzal fungi can use these under ground connections to produce and receive warning signals (Porcel *et al.,* 2012; Babikova *et al.,* 2013). Specifically, when a host plant is attacked by an aphid the plant signals surrounding connected plants of its condition. The host plant releases volatile organic compounds (VOCs) that attract the insect's predators. The plants connected by mycorrhizal fungi are also prompted to produce identical VOCs that protect the uninfected plants from being targeted by the insect. Additionally, this assists the mycorrhizal fungi by preventing the plant's carbon relocation which negatively affects the fungi's growth and occurs when the plant is attacked by herbivores.

(e) Resistance to Toxicity

AM fungi have been found to have a protective role for plants rooted in soils with high metal concentrations, such as acidic and contaminated soils.

3. Application of AMF in Apple Trees

Apple trees (*Malus domestica* Borkh.) are cultivated over a wider geographic range than any other temperate zone fruit tree. Apple cultivars are propagated as scions grafted to clonal rootstocks which have been produced by stooling or mound layering. These rootstocks have been selected for their effects on tree form and vigor. Clones that induce dwarfing are becoming increasingly important, since the trees thus formed facilitate more economic production methods. Although much is known about the effects of rootstock genotype on fruit production and tree form few studies exist in which their mycorrhizal relationships have been examined. Indeed, since early examinations showed the species to be endomycorrhizal, there have been few studies until recently, when some examinations of the interaction of mycorrhizae and nutrient regimes have been made.

The roots of apple trees can support two very different types of mycorrhiza: arbuscular mycorrhizal fungi belong to the kingdom Glomeromycota and transport water and mainly phosphate to the plant. On the other hand, ecotmycorrhizal fungi are either Asco- or Basidiomycota providing the plant with water and mainly nitrogen. Thus, both types of mycorrhiza are important for adaptation of trees to local conditions, but, being fungi, are potentially threatened by the expression of enzymes that attack fungi. Under unsterilized low P field conditions, inoculation of greenhouse-produced apple seedlings by mycorrhizae before field planting can significantly increase the growth at least in the first season after planting as compared to both phosphorus fertilized and unfertilized. These results indicate the benefit of planting apple plants pre-colonized with a highly compatible AMF fungus, even when the plants become colonized by the indigenous AMF flora once they are in the field. Apple trees show a strong dependency on mycorrhizae, and in orchards they form symbiosis with the naturally occurring arbuscular mycorrhizal (AMF) flora. Mycorrhiza benefit apple plants by improving growth and nutrition, involving mainly P and, in some cases, other immobile nutrients such as Zn and Cu. There are numerous reports in the literature on the effects of inoculation with AMF on the assimilation and accumulation of phosphorus and other nutrients by the roots of apple trees. There have been many studies on the use of mycorrhizal fungi to inoculate apple plants grown *in vitro* cultures, which can help in the acclimatization of these plants to field conditions.

Several greenhouse studies have reported growth stimulation, with or without associated increases in plant nutrient concentrations, of mycorrhizal apple seedlings grown under low phosphorus regimes. The AMF-induced growth response in low zinc soil has been noted with apple seedlings.

Apple root systems are composed of roots of a variety of ages, diameters and distributions. All young apple roots are white and succulent with very short (0.025 to 0.075 mm) root hairs. After 1 to 4 weeks the root cortex begins to turn brown and root hairs shrivel. The cortex is sloughed and either the root dies and disintegrates or suberization and secondary thickening occur and the root becomes part of the perennial root system bearing new roots as laterals. An estimated one-half of the dry matter produced by extension roots is shed soon after it is produced. Phosphorus is the element most limiting to plant growth in natural ecosystems due to its low availability. It readily complexes with iron and aluminum at low pH and with calcium at high pH. It is also the element whose uptake is most typically increased by mycorrhizae.

(a) AMF Fungi, Pollutants, Herbicides and Pesticides

Work with AMF strains tolerant to heavy-metal has provided evidence for their rapid adaptation to contaminated soils. It was found that cadmium-tolerant *Glomus mosseae* AMF isolates were responsible for uptake, transport and immobilization of cadmium. Copper (Cu) was absorbed and accumulated in the extraradical mycelium of three AMF isolates, as observed in a study with *Glomus* spp. (Gonzalez-Chávez *et al.*, 2002). Other references indicated resistance of arbuscular mycorrhizal fungi to aluminum. Soil aluminum normally causes significant reduction in tissue Calcium and Magnesium concentrations (Cumming and Ning, 2003). *Glomus caledonicum* seems to be a promising mycorrhizal fungus for bioremediation of heavy metal contaminated soil (Liao *et al.*, 2003). Rufykiri *et al.* (2002) found that AM fungus could uptake and translocate uranium towards the roots. At varying zinc levels, mycorrhizal colonization increases zinc absorption and accumulation in the roots. This may help to explain the alleviation of zinc toxicity at high concentrations. Mycorrhizae were found to ameliorate the toxicity of trace metals in polluted soils growing in soybean and lentil plants. Mycorrhizal colonization, however, was reduced in field plots through applications of the fungicide benomyl as a soil drench.

(b) Restoration of Degraded Areas using AMF Fungi

The soils of disturbed sites are frequently low in available nutrients and lack the nitrogen-fixing bacteria and mycorrhizal fungi usually associated with root rhizospheres. As such, land restoration in semi-arid areas faces a number of constrains related to soil degradation and water shortage. As mycorrhizae may enhance the ability of the plant to scope with water stress situations associated to nutrient deficiency and drought, mycorrhizal inoculation with suitable fungi has been proposed as a promising tool for improving restoration success in semi-arid degraded areas. By stimulating the development of beneficial microorganisms in the rhizosphere, the use of AMF-infected plants could reduce the amount of fertilizer needed for the establishment of vegetation and could also increase the rate at which the desired vegetation becomes established by stimulating the development of beneficial microorganisms in the rhizosphere. Degraded soils are common targets of re-vegetation efforts in the tropics, but they often exhibit low densities of AMF fungi. This may limit the degree of mycorrhizal colonization in transplanted seedlings and consequently hamper their seedling establishment and growth in those areas. Soil inoculation with *G. mosseae* has significantly enhanced plant growth and biomass production in limestone mine spoils (Rao and Tak, 2002).

(c) Root Pathogens

The phenomenon of AMF protecting plants from root pathogens is known from studies involving root-infecting pathogens *e.g. Phytophthora parasitica* or *Fusarium* sp., root-invading nematodes and horticulutural and agricultural species such as tomato (*Lycopersicum esculentum* Mill.), alfalfa (*Medicago sativa* L.), and in grasses. *G. mosseae* induced local and systemic resistance to *P. parasitica* and was effective in reducing symptoms produced by this pathogen (Pozo *et al.,* 2002). Larsen and Bodker (2001), however, found that in severely infected root cortical tissue *G. mosseae* had reduced energy reserves and biomass and did not protect the plant from the biotrophic pathogen, *Aphanomyces euteiches*. In wheat, high levels of colonization by AMF did not protect crop roots from damage by root pathogens Ryan *et al.* (2002).

4. Conclusion

Arbuscular mycorrhizal fungi are ubiquitous in soil habitats and form beneficial symbiosis with the roots of angiosperms and other plants. Most terrestrial plants associate with root colonizing mycorrhizal fungi, which improve the fitness of both the fungal and plant associates. Ubiquitous occurrence and importance of AMF for plant growth is now a well-established fact. Distribution and abundance of AMF vary greatly among different sites *i.e.* natural and man made ecosystems. Natural soil offers consortium of indigenous mycorrhizal fungi and often used as source of inoculum. AMF can be produced on a large scale by pot culture technique. The beneficial use of AMF inoculum in agriculture and raising nurseries has been reported. Fungi play a key role in crop field management with biofertilizers. They are also environment friendly fertilizers and do not cause the pollution of any sort.

References

Babikova, Z., Gilibert, L., Bruce, T.J.A., Birkett, M., Caulfield, J.C., Woodcock, C., Pickett J. A. and Johnson, D. 2013. Underground signals carried through common mycelial networks warn neighbouring plants of aphid attack. *Ecology Letters.* 16 (7): 835–43.

Bucking, H., Liepold, E. and Ambilwade, P. 2012. *The role of the mycorrhizal symbiosis in nutrient uptake of plants and the regulatory mechanisms underlying these transport processes.* In: Dhal, N.K. and Sahu, S.C. (Eds.) Plant Science Intech, Rijeka, pp. 107-138; DOI: 10.5772/52570.

Cumming, Jr. and Ning, J. 2003. Arbuscular mycorrhizal fungi enhance aluminum resistance of broomsedge (L). *J.Exp. Bot.* 54(386):1477-1459.

Gonzalez-Chavez, C., D'Haen J., Vangronsveld, J. and Dodd, J.C. 2002.Copper absorption and accumulation by the extraradical mycelium of different *Glomus* spp (arbuscularmycorrhizal fungi) isolated from the same pollued soil. *Andrpongonvirginicus. Plant Soil* 240(2):287-297.

Liao, J.P., Lin, X.G., Cao, Z.H., Shi, Y.Q. and Wong, M.H. 2003. Interactions between arbuscular mycorrhizae and heavy metals under sand culture experiment. *Chemosphere* 50(6): 847-853.

Linderman, R. G. 1994. Role of AM fungi in biocontrol. Pages 1-25 in: *Mycorrhizae and Plant Health.* F. L. Pfleger and R. G. Linderman, eds. APS Press, St. Paul, MN.

Morton, J. B. and Benny, G. L. 1990. Revised classification of arbuscular mycorrhizal fungi (zygomycetes): a new order glomales, two new suborders, glomineae and gigasporineaeand gigasporaceae, with an amendation of glomaceae. *Mycotaxon* 37:471-491.

Nikolaou, N., Angelopoulos, K. and Karagiannidis, N. 2003. Effects of drought stress on mycorrhizal and non-mycorrhizal Cabernet Sauvignon Grapevine, grafted onto various rootstocks. *Experimental Agriculture* 39 (3): 241–252.

Porcel, R., Aroca, R. and Ruiz-Lozano, J.M. 2012. Salinity stress alleviation using arbuscular mycorrhizal fungi. A review. *Agronomy for Sustainable Development* 32: 181-200.

Pozo, M.J., Cordier, C. and Dumas-Gaudot, E. 2002. Lozalized versus systemic effect of arbuscular Mycorrhizal fungi on defense responses to *Phytophthora* infection in tomato plants. J. *Exp. Bot.* 53(368):525-534.

Rao, A.V. and Tak, R. 2002. Growth of different tree species and their nutrient uptake in limestone mine spoil as influenced by arbuscular mycorrhizal (AM)-fungi in Indian arid zone. *J. Arid Environ.* 51(1):113-119.

Rufykiri, G., Thiry, Y., Wang, L., Delvaux, B. and Declerck, S. 2002. Uranium uptake and translocation by the arbuscular fungus, *Glomus intraradices*, under root-organ culture conditions. *New Phytol.* 156(2):275-281.

Ryan, M.H., Norton, R.M., Kirkegaard, J.A., McCormick, K.M., Knights, S.E. and Angus, J.F. 2002. Increasing mycorrhizal colonisation does not improve growth and nutrition of wheat on vertsols in south-eastern Australia. *J. Agric. Res.* 53(10):1173-1181.

Selosse, M.A., Richard, F., He, X. and Simard, S.W. 2006. Mycorrhizal networks: des liaisons dangereuses.*Trends Ecol Evol.*21 (11): 621–628.

Sharma, D., Kapoor, R. and Bhatnagar, A. R. 2009. Differential growth response of *Curculigoorchoides* to native AMF communities varying in number and fungal components. *European Journal of Soil Biology* 45 (4): 328-333.

Simard, S.W., Beiler, K.J., Bingham, M.A., Deslippe, J.R., Philip, L.J. and Teste, F.P. 2012. Mycorrhizal networks: Mechanisms, ecology and modeling. *Fungal Biology Review* 26: 39-60.

Wang, F. Y. and Zao, Y. S. 2008. Biodiversity of arbuscular fungi in China: A review. *Advances in Environmental Biology* 2: 31-39.

Chapter 14

Crop Weather Relationship of Apple

Raihana Habib Kanth, Latief Ahmad
and Sabah Parvaze

Division of Agronomy,
Sher-e-Kashmir University of Agricultural Sciences and
Technology of Kashmir, Srinagar
E-mail: raihana_k@rediffmail.com

1. Introduction

(a) Apple

Apple is a rosaceous fruit tree that belongs to genus *Malus*. It is the most widely grown fruit tree in the world and is propagated in temperate regions of both northern and southern hemispheres of the world owing its high economic value. The genus has five sections including 122 species and subspecies. Natural varieties of cultivated apple belong to *Malus pumila* Mill,whileas its hybrid varieties belong to *Malus domestica* Bork. Apple is a dicotyledonous tree belonging to family Rosaceae with subfamily Maloideae (former Pomideae) having basic chromosome number 17. Cross breeding of various species of *Malus* led to the development of cultivated apple.

Historically, apple tree owes its origin in South-eastern Europe and Tien Shan mountains of Kazakhstan in Asia. The wild apple of ancient Asia, *Malus pumila var. mitris,* produced hundreds of tiny fruits that were sour having numerous, small, dark brown seeds, hardly a fruit that anyone would anticipate eating. The wild apple of Europe, the main ancestor of the domestic apple, is classified as *Malus sylvestris.*

From economic point of view, apple is the most important fruit grown all over the world. Nearly half of the production is consumed as fresh fruit and most of the rest is processed into apple juice, canned apple sauce, apple jam and apple butter. Dehydrated apples, apple flour, apple dumpling, charoset (apple relish), apple haystacks are other important commercial products.

Apple is grown throughout the temperate zones of the world. In India, the major apple producing regions include Kashmir, Himachal Pradesh, Uttar Pradesh, Kumaon, Assam and Nilgiri Hills. Apple is the principle fruit crop of Jammu and Kashmir and accounts for (154720 hectares) of total area of 342791 hectares of total area under all temperate fruits grown in this state. The annual apple production in the state is 1.3 million metric tons (2013). Average yield of commercially important apple cultivars per unit area is the highest in the country ranging between 10-12 tonnes/ha. About 330 varieties of apple are known to have been under cultivation in Kashmir valley in around 1972 but only a dozen are propagated at present on commercial scale. Apple is now available throughout the country and RTS apple juice has become a popular drink.

(b) Climate

Most of the apple varieties require about 1500 hours below 7 °C to break the rest period. The average temperature should be 21 °C to 24 °C during growing season of the crop. Locations having frequent frost and hail storms should be avoided for cultivation of apple. In general, a cool climate with low winter temperature and little rainfall in summer are most suitable for apple cultivation.

(c) Varieties

The most important factor in the cultivation of apple is proper selection of varieties which have commercial value and are also suitable for effective cross-pollination. A large number varieties of apple are found in Jammu and Kashmir. There are 113 varieties of apple in the state. The important varieties being Delicious, American, *Ambri, Maharaji, Kesari, Hazaratbali.* Some of the important varieties of apple cultivated in Kashmir valley with their approximate time of harvest are listed in Table 14.1.

Table 14.1: Important Varieties of Apple Cultivated in Kashmir Valley with their Approximate Time of Harvest

Varieties	Approximate Harvest Time
Irish Peach (*Sur Saharanpuri*)	2nd-3rd week of July
Benoni (*Hazratbali Saharanpuri*)	3rd-4th week of July
American Mother (*Shand Gund*)	2nd-3rd ',seek of August
Shireen	3rd week of August
Johanthan (*Janathan*)	4th of August
Firdous	4th week of August
Shalimar Apple–1	4th week of August to Ist week of September

Varieties	Approximate Harvest Time
Cox's Orange Pippin (*Kesri Trel*)	5th week of August to lst week of September
Shalimar Apple–2	6th week of August to 2nd week of September
Red Gold	2nd-3rd week of September
Queen's Apple/Quince Apple (*Behi Chont*)	3rd week of September
Rome Beauty	3rd week of September
Scarlet Siberian (*Kichhami Trel*)	4th week of September
King of Pippin	3rd week of September to lst week of October
Golden Delicious	4th week of September to 2nd week of October
American Apirouge(*American Trel*)	lst week of October
Kerry Pippin (*Phokla*)	lst week of October
Lal Ambri(*Lal Ambur*)	lst week of October
Sunhari(Sona Ambur)	lst week of October
Chamure(*Chamur*)	lst week of October
Royal Delicious (*Starking Delicious*)	lst week of October to 2nd week of October
Red Delicious	lst week of October to 2nd week of October
Ambri (*Ambur*)	lst week of October to 2nd week of October
Baldwin(*Lal Farashi*)	2nd week of October to 3rd week of October
Yellow Newton (*Khour*)	3rd week of October to 3rd week of October
White Dotted Red (*Versified, Maharaji*)	lst week of november to 2nd week of november

2. Crop Weather Relationship for Apple

Weather is one of the most important factors influencing the growth and development of crops. The annual departures and variations in weather have a detrimental effect on productivity and thus effect food supply. Moreover, the impact of weather is more pronounced on perennial crops as they are influenced by weather throughout the year. In order to maintain a continuous demand, apart from other factors weather should also be taken into consideration for sustainable production. The weather parameters which play a significant role in the growth and development of crops/plants are temperature, humidity, light, wind, solar duration and precipitation. However, five weather parameters have been identified which significantly affect the apple yield. These are rainfall, maximum temperature, minimum temperature, morning relative humidity and evening relative humidity.

(a) Temperature

Temperature is a primary factor affecting the rate of plant development. Phenological events are highly responsive to temperature. Rate of plant growth and development is dependent upon the temperature surrounding the plant and each species has a specific temperature range represented by a minimum, maximum, and optimum. Temperature directly affects the processes of photosynthesis, respiration and transpiration. The rate of these processes increases with an increase in temperature. Temperature plays an important role in various developmental stages

Table 14.2: Critical Spring Temperatures (°C) forApple Tree Fruit Bud Development Stages

Apples									
Apples	Silver tip	Green Tip	Half inch green	Tight Cluster	First Pink	Full Pink	First Bloom	Full Bloom	Post Bloom
Old temp	16	16	22	27	27	28	28	29	29
10% kill	15	18	23	27	28	28	28	28	28
90% kill	2	10	15	21	24	25	25	25	25

of the apple. Critical spring temperatures for apple tree fruit bud development stages is given in Table 14.2.

Old standard temperature is the lowest temperature that can be endured for 30 minutes without damage. This chart also shows the temperature that will kill 10 per cent and 90 per cent of normal fruit buds (WSU EB0913).

(b) Chilling Requirements

The amount of cold needed by a plant to resume normal spring growth following the winter period is commonly referred to as its "chilling requirement." During the fall and winter, deciduous fruit plants enter a dormant period which is generally referred to as the plants' "rest period." Plants enter the rest period in the fall as air temperatures begin to drop below 50 °F, leaf fall occurs, and visible growth ceases. Plants enter the dormant, or rest, period as the level of growth-regulating chemicals in buds changes. In other words, as the growth-regulating inhibitors increase and the growth-regulating promoters decrease, plants begin their dormant period.

As the chilling requirement of a plant is being satisfied by cold temperatures, the level of promoters begins increasing while the level of inhibitors decreases. The higher levels of promoters in the buds allow normal resumption of growth and flowering in the spring as the chilling requirement is met. The standard chill requirements of apple are 400-1000 hours. However, certain low chill varieties have been developed which require comparatively low chill hours of 100-500 hours.

(c) Growing Degree Days

Plant development depends on temperature. Plants require a specific amount of heat to develop from one point in their life-cycle to another.

GDD are calculated by determining the mean daily temperature and subtracting it from the base temperature needed for growth of the organism. The GDD value for one day is represented by the following equation:

$$GDD = \frac{(T_{max} + T_{min})}{2} - T_b$$

Where,

T_{max} = Daily Maximum Temperature

T_{min} = Daily Minimum Temperature

T_b = Base Temperature

"Growing degree days" (GDD or DD) is a way of assigning a heat value to each day. The values are added together to give an estimate of the amount of seasonal growth plants have achieved. The GDD requirement of apple during different phenophases from dormancy to bloom are given below.

Stage	GDD Requirement
Dormant	-
Silver tip	24.4
Green tip	20.5
Half inch green	35.45
Tight cluster	39.75
Pink	49.5
Bloom	57.3

Approximate values of total GDD of different varieties of apple during the stages from dormancy to harvest are given in Table 14.3.

Table 14.3: Growing Degree Days of different Varieties of Apple

Variety	No. of Days from Bloom to Harvest	Total GDD
Irish Peach (*Sur Saharanpuri*)	90-97	1540
Benoni (*Hazratbali Saharanpuri*)	99-100	1655
Firdous	119-122	2065
American Mother (*Shand Gund*)	116-126	2065
Johanthan (Janathan)	127-132	2239
Shalimar Apple–1	130-132	2257
Cox's Orange Pippin (*Kesri Trel*)	135-147	2445
Shireen	139-144	2463
Shalimar Apple–2	138-151	2516
Red Gold	150-156	2654
Queen's Apple/Quince Apple	155-161	2737
Rome Beauty	158-163	2785
Scarlet Siberian (*Kichhami Trel*)	160-165	2817
King of Pippin	164-174	2907
American Apirouge (*American Trel*)	164-174	2907
Kerry Pippin (*Phokla*)	168-175	2950
Lal Ambri (*Lal Ambur*)	170-176	2963
Sunhari (*Sona Ambur*)	172-177	2991
Chamure (*Chamur*)	174-178	3004
Golden Delicious	174-180	3017
Royal Delicious (Starking Delicious)	175-181	3031
Red Delicious	175-182	3044
Ambri (*Ambur*)	177-182	3057
Baldwin (*Lal Farashi*)	180-187	3105
Yellow Newton (*Khour*)	184-190	3139
White Dotted Red (*Versified, Maharaji*)	204-211	3287

(d) Sunshine

Apple trees need sunlight to produce fruit. Branches exposed to sunlight remain fruitful and produce bigger fruit than branches exposed to less sunlight. When branches are shaded, they stop setting fruit. Light is one of the most important factors in the development of anthocyanin pigments, not only in the apple, but in other kinds of fruit as well, such as pears, peaches,plums, nectarines, *etc.* (Overholser, 1917). Apples must reach a certain state of maturity before red color will develop on exposure to light (Magness, 1928). In other words, certain substances must be present in the fruit before the anthocyanin pigments can be produced by the action of light. This pigment has been identified as flavonol (Pearce, 1931).

3. Apple Scab

Apple scab is caused by a fungus *Venturia inaqequalis*. Apple scab results in symptoms on the aerial parts of the apple tree, including leaves, petioles, flowers, sepals, fruit, pedicels, young shoots, and bud scales. Development of apple scab is favored by rainy, humid and cool spring weather conditions. Scab is characterized by olive green lesions on leaves. Early season infections frequently occur on the lower leaf surface, but lesions can be found on the upper surface as well. Extensive infections can cause early defoliation and may reduce the next year's crop. Small, dark lesions occur on the fruit, often on the sepals or near the calyx end. The spots soon become black. Apple scab infections do not rot the fruit but may cause cracking as the fruit enlarge.

Apple scab over winters in infected apple leaves on the orchard floor. During the winter and early spring, small black pseudothecia develop in the infected leaves on the orchard floor. By early spring, ascospores which serve as the primary inoculum for early season infections are formed inside the pseudothecia.Maturation of the ascospores in the dead leaves on the orchard floor usually occurs at the same time the apple tree is emerging from dormancy. Mature ascospores are present and ready to infect the first green tissue in spring. The percentage of mature ascospores in the orchard generally peaks when apples are at the late pink to early bloom stages of bud development. Mature ascospores are discharged from the pseudothecia by rain and carried up to emerging green tissue in the trees by wind currents. Moisture - dew or rain - is necessary for ascospore discharge and germination, as well as subsequent infection of apple tissue. Olive green, velvety lesions appear 10-28 days after infection by an ascospore. The lesions initiated by ascospores result in primary infections, and in turn, produce spores called conidia.

Conidia are spread from primary lesions by splashing rain drops and wind, and initiate further infections when the combination of temperature and leaf wetness enables them to germinate and become established. These are called secondary infections, and generally occur within a tree or between adjacent trees rather than at a long distance. The secondary cycle can be repeated many times during the growing season. With frequent rainfall, the control of apple scab becomes extremely difficult, particularly if the disease becomes established from primary infections in the spring.

Leaves are most susceptible to infection until they are fully expanded. Old leaves may again become susceptible to the fungus in late season, and previously inhibited mycelia inside the leaf tissues may resume growth, resulting in new visible lesions. This phase of epidemics in autumn has significant implications for disease management because it provides additional primary (ascospores) inoculum next spring.The Mills table relating leaf wetness duration and temperature is used to determine the likelihood that infection will occur if conidia are present (Table 14.4). The Mills table continues to be revised as more data are gathered from different regions.

Table 14.4: Temperature and Moisture Requirement for Apple Scab Infection Periods

		Wetting Period (Hours)			
Average Temperature (F)	Average Temperature (C)	Light Infection	Moderate Infection	Heavy Infection	Incubation Period (days)
78	25.6	13	17	26	...
77	25	11	14	21	...
76	24.4	9.5	12	19	...
63-75	17.2-23.9	9	12	18	9
62	16.7	9	12	19	10
61	16.1	9	13	20	10
60	15.6	9.5	13	20	11
59	15	10	13	21	12
58	14.4	10	14	21	12
57	13.9	10	14	22	13
56	13.3	11	15	22	13
55	12.8	11	16	24	14
54	12.2	11.5	16	24	14
53	11.7	12	17	25	15
52	11.1	12	18	26	15
51	10.6	13	18	27	16
50	10	14	19	29	16
49	9.4	14.5	20	30	17
48	8.9	15	20	30	17
47	8.3	17	23	35	17
46	7.8	19	25	38	17
45	7.2	20	27	41	17
44	6.6	22	30	45	17
43	6.1	25	34	51	17
42	5.5	30	40	60	17

Source: "Mills' Chart": Determined by Mills and modified by A.L. Jones.

References

Magness, J. R., 1928. PTOC. Am. Sot. Hart. SC., 289

Mills, W.D., 1944. *Efficient use of sulfur dusts and sprays during rain to control apple scab*. New York State College of Agriculture. Extension Service.

Overholser, E. L., 1917. Proc. Am. Sot. Hod. SC., 73

Pearce, G. and Streeter, L.R., 1931. A report on the effect of light on pigment formation in apples. *Journal of Biological Chemistry*, *92*(3), pp.743-749.

WSU EB0913: *Washington State University* (WSU) Extension bulletin for apple.

Chapter 15

Pollination Management Studies in Apple Orchards of Himalayas

M. A. Paray, Shoukat Ara, Munazah Yaqoob,
Rizwana Khurshid, G. H. Rather, S. H. Paray,
Waseem A. Bhat, Zubair A. Rather, and Farhana Majid

Research and Training Centre for Pollinators, Pollinizers and Pollination Management, Sher-e-Kashmir University of Agricultural Sciences and Technology of Kashmir, Srinagar
E-mail: manzoor_paray@yahoo.in

1. Introduction

Pollination is the transference of effective pollen grains from the anther of a flower to the receptive stigma of the same flower or of another flower of the same or sometimes allied species. Pollinations are of two kinds and both the methods are widespread in nature.

(a) Self-Pollination or Autogamy

Self-pollination is the transference of pollen grains from the anther of a flower to the stigma of the same flower evidently bisexual. In self-pollination only one flower is concerned to produce the offspring. The self-pollination crop species occupy less than15 per cent. The self-pollination crop species also benefit from cross pollination and hybrids grow these days require pollination in order to bear satisfactory marketable crops.

(b) Cross-Pollination or Allogamy

Cross-pollination is the transference of pollen grains from one flower to another flower borne by the same plant or by two separate plants of the same or allied species irrespective of whether the flowers are bisexual or unisexual. Here two flowers are involved and therefore a mingling of two sets of parental characters takes place resulting in better offspring. The cross pollination crop species occupy more than 85 per cent. Some crops also exhibit often cross pollinated nature. The genetic architecture of such crops is intermediate between self- and cross-pollinated species.

(c) Why is Cross-Pollination Important?

Pollination is crucial for the production of fruit and seed. There are many plants that cannot produce fruit and seed if pollinated by their own pollen and so require cross-pollination. Such plants include those where male and female parts are either borne on separate plants or on separate parts/flowers of the same plant. Cross-Pollination is also essential in those crops where male and female parts are borne on the same flower but they are physically excluded from each other. Cross-Pollination in normally self- Pollinated crops also results in higher yields and better- quality fruit and seed.

☆ Cross-Pollination is important in many partially or fully self-Incompatible/ self-sterile varieties of agricultural and horticultural crops; commercial varieties of cabbage, cauliflower, broccoli, radish, apple, almond, peach, pear, plum, *etc.*

☆ Cross-pollination is also important for fruit and seed production in plants that produce unisexual flowers; *e.g.*, species belonging to families Actinidiaceae, Anacardiaceaae, Cucurbitaceae.

☆ Cross-Pollination enhances the yield and quality of many self-Pollinated crops.

There are several factors that affect the productivity of mountain crops. They include soil fertility; poor quality of planting material; agronomic inputs including irrigation, fertilizer, and/or manuring; and use of pesticides- but pollination plays possibly the single most significant role. Pollination is an essential pre- requisite for fertilization and fruit and seed set. If there is no pollination, no fruits or seeds will be formed, and there will be nothing to harvest. Crops can be divided into two categories; self-fertile (self -Compatible) and self- sterile (self-incompatible). Self –fertile crops include such plants as wheat, rice and maize. They are largely self-pollinated and farmers rarely have any pollination problems with them. In contrast, many commercial varieties of fruit and vegetables are partially or fully self- incompatible. Successful pollination requires the presence of another appropriate compatible plant (*i.e.* a Pollinizer), conditions that ensure synchronized or overlapping flowering in the two plants (the stigma of the commercial variety flowers must be receptive and the same time as the anthers ripen in the Pollinizer flowers), and a pollinizing agent like a bee. To obtain good yields, farmers must ensure that these conditions are met-this is crop pollination management (Partap and Partap, 2002).

Jammu and Kashmir is known as the apple bowl of India both interms of area under plantation and production. It provides livelihood to lakhs of people in the valley, both on farm and off farm sector. The total area under apple cultivation in Jammu and Kashmir is 157280 hectares The Kashmir valley alone has 140150 hectares for apple cultivation while the Jammu region has 17124 hectares under apple cultivation. Apples have emerged as the leading cash crop in Kashmir Himalayan region, assuring great importance in helping many farmers move out of poverty trap. The average yield of apples in Hindkush Himalayas (HKH) region varies from 2.5 to 12.9 tonnes per ha (Partap and Partap, 2002). The figures are very low as compared with the average yield of 25-30 tonnes per ha in Europe and other countries advanced in horticulture. Several factors have been identified that affect apple productivity in Kashmir Himalayans. The social, physiographic and physiological factors are mostly well understood and have been studied in some detail. However, the importance of pollination for maintaining apple productivity and quantity is generally less well recognized, however and the problems of inadequate pollination and poor fertilization due to lack of pollinating insects and inclement weather conditions have received little attention in the region in comparison with other factors, although there is indication that this is the single most important factor for the decline in productivity. For the successful pollination of apple it needs the presence of the appropriate compatible plant known as pollinizer and the carrier of pollen grains known as pollinator. Pollination is one of the most important mechanisms in the maintenance and promotion of biodiversity, and in general, life on earth. Many ecosystems, including agro ecosystems, depend on pollinator diversity to maintain overall biological diversity. Pollinators provide pollination services that are crucial for the productivity of agricultural and natural ecosystems. It has been estimated that over three quarters of the world's crops and over 80 per cent of all flowering plants depend on animal pollinators, especially bees. However, pollinators are currently under threat with declines in pollinator populations and diversity occurring worldwide. This presents a serious threat to agricultural production affecting the livelihood of farmers, national agricultural economies, and food security. Key factors behind this are loss of pollinator habitats and modern agricultural practices, which are dominated by the excessive and indiscriminate use of pesticides and other agro chemicals. The most widely used insect for fruit pollination is the European honey bee *Apis mellifera*. Honey bees are ideal pollinators in many crop systems; each colony produces thousands of foraging workers and colonies can be moved into orchards and fields during the flowering period, their body parts are especially modified to pick up pollen grains, they can work for long hours, show flower consistency and are adopted to different climates (Free, 1993 and Mc Gregor, 1976). Other pollinators like native bees and Syrphid flies play an important andappreciated role in apple pollination of Kashmir Himalaya (Paray *et al*, 2014). Bees are an enormously diverse group.

There are over 20,000 species of bees in the world (Michener, 2007). Approximately 679 species of bees are in India (Ascher and Pickering, 2013). The ecological relationship of the pollinators was recognized long before by Knutson *et al.,* 1990 that cross pollination is the only means of maintaining the ecological diversity. Good pollination improves both yield, and size of fruit (Gautier Hion and Maisels, 1994; Free, 1993). Yields of fruit, legumes and vegetable seeds often have been doubled or tripled by providing adequate number of bees for pollination (Mc Gregor, 1976). Globally the annual contribution of pollinators to the agricultural crop has been estimated at about US \$54 billion (Buchmann and Nabhan, 1996; Kenmore and Kreller, 1998). Among the dipterans pollinators the syrphidae is a large (6000 ± world spp.)family with striking flies of much diversity, and about 500 species have been found in Indian sub-continent so far (Ghorpede *et al.,* 2011) and under the right conditions numerous species can be very abundant and very little work has been done on fly (Diptera) Pollinators (Ssymank *et al.,* 2008). In Jammu and Kashmir this family is represented by 90 species belonging to 40 genera (Ghorpede personal communication).A Number of insect pollinators visiting to various cultivated/wild plants in different agro climatic zones of J&K State has been reported earlier however, negligible amount of work has been carried on enlisting of native bees of Kashmir Himalaya. Whatever work is done, it is by foreign authors who either got the material from different museums or collected them during their expeditions to India. The published work of native bees by these foreigners on the Kashmir Himalaya is scattered throughout the world and hence a through taxonomic review of native bees is the need of an hour, as they play an important role in enhancing production and productivity of apple and other fruit crops. Therefore, the present work was undertaken to explore the diversity and abundance of insect pollinators in apple orchards of Kashmir Himalayas.

2. Estimation Procedure

Extensive field surveys were carried out during the year 2012-2013 and 2014-2015 in approximately 60 commercial apple growing villages of Kashmir valley covering the six important apple producing districts, Baramulla, Shopian, Pulwama, Budgam, Kulgam and Kupwara. The survey included 36 questions related to grower's practices and perceptions about honey bees and native bees as pollinators. A total of 1986 growers in six districts responded to all parts of the survey. The survey included the statistics on the size of orchard, the management practices used (conventional, IPM or organic), the number of apple varieties grown, and the type of flora in and around the apple orchards. The initial survey of apple growers provided baseline information on current management practices, perceptions about the importance of honey bees and native bees in pollination and willingness to adapt practices that would enhance wild bee pollination in apple orchards. During the bloom period (March –April) of apple, the collection of bees in 102 orchards from the six apple growing districts Baramulla, Shopian, Pulwama, Budgam, Kulgam and Kupwara on warm sunny days between 10 am and 2 pm were carried out. The bees were netted visiting apple blossoms using the following two methods "General collecting method" consisted of walking along the rows of apple trees and netting any native bees observed landing on or flying around apple blossoms. The aim of this method was to characterize the diversity of native bee species present in each orchard. The second method used was the "time trial collecting method, which consisting of collecting all the bees (honey bees and native bees) during 15 minute intervals. For 15 minutes bees were observed and were collected along a row of apple trees. The 15 minute timed collections gave information on the abundance (No. of individuals of different species) per unit time.

In order to study the proportion of each species within the local community, species diversity were computed after completing the identification of specimens up to species level with the help of literature and with Taxonomists working in India or abroad. Identified species of insect pollinators is given in the Table 1. Species diversity (H) was calculated by the formula given by Margalef (1958) based on Shannon- Wiener functions as:

$$H= OPi\ (LnPI)$$

where,

Pi =Ni [N], Ni = total number of individuals in a species,

N = Total number of individuals in all species.

Evenness (J) was calculated to estimate the equitability component of diversity using the formula:

J = H/log 10s (Pielou, 1975)

Richness was computed by using formula:

ma= S-1/log 10N (Pielou, 1966)

where,

s = total number of species collected.

The present study provides quantitative evidence in the form of species richness, evenness and characterization of species diversity and abundance of insect pollinators that play an important role in the pollination of apple during flowering season in Kashmir valley.

3. Pollinators in Orchard Ecosystems

(a) Occurrence of Hymenopterous Bee EMS

Field surveys for collection of the bees were carried out during 2012-2016 (March –April). The specimens were identified up to species level. A total of 44 species belonging to 25 genera that fall in 11families, were identified. The species diversity was found highest in the district Shopian (0.24873) followed by district Baramulla (0.24253) and district Kulgam (0.24114). Species richness was highest in district Shopian (42.65017) followed by district Pulwama (37.63) and lowest among six districts was district Kulgam (27.6086). Among the wild bees the *Lassioglossum*

**Table 15.1: Pollinators/Insect Visitors Identified from Apple
Orchards of Kashmir Valley**

Sl.No.	Species	Family	Order
1.	*Xylocopa valga* Gerstaecker	Apidae	Hymenoptera
2.	*Xylocopa violacea* Linn	Apidae	Hymenoptera
3.	*Bombus simillmus* Smith	Apidae	Hymenoptera
4.	*Bombus tunicatus* Smith	Apidae	Hymenoptera
5.	*Bombus trifasciatus* Smith	Apidae	Hymenoptera
6.	*Amegilla fallax* (Smith)	Apidae	Hymenoptera
7.	*Apis cerana* Fabricius	Apidae	Hymenoptera
8.	*Apis mellifera* Linn.	Apidae	Hymenoptera
9.	*Mellitina harrietae* Bingham	Apidae	Hymenoptera
10.	*Lassioglossum himalayense* Bingham	Halictidae	Hymenoptera
11.	*Lassioglossum nursei* Blüthgen	Halictidae	Hymenoptera
12.	*Lassioglossum rugolatum* Smith	Halictidae	Hymenoptera
13.	*Lassioglossum polyctor* Bingham	Halictidae	Hymenoptera
14.	*Lasioglossum marginatum* Brullé	Halictidae	Hymenoptera
15.	*Lasioglossum sublaterale* Blüthgen	Halictidae	Hymenoptera
16.	*Lasioglossum leucozonium* Schrank	Halictidae	Hymenoptera
17.	*Halictus constrictus* Smith	Halictidae	Hymenoptera
18.	*Halictus (Seladonia) propinquus* Smith	Halictidae	Hymenoptera
19.	*Sphecodes tantalus* Nurse	Halictidae	Hymenoptera
20.	*Sphecodes lasimensis* Blüthgen	Halictidae	Hymenoptera
21.	*Andrena patella* Nurse	Andrenidae	Hymenoptera
22.	*Andrena cineraria*Linn.	Andrenidae	Hymenoptera
23.	*Andrena floridula* Smith	Andrenidae	Hymenoptera
24.	*Andrena flavipes* Panzer	Andrenidae	Hymenoptera
25.	*Ceratina hieroglyphica* Smith	Ceratidae	Hymenoptera
26.	*Ceratina propinqua* Cameron	Ceratidae	Hymenoptera
27.	*Ceratina lepida* Smith	Ceratidae	Hymenoptera
28.	*Anthidium conciliatum* Nurse	Megachalidae	Hymenoptera
29.	*Megachile conjuncta* Smith	Megachalidae	Hymenoptera
30.	Megachile sp.	Megachalidae	Hymenoptera
31.	Heriades spp.	Megachalidae	Hymenoptera
32.	*Athalia proxima* Klug	Tenthredinidae	Hymenoptera
33.	*Metasyrphusbucculatus* Rondani	Syrphidae	Diptera
34.	*Sphaerophoria bengalensis* Macqaurt	Syrphidae	Diptera
35.	*Episyrphus balteatus* (Degeer)	Syrphidae	Diptera
36.	*Eristalodes paria* (Bigot)	Syrphidae	Diptera
37.	*Eristalis tenax* Linn	Syrphidae	Diptera
38.	*Eoseristalis cerealis* Fabricius	Syrphidae	Diptera

Sl.No.	Species	Family	Order
39.	*Bibio* sp.	Bibionidae	Diptera
40	*Plecia* sp.	Bibionidae	Diptera
41.	*Scathophaga* sp.	Scathophagidea	Diptera
42.	*Pieris brassicae* Linn	Pieridae	Lepidoptera
43.	*Vanessa cashmirensis* Kollar	Nymphalidae	Lepidoptera
44.	*Osmia* sp.	Megachalidae	Hymenoptera

nursei is highly diversified species in all the six districts. The least diversified species in all the six districts was *Andrena patella* with species diversity index (1.8). The abundance, richness and evenness of insect pollinators of the order Hymenoptera that were found in the sampled commercial fruit orchards particularly on apple are in accordance with Hussain *et al.* (2012), Jasara and Rafi, (2008) and Jasara *et al.* (2000). Abundance, richness and evenness of the hymenopterans pollinator bees was found at peak during the flowering period of apple in the months of April to May. Verma and Partap, (1993) also highlighted the impact of mountain pollinators during spring from the Himalayan region. The list of species diversity index (H) for the insect order Hymenoptera, Diptera and Lepidoptera that were collected during the collection.

(b) Occurrence of Dipteran Pollinators

A total of 6 species belonging to 6 genera that fall in single main important pollinator family Syrphidae in the insect order Diptera were identified during the flowering period of apple in three main apple growing districts of Kashmir valley. The species diversity was found highest in the district Shopian (0.24873) followed by district Baramulla (0.24253) and district Kulgam (0.24114). Species richness was almost equal in district Pulwama (2.55) and in district Baramulla (2.00) followed by district Shopian (1.94). Species evenness was highest in district Shopian (1.97) and equally followed by district Pulwama (0.97) and district Shopian (0.97). Among Syrphids the highly diversified species in district Pulwama is *Metasyrphus bucculatus* (0.15), in district Shopian that of *Eristalis tenax* (0.15) and in district Baramulla *Episyrphus balteatus* (0.15). Abrol, (1993) listed various species of Syrphid flies that play an important role in pollination ecology of apple and other fruits in Jammu and Kashmir. In Europe, Ssymank's, (2001) regional studies in Germany, conducted over 10 years with more than 21,000 flower observations, showed that 80 per cent of the total flowering plants in the area were visited by flower flies, including a number of plants previously thought to be visited by bees only. Jennifer and Gibbert, (1979) observed that Syrphid flies are less abundant in harsh months due to unfavorable weather conditions and their number increases when the conditions become favorable. The list of species diversity index (H) for the insect order Diptera particularly the family Syrphidae that were collected during the collection surveys from different apple orchards of six districts are given in the Table 15.2.

Table 15.2: Rank List Along with the List of Taxa of Order Hymenoptera, Diptera and Lepidoptera Collected from Six Districts of Kashmir Valley

Species	Pul-wama	Species Diversity (H)	Shopian	Species Diversity (H)	Bara-mulla	Species Diversity (H)	Budgam	Species Diversity (H)	Kulgam	Species Diversity (H)	Kup-wara	Species Diversity (H)
Xylocopa valga Gerstaecker	49	0.20771	59	0.20466	39	0.19874	57	0.20835	38	0.23771	27	0.18881
Xylocopa violacea Linn	32	0.15895	41	0.16289	29	0.16493	35	0.15315	19	0.15554	24	0.17533
Bombus simillimus Smith	0	0	13	0.07233	0	0	03	0.02401	02	0.02891	3	0.03846
Bombus tunicatus Smith	25	0.13473	29	0.12912	26	0.15354	26	0.12518	11	0.1068	8	0.08176
Bombus trifasciatus Smith	0	0	33	0.14103	0	0	22	0.11135	09	0.09241	7	0.07402
Amegilla fallax (Smith)	23	0.12723	29	0.12912	08	0.06606	07	0.04727	0	0	3	0.03846
Apis cerana Fabricius	35	0.16849	38	0.15497	34	0.18257	31	0.1412	27	0.1946	22	0.1658
Apis mellifera Linn.	46	0.19996	71	0.22808	52	0.23513	39	0.16442	09	0.09241	9	0.08916
Mellitina harrietae Bingham	18	0.10712	34	0.1439	20	0.12858	22	0.11135	0	0	8	0.08176
Lassioglossum himalayense Bingham	35	0.16849	55	0.19613	32	0.1757	41	0.16982	23	0.17604	23	0.17062
Lassioglossum nursei Blüthgen	21	0.11943	54	0.19394	31	0.17218	45	0.1802	32	0.2155	27	0.18881
Lassioglossum rugolatum Smith	28	0.14548	18	0.09204	13	0.09475	12	0.07148	11	0.1068	9	0.08916
Lassioglossum polyctor Bingham	17	0.10283	10	0.05927	12	0.08938	22	0.11135	18	0.15006	25	0.17993
Lasioglossum marginatum Brullé	56	0.2246	87	**0.24873**	55	0.24253	78	**0.23897**	39	0.24114	32	0.20936

Species	Pul-wama	Species Diversity (H)	Shopian	Species Diversity (H)	Bara-mulla	Species Diversity (H)	Budgam	Species Diversity (H)	Kulgam	Species Diversity (H)	Kup-wara	Species Diversity (H)
Lasioglossum sublaterale Blüthgen	17	0.10283	13	0.07233	17	0.11481	27	0.12849	13	0.12016	33	0.21321
Lasioglossum leucozonium Schrank	10	0.06956	08	0.04989	27	0.15741	32	0.14426	08	0.08477	13	0.11611
Halictus constrictus Smith	12	0.07973	02	0.01631	05	0.04598	15	0.08441	0	0	0	0
Halictus (Seladonia) propinquus Smith	13	0.08459	05	0.03444	0	0	08	0.05245	0	0	0	0
Sphecodes tantalus Nurse	17	0.10283	27	0.12289	35	0.18592	26	0.12518	18	0.15006	11	0.10312
Sphecodes lasimensis Blüthgen	0	0	09	0.05466	03	0.03065	12	0.07148	02	0.02891	3	0.03846
Andrena patella Nurse	12	0.07973	08	0.04989	09	0.0722	21	0.10773	04	0.05011	6	0.0659
Andrena cinerariaLinn.	04	0.03409	07	0.04495	04	0.03857	17	0.09252	0	0	2	0.02779
Andrena floridula Smith	03	0.02704	05	0.03444	02	0.02205	12	0.07148	0	0	1	0.01574
Andrena flavipesPanzer	06	0.0469	02	0.01631	01	0.01241	11	0.06694	0	0	0	0
Ceratina hieroglyphica Smith	0	0	05	0.03444	0	0	05	0.03625	0	0	0	0
Ceratina propinqua Cameron	10	0.06956	02	0.01631	01	0.01241	06	0.04188	0	0	3	0.03846
Ceratina lepida Smith	06	0.04697	03	0.02278	0	0	02	0.01721	0	0	4	0.04823
Anthidium conciliatum Nurse	08	0.05587	02	0.01631	04	0.03857	05	0.03625	0	0	0	0

Species	Pul-wama	Species Diversity (H)	Shopian	Species Diversity (H)	Bara-mulla	Species Diversity (H)	Budgam	Species Diversity (H)	Kulgam	Species Diversity (H)	Kup-wara	Species Diversity (H)
Megachile conjuncta Smith	04	0.03409	07	0.04495	07	0.05967	02	0.01721	12	0.1136	4	0.04823
Megachile rotundata	06	0.04697	05	0.03444	04	0.03857	06	0.04188	10	0.09974	1	0.01574
Heriades spp.	06	0.04697	04	0.02879	02	0.02205	03	0.02401	01	0.01639	0	0
Athalia proxima Klug	03	0.02704	04	0.02879	02	0.02205	05	0.03625	0	0	5	0.05733
Metasyrphus bucculatus Rondani	06	0.04697	03	0.02278	01	0.01241	04	0.03032	0	0	2	0.02779
Sphaerophoria bengalensis Macqaurt	05	0.0407	03	0.02278	0	0	05	0.03625	0	0	2	0.02779
Episyrphus balteatus (Degeer)	07	0.05296	05	0.03444	02	0.02205	02	0.01721	03	0.03998	0	0
Eristalodes paria (Bigot)	05	0.0407	04	0.02879	04	0.03857	03	0.02401	02	0.02891	0	0
Eristalis tenax Linn	05	0.0407	02	0.01631	01	0.01241	04	0.03032	04	0.05011	4	0.04823
Eoseristalis cerealis Fabricius	0	0	03	0.02278	02	0.02205	02	0.01721	01	0.01639	5	0.05733
Bibio sp.	05	0.0407	02	0.01631	01	0.01241	01	0.00963	01	0.01639	2	0.02779
Plecia sp.	11	0.07472	04	0.02879	06	0.05299	01	0.00963	05	0.05952	7	0.07402
Scathophaga sp.	10	0.06956	03	0.02278	07	0.05967	0	0	03	0.03998	16	0.1341
Pieris brassicae Linn	04	0.03409	02	0.01631	02	0.02205	0	0	0	0	4	0.04823
Vanessa cashmirensis Kollar	05	0.0407	02	0.01631	01	0.01241	0	0	01	0.01639	2	0.005305
Osmia sp.	0	0	0		0	0	0	0	0	0		
Total No. of Individuals	585		722		501		677		359		377	
Total No. of species	38		43		37		40		28		35	

Lassioglossum polyctor Bingham   Halictidae   *Sphecodes lasimensis* Blüthgen

Sphecodes tantalus Nurse   Halictidae   *Halictus (Seladonia) propinquus* Smith

Megachile conjuncta Smith
Megachalidae

Mellita harrietae Bingham
Apidae

Andrena patella Nurse
(Andrenidae)

Andrena cineraria Linn.
(Andrenidae)

Bombus simillmus Smith
(Apidae)

Bombus tunicatus Smith
(Apidae)

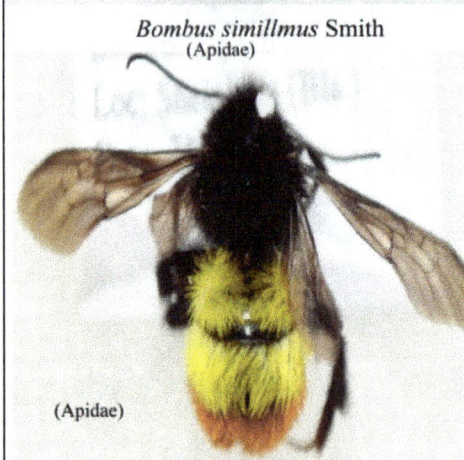

(Apidae)

(Nymphalidae)

Bombus trifasciatus Smith

Vanessa cashmirensis Kollar

Xylocopa valga Gerstaecker
(Apidae)

Xylocopa violacea Linn
(Apidae)

Euglossa sp. (Apidae)

Ceratina hieroglyphica Smith
(Ceratidae)

Ceratina propinqua Cameron
Ceratidae

Ceratina lepida Smith

Lassioglossum marginatum Brullé

Lassioglossum himalayense Bingham

Lassioglossum leucozonium Schrank

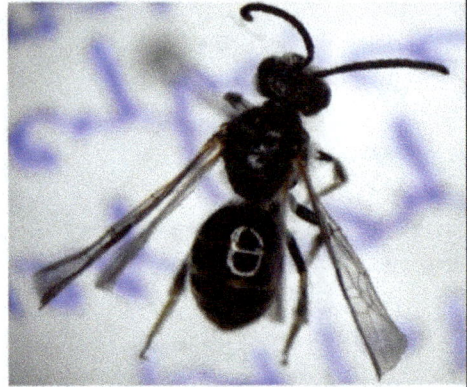

Lassioglossum nursei Blüthgen

Lassioglossum sublaterale Blüthgen

Lassioglossum rugolatum Smith

Family: Halictidae

(c) Occurrence of Lipdopteran Pollinators

Among lipdopterans two species *Pieris brassicae* and *Vanessa cashmirensis* belonging to two families Pieridae and Nymphalidae respectively were collected and identified during the bloom period of apple in Kashmir valley. Among Pieridae the highest species diversity was found in district Shopian (0.01631) and lowest in district Pulwama (0.03409), while in the Nymphalidae, species diversity was found highest in district Shopian (0.01631) and lowest in district Pulwama (0.0407), respectively.

Table 15.3: Species Diversity, Richness and Evenness of Insect Order Hymenoptera, Lepidoptera and Diptera

Sl.No.	District	No. of Species	No. of Individuals	Species Diversity (H)	Species Richness (ma)	Species Evenness (J)
1.	Pulwama	38	585	0.2246	37.63	0.14217
2.	Shopian	43	722	0.24873	42.65017	0.8916
3.	Baramulla	37	501	0.24253	35.62961	0.15584
4.	Budgam	40	677	0.23897	36.467	0.15541
5.	Kulgam	28	359	0.24114	27.6086	0.16663
6.	Kupwara	35	377	0.20936	34.61185	0.13559

4. Synchronized Species of Plants Occurring in Kashmir Valley during Bloom Period of Apple

Native bees respond positively to the abundance of wild flowering plants surrounding the apple orchards in Kashmir valley. During the surveys to different apple orchards of Kashmir valley a total of 29 families containing 64 species of flowering plants were identified that provide a continuous supply of pollen and nectar to the bees before and after apple bloom. Floral resource around the crop peripheries is important to pollinators during the bloom period of fruit plants of the family Rosaceae and other tree fruits (Scott-Dupree and Watson, 1987). Most bees have flight period that extends beyond the bloom and requires a continuous source of nectar and pollen throughout the season (Goulson, 2003; Saure, 1996; Westrich, 1996). Macfarlane and Patten, (1997) examined the relative attractiveness of perennial floral resources to bumble bees around the cranberry beds in the Pacific Northwest throughout the summer. They found a shortage of bumble bees forage resources early and late in the season and suggested that planting for early and late-blooming wild flowering plants may boost bumble bee abundance. This highlighted the necessity of a constant source of nectar and pollen from early spring through the fall. A complete list of flowering plants that constantly supply pollen and nectar to the native bees before and after flowering period of apple is given in the Table 15.4.

Table 15.4: Blooming Months and Availability of Pollens on different Flowering Source to Pollinators in Apple Orchards of Kashmir Valley

Family	Genus/Species	Source	Flowering Time	Pollinators
Amaranthaceae	Chenopodium album	Pollen and nectar	June-Sep	Apis cerana, A. mellifera.
Asteraceae	Dahlia daenranthema	Pollen	Aug- Nov	Apis cerana, A. mellifera, Bombus tunicates, B. trifeciatus, B. simmillmus
	Cicerbita hortons	Pollen and nectar	May- Aug	Apis cerana, A. mellifera
	Chrysanthemum coronarium	Pollen	May- July	Apis cerana, A. mellifera, Bombus tunicates,
	Aster thomsonii	Pollen		Apis cerana, A. mellifera, and xylocpa spp.
	Parthenium hysterophorus	Pollen and nectar	July-Nov	Formica spp.
Apocyanaceae	Vinca major	Nectar	May	Vaneesa, Pieris, syphingidae
Bignonaceae	Capsid grandiflora	Pollen	June-Aug	Apis cerana, A. mellifera, Bombus tunicates, B.trifeciatus, B. simmillmus and xylocopa spp.
Brassicaceae	Rorippa islandica	Nectar	May-Aug	Apis cerana, A. mellifera and dipteral
	Descuriania sophia	Pollen	July-Aug	Apis cerana, A. mellifera, and Bombus.
	Iberis amara	Pollen and nectar	May-July	Apis cerana, A. mellifera
	Alliaria etiolates	Pollen and nectar	March- May	Apis cerana, A. mellifera
Caprifolaceae	Lonicera japonica	Pollen and nectar	May-Sep	Apis cerana, A. mellifera, Bombus tunicates,
Caryophylaceae	Dianthus anatolicus	Pollen and nectar	May-July	Apis cerana, A. mellifera, and n dipteral
	Menuartia kashmirica	Nectar	May-August	Apis cerana, A. mellifera,
Fabaceae	Astragalus candolleanus	Nectar	May-June	Apis cerana, A. mellifera, Bombus spp.and certain spp.
	Melilotus alba	Pollen and nectar	May-June	Apis cerana, A. mellifera,
	Lathyrus emodi	Pollen and nectar	June-July	Apis cerana, A. mellifera,and dipteral
	Circis siliquastrum	Nectar	March-April	Apis cerana, A. mellifera, Bombus.and Xylocopa

Family	Genus/Species	Source	Flowering Time	Pollinators
Fumariaceae	*Fumaria indica*	Nectar	May-Oct	*Apis cerana, A. mellifera,*
Iridaceae	*Gladiolus sp.*	Pollen and nectar	July-Oct	*Apis cerana, A. mellifera,*
	Crocus sativa	Pollen and nectar	Oct-Nov	*Apis cerana, A. mellifera,*
Juglandaceae	*Carya illinoinensis*	Nectar	May-June	*Apis cerana, A. mellifera,and dipteral*
Liliaceae	*Lilium album*	Nectar	May-July	*Apis cerana, A. mellifera,*
	Fritillaria cirrhosa	Nectar	March-May	*Apis cerana, A. Mellifera*
Lamiaceae	*Salvia hians*	Nectar	June-Sept	*Apis cerana, A. mellifera and Bombus spp*
	Nepeta eriostachya	Nectar	June –Aug	*Apis cerana, A. mellifera*
	Lavandula officinale	Nectar		*Apis cerana, A. mellifera, and Bombus.*
Loganiaceae	*Buggleja crisped*	Nectar	June-Aug	*Apis cerana, A. mellifera, and*
Mimosaceae	*Albia julibrissin*	Nectar	June-Aug	*Apis cerana, A. mellifera,*
Oleraceae	*Ligustrum ovaliolium*	Pollen and nectar	May-July	*Apis cerana, A. mellifera,*
Papaveraceae	*Eschscholzia Californian*	Nectar	May-June	*Apis cerana, A. mellifera*
Rosaceae	*Sorbus cashmeriana*	Nectar	June-July	*Apis cerana, A. mellifera,and dipteral*
Ranunculaceae	*Delphinium royle*	Pollen and nectar	July-Sept	*Apis cerana, A. mellifera, and Bombus.*
Schro-phulariaceae	*Antirrhinum majus*	Nectar	June-Nov	*Apis cerana, A. mellifera,and xylocpa*
	Veronica laxa	Pollen and nectar	Jan-May Nov-Dec	*Apis cerana, A. mellifera,*
Solanaceae	*Atropa acuminata*	Pollen	June-Aug	*Apis cerana, A. mellifera,*
	Petunia alba	Pollen	July-Nov	*Apis cerana, A. mellifera,*
	Physalis philadelphica	Pollen and nectar	July-Sep	*Apis cerana, A. mellifera,and Bombus spp.*
	P. peruviana	Pollen and nectar	July- Sep	*Apis cerana, A. mellifera,and Bombus spp.*

5. Acknowledgement

The authors are thankful to the Indian Council of Agricultural Research (ICAR), New Delhi for providing the financial grant under Niche Area of Excellence (NAE-II) project no. F. No. 10(5)/2012-EPD dated: 26-03-2012. The help rendered by the orchardists during the collection survey is highly acknowledged.

References

Abrol, D. P. 1993. Insect pollination and crop production in Jammu and Kashmir. *Current Science* 65(3): 265-269.

Ascher, J. S., Pickering, J. 2013. Discover life bee species guide and world checklist (Hymenoptera: Apoidea: Anthophila). – Online. [Accessed on November, 25, 2013. http://www. Discoverlife. Org/mp/20q? Guide=Apoidea species.

Buchmann, S. L., Nabhan, G. P. 1996.*The forgotten Pollinators*. Island Press, Washington, D.C., U. S. A. p. 292.

Free, J. B. 1993. *Insect pollination of crops*. 2nd edition, Academic press, London.

Gautier-Hion, A and Maisels, F. 1994. Mutualism between a leguminous tree and large African monkeys as pollinators. *Behavioural Ecology* 34: 203-210.

Ghorpede, H., Prasad, K. D., Pavan, S. 2011.Hover-flies (Diptera: Syrphidae) of the Coromandel Coast in Andhra Carnatic, Peninsular India.*Bionotes* 13(2): 78-86.

Goulson, D. 2003.*Bumblebees: Their behavior and ecology*. Oxford University Press, New York.

Hussain, A., Khan, M. R., Ghaffar, A., Hayat, A., Jamil, A. 2012.The hymenopterous Pollinators of Himalayan foot hills of Pakistan (distributional diversity).*African Journal of Agriculture Research*11 (28): 7263-7269. DOI: 10.5897/AJ11.470

Jasara, A. W., Ashfaq, S. and Kasi, A. M. 2000.*Apple pollination in Baluchistan*, Pakistan National Arid land Development and Research Institute, Ministry of Food, Agricultural and Livestock, Islamabad, p. 33.

Jasara, A. W. and Rafi, M. A. 2008. Pollination management of apricot as a livelihood source in northern areas. *Pakistan Journal of Agriculture* 24: 34-40.

Kenmore, P. and Krell, R. 1998.Global perspectives on pollination in agriculture and agro-ecosystem management. *International workshop on the conservation and sustainable use of pollinators in agriculture with emphasis on bees*, October 7-9 Sao Paulo, Brazil.

Knutson, R. D., Taylor, R. G., Penson, B. J. and Smith, G. E. 1990.*Economic impacts of reduced chemical use*. Knutson and Associates, College Station, Texas pp. 30-31.

Macfarlane, R. P. and Patten, K. D. 1997. Food sources in the management of bumble bee population round cranberry marshes. In: Proceedings of International Symposium on Pollination. Ed. K.W. Richards. *Acta Horticulture* 437: 239-244

Margalef, R. 1958.In*Perspectives in Marine biology*. – Buzzati-Traverso (Ed.). Berkeley: University California Press. p. 323.

McGregor, S. E. 1976.*Insect pollinators of cultivated crop plants*. United State Department of Agriculture, Agriculture Hand book, p. 496.

Michener, C. D. 2007.*Bees of the World. Johns Hopkins Press*, Baltimore, MD.

Partap, T. and Partap,U.2002. *How to manage crop Pollination using honey bees as pollinators*, Training Manual-Westville Publishing House, New Delhi.

Pielou, E. C. 1966.The measurement of diversity in different types of biological collections. *Journal of Theoretical Biology* 13: 131-144.

Pielou, E. C. 1975. *Ecological diversity*. New York: John Willey.

Saure, C. 1996. Urban habitats for bees: the example of the city of Berlin. In: *The conservation of Bees,* ed. A. Matheson, S. Buchmann, C. O'Toole, P. Westrich, and I. H. Williams.

Scott-Dupree, C. D., Watson, M. L. 1987. Society Linnean of London and the International Bee Research Associates, Academic Press, San Diego. Wild bee pollinator diversity and abundance in orchard and uncultivated habitats in the Okanagan Valley, British Cloumbia. *Canadian Entomologist* 119: 735-745.

Ssymank, A., Kearns, C. A., Pape, T. and Thompson, F. C. 2008. Pollinating flies (Diptera): A major contribution to plant diversity and agricultural production. *Tropical Conservation* 9 (1 and 2): 86-89.

Verma, L. R., and Partap, U. 1993. The Asian hive bee *Apis cerana* as a pollinator in vegetable seed production. *International Centre for Integrated Mountain Development*, Katmandu Nepal. p183.

Westrich, P. 1996. Habitat requirements of central European bees and the problems of partial habitats. In: *The conservation of Bees*, ed. A. Matheson, S. Buchman, C. O'Toole, P. Westrich, and I. H. Williams. Linnean Society of London and the International Bee Research Associates, *Academic Press*, San Diego, CA.

Chapter 16

Fruit and Foliar Diseases of Apple and their Management

Qazi Nissar Ahmad, S. Rovida,
Kursheed A. Hakim and Nisar A. Khan

Division of Plant Pathology,
Sher-e-Kashmir University of Agricultural Sciences and
Technology of Kashmir, Srinagar
E-mail: qazinissar11@gmail.com

1. Introduction

Pome and stone fruits are important temperate horticultural crops. Apple (*Malus domestica* Borkh.) is the predominant pome fruit cultivated globally. In India, apple farming has attained the status of an industry in Jammu and Kashmir (J&K) and Himachal Pradesh states. In J&K, apple farming is spread over an area of 1, 57,280 ha with an annual fruit production of 13, 48,149MT, representing 45.3 per cent of total area under fruits and 77.4 per cent of total fruit production (Anonymous, 2013). Apple industry of J&K, especially Kashmir valley, has been significantly contributing to the economy of the state through its cultivation and allied trade. The monoculturing of Red Delicious cultivar on seedling rootstocks predominates most of the apple orchards under existing apple farming scenario in the valley. The process of establishing high density orchard systems on clonal rootstocks has been initiated by the state government in order to increase the production and productivity of apple. Apple, in general, and Red Delicious cultivar, in particular, is prone to a number of diseases, affecting leaves, fruits, twigs, blossoms and roots which inflict huge economic losses to fruit growers and also increase the cost of management. The success of any plant disease control programme is primarily governed by a crop production system which fulfils the essential goals of pest management (Qazi, 2015). Economic losses due to the disease vary with climatic conditions, cultivar susceptibility and orchard or nursery management practices. The monoculturing of susceptible apple cultivar especially under temperate environmental conditions promote severe disease development. Under such situations, any lapses or interruptions in the disease management process, sometimes, facilitate epiphytotics. The recent examples are 2013-Alternaria leaf blotch and 2014-Scab epipyhtotics of apple fruit crop in Kashmir valley triggered by highly un- favourable weather conditions. Thus, adoption of appropriate and effective protection strategies to restrict such losses becomes unavoidable. Therefore, cultivation of apple fruit crop for optimal yield and quality demands improved technical support for management of various diseases (MacHardy, 1996).

2. Prevelent Diseases

The major diseases of apple are mainly caused by fungi, and to some extent, by bacteria and viruses (Verma and Sharma, 1999; Zaki *et al.,* 2010; Qazi, 2015). Amongst them, the most common and devastating foliar and fruit diseases found in almost all apple growing countries of the world, which reduce yield and fruit quality or even render fruit unmarketable or weaken the plant vigour, are elaborated under the following heads (MacHardy, 1996; Qazi, 2015).

(a) Apple Scab

Scab is the most important and destructive disease of apple occurring predominantly all over the world. Scab is particularly severe in areas having high relative humidity and rain fall. Apple scab inflicts huge losses through deteriorating quality fruit yield and potential. The disease was first reported in 1819 from Sweden by Fries and is one of the earliest reported diseases. In India, the disease was declared as one of the national diseases of plants in 1974, soon after its first devastating epiphytotic of 1973 in Kashmir valley. The endemic status of scab has resulted in several ephiphytotics all over the world.

(i) Causal Organism

The disease is caused by *Venturia inaequalis* (Cooke) Wint. (Anamorph: *Spilocaeapomi* Fr.).

(ii) Symptoms

The most obvious symptoms occur on leaves and fruits. The fungus also attacks blossoms, young shoots and bud scales. During early spring, the symptoms first appear as velvety brown or olive green spots with feathery, indistinct margins on adaxial surface of leaves. After unfolding of foliar and floral buds, they appear on both surfaces of leaves and other susceptible plant parts. With age, the lesions turn brown to black in colour and have distinct margins, which later on coalesce. The

Plate 16.1: Initial Velvety Brown Scab Lesion on under Surface of Leaf.

Plate 16.2: Scab Lesions on Leaf Petiole.

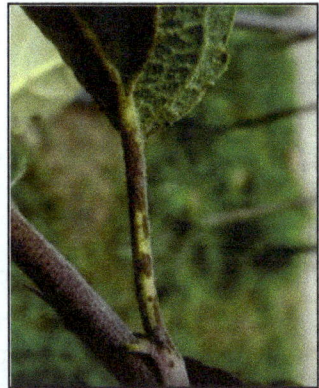

Plate 16.3: Sheet Scab Covering the Entire Leaf Surface.

lesions on leaves from a convex surface with corresponding concave area on the opposite side. Sometimes, the entire surface of leaf is covered with scab lesions, which is referred as sheath scab. Infection of petioles and pedicels result in premature abscission of leaves and fruits, respectively. On young fruits, the lesions are usually brownish in colour initially and with age, they turn black having brownish centre. The infected fruits become cracked and deformed.

(iii) Disease Development

Apple scab pathogen overwinters mainly sexually as pseudothecia on fallen infected apple leaves, which upon soaking forcibly liberate ascospores into the air. Major proportion of ascospores is discharged during day time. The ascospores are

Plate 16.4: Light Olive Green Scab Lesions on Young Fruit.

Plate 16.5: Black Corky Lesions on Mature Fruit.

dispersed by wind and act as primary source of infection for unfolding buds, leaves, flowers and fruits in the spring, which in turn produce several generations of conidia for secondary infection. Sometimes, mycelium over wintering on twigs and buds produce conidia, which on dissemination also act as primary source. Maturation, production and discharge of ascospores and conidia as well as infection process are highly influenced by environmental factors particularly high humidity, rain fall and moisture, besides temperature and wind.

(b) Alternaria Leaf Blotch

Alternaria leaf blotch is one of the important foliar diseases that cause premature defoliation, thereby weakening apple plant vigour leading to poor fruit yield. The disease occurs in many apple growing countries of the world especially on susceptible apple cultivars like Red Delicious.

(i) Causal Organism

The disease is caused by *Alternaria mali* Robert.

(ii) Symptoms

The pathogen mainly affects the leaves. The disease initially appears on leaves as small, circular, light brown non-sporulating lesions measuring about 0.5 mm in size. Later on, lesions enlarge, coalesce and turn silvery grey with black tiny mass of spores. Under severe infection, large irregular necrotic patches develop on coalesced areas. Infection on petioles often leads to chlorosis and premature leaf fall. Severe infection causes pre-mature leaf fall even upto an extent of 60-80 per cent. Sometimes, brownish black circular lesions appear on growing twigs and may get killed under severe infection.

(iii) Disease Development

The pathogen perpetuates asexually on overwintered fallen diseased leaves, infected buds of intact twigs and pruned snags in the form of mycelium and conidia. Peak period of conidial production for primary infection usually occurs during first fortnight of May. Primary infection occurs in late spring usually after petal fall

Plate 16.6: Circular to Irregular Light Brown Non-sporulating Lesions.

Plate 16.7: *A. mali* Outbreak Leading to Severe Pre-mature Leaf Fall.

and several generations of conidia are produced which act as secondary inoculum throughout the season. Disease severity increases rapidly during the rainy season. Pre-mature defoliation is many-fold increased by leaf injuries due to European red mite (Zaki *et al.,* 2010). Optimum temperature for infection, mycelial growth, sporulation and spore germination ranges between 25 to 30 °C.

(c) Powdery Mildew

Powdery mildew is one of the serious foliar diseases of apple world over. The disease occurs in nurseries on seedling plantation and in orchards on young shoots of many apple cultivars.

(i) Causal Organism

The disease is caused by *Podosphaera leucotricha* (Ell. and Ever.) Salmon

(ii) Symptoms

The disease affects buds, leaves, flowers, young shoots and fruits. In spring, symptoms are first noticed on under surface of leaves as small greyish or whitish irregular or felt-like patches of mycelium. Soon after, the entire leaf becomes covered and coated with white mycelium and powdery mass of spores. Severely infected leaves become inrolled or folded longitudinally, hard and brittle, giving a scorched appearance. The young shoots also start exhibiting the typical symptoms and their growth is stunted with shortened internodes. Late on, powdery coating disappears giving brown felt-like cover with dark brown fruiting bodies. Unfolding of infected flower buds is usually delayed and flowers become blighted, which rarely set fruit. Fruit infection is common on severely infected trees. Infected fruits are stunted and typical russetting on infected areas appear as a very closely interwoven network of fine lines. Severely infected fruits often crack.

(iii) Disease Development

The causal fungus is anobligate parasite and overwinters asexually as mycelium in buds and twigs. In early spring, it becomes active to produce conidia for primary infection of emerging foliage. The pathogen reportedly also perpetuates sexually as cleistothecia on infected plant parts to produce ascospores (Marine *et al.,* 2010). Relatively cool and humid conditions favour disease development but severe infection, sometimes, takes place under hot and dry as well as warm and moist conditions over a temperature range of 5-30 °C with an optimum temperature of 20 °C.

(d) Marssonina Blotch

Marssonina blotch is one of the most destructive diseases of apple causing severe premature defoliation and reducing the yield and quality of fruit (Sharma *et al.,* 2004).

(i) Causal Organism

The disease is caused by *Diplocar ponmali* Harada and Sawamura (Anamorph: *Marssonina coronaria* (Ell. and Davis).

(ii) Symptoms

The disease appears as dark green patches or grey-black spots interspersed with lighter or yellow portions on upper surface of mature leaves, sometimes giving a mosaic-like appearance. The symptoms develop as brown to dark brown spots of 5 to 10 mm in diameter. Areas around these spots become pale and entire leaf soon turns yellow and prematurely falls off the tree by mid-summer. Under highly humid conditions, the lower portion of the tree becomes defoliated within few weeks and fruits hanging on the defoliated or naked branches are exposed to direct sunlight and become prone to sun scald. Thus, affecting the yield and quality of fruit adversely. Sometime, flower initiation occurs in autumn, leading to reduction or failure in fruit set or fruit bearing capacity of affected trees in the subsequent years. Symptoms on fruits, although not observed in Kashmir, are characterized by the appearance of circular to oval brown spots of usually 3-5 mm in diameter which later become depressed and turn dark brown to black in colour. As the disease progresses, these spots develop into larger areas surrounded by red edges and dark coloured dot- or pinhead- like asexual fruiting bodies (acervuli) develop on the affected surfaces.

(iii) Disease Development

The fungus overwinters sexually develop in the form of apothecia on infected fallen leaves (Lee *et al.,* 2011). Primary infection has been reported to initiate by

Plate 16.8: Powdery Mildew Symptoms on Terminal Growth of Apple.

Plate 16.9: Marssonina Blotch on Apple Leaves.

ascospores produced in apothecia on over wintered leaves (Harada *et al.,* 1974). Rain is necessary for spore dispersal and leaf infection. The initial disease symptoms under valley conditions appear in July-August, usually in rainy summers. The secondary infection occurs more frequently through conidia. Optimum temperature ranging from 20 to 25 °C favours disease development.

(e) Sooty Blotch and Flyspeck

Sooty blotch and flyspeck are two economically important diseases of apple and reduce market acceptability by deteriorating fruit quality. Both the diseases are prevalent in all apples growing countries of the world particularly in areas with high relative humidity; both are superficial usually appearing simultaneously and mistaken as single disease.

(i) Causal organism

Sooty blotch is a disease complex caused by more than one fungal pathogen *viz., Peltaster fruticola, Leptodontium elatius* and *Geastrumia polystigmatis.* All three fungi are not necessarily present in all sooty blotch lesions. Flyspeck is caused by Zygophiala *jamaicensis* Mason.

(ii) Symptoms

Sooty blotch and flyspeck appear simultaneously on fruits in the summer after walnut-size fruit development stage. Sooty blotch is characterized by superficial dark filmy smudge or shades of olive green to black or cloudy areas on fruit surface. The individual colonies are initially circular and variable in size which coalesce later exhibiting indefinite or diffuse margin and may even completely cover the fruit surface.

Flyspeck often associated with sooty blotch appears as superficial small circular black glistering specks or dots with definite outlines, usually few to more than 50 in number. The symptoms resemble with flyspecks in size and colour. These specks are fruiting bodies that develop on mycelium appearing as rather loose feathery or fern-like strands, or clumps or dots scattered within the thin sooty area.

Plate 16.10: Brown Spots with Greyish Centre of Frog Eye Spot on Apple Leaf.

Plate 16.11: Sooty Blotch as Dark Filmy Smudge/Shades, and Fly Speck as Small Circular Black Glistering Dots.

(iii) Disease Development

The pathogens of both the diseases usually perpetuate asexually as mycelium or conidia or chlamydospores on wild hosts or apple twigs. The spores are dispersed or washed on to developing fruits by rains and require water for their germination (Jones and Sutton, 2001).

(f) Frog Eye Leaf Spot, Black Rot and Smoky Canker

Frog eye leaf spot is an important disease of apple affecting leaves, fruits and shoots. The disease is referred to as frog eye spot on leaf, black rot on fruit and smoky canker on twig and trunk. The disease occurs in all apple growing areas of the world including Kashmir valley.

(i) Causal Organism

The disease is caused by *Physalospora obtuse* and *Botryosphaeria obtuse.*

(ii) Symptoms

The symptoms initially appear on leaves as purplish flecks and then develop as brown spots with greyish centre, resembling a frog's eye. On young fruits, symptoms are characterized by irregular light green spots with brown halo. The spots later turn brown with alternate light green and brown margins and become depressed. Severe infection leads to complete decay of mature fruits which finally turn black and numerous black pimple-like asexual fruiting bodies (pycnidia) of the fungus appear on fruit surface. Infection of limbs, branches and trunks develop as small, slightly sunken, reddish-brown areas in the bark. These areas slowly enlarge and darken leading to canker formation.

(iii) Disease Development

The causal pathogen overwinters sexually as perithecia and asexually as pycnidia on infected limbs or shoots usually underneath the bark or on pruned sites and produce ascospores and conidia, respectively, for primary infection (Venkatasubbaiah *et al.,* 1991). Previous years mummified fruits also provide primary inoculum. The cultural practices like improper pruning, soil moisture and nutrients *etc.* favour the disease development while as unprotected wounds and injuries provide the sites for infection.

(g) Other Cankers

Various types of cankers occur in apple orchards all over the world. They are posing a serious threat to apple plantation in Kashmir valley. Some of the important cankers include stem bark or white rot canker, smoky canker, Cytospora canker, anthracnose canker, Fusicocum canker, *etc.* They are caused by various fungi *viz.,* Botryosphaeria spp., Nectria spp., Cytospora spp., Cryptosporiopsis spp. *etc.* Cankers usually develop on twigs, limbs, branches and tree trunk. Various symptoms include depressed lesions of different colours which slowly increase in size particularly around injured sites. Cankers may become dry, hard and tough. Sometimes, blisters appear on the bark exuding some watery liquid. Some cankers remain superficial causing roughening of the skin. The infected bark may also get loosened and papery. Wood below the bark appears stained, showing cracks and may emitting alcoholic smell. Cankers later develop rapidly lengthwise and coalesce to become elliptical extending up to more than one meter and may completely girdle the limb. Small shoots above the cankered lesions show die back symptoms. The predisposing factors for disease development are more or less similar as mentioned for smoky canker.

3. Management of Fungal Diseases

In Kashmir, SKUAST-K issues advisory to apple orchardists through print and electronic media and devises an apple plant protection spray schedule annually for recommendation under normal weather conditions (Table 16.1). For such guidance, SKUAST-K evaluates different molecules for their bio-efficacy against apple pathogens and pests every year. In addition, trainings are organized for extension specialists and field functionaries of state development departments and *Krishi Vigyan Kendras* (KVK's) as well as fruit growers with regard to importance of various plant protection strategies. However, the use of computerised forewarning devices to record or predict the occurrence of infection periods based on temperature, relative humidity, rainfall, hours of leaf wetness, *etc.* are also being evaluated for scheduling the fungi toxicant spraying (Qazi, 2015). The important strategies include cultural and canopy management, orchard sanitation, monitoring and scouting of apple orchards and chemical sprays.

(a) Monitoring and Scouting

Apple orchardists and field functionaries are educated to closely monitor orchards frequently, which is important for effective management and advised to consult extension specialists and KVK's of respective localities for any advisory, as and when required. For location specific decisions and under abnormal weather conditions, orchardists are advised to consult the extension specialists of state Development Departments/*Krishi Vigyan Kendra* (KVK) scientists for advice.

(b) Orchard Sanitation

The success and effectiveness of sprays particularly during spring months depends on the reduction of primary inocula which could be achieved by ensuring collection and complete destruction of fallen leaves, pruned, diseased and infected twigs/snags and fallen/mummified fruits (Hardwick, 2006; Qazi *et al.,* 2008).

(c) Canopy Management

The micro climate within an orchard plays a significant role in the disease development, which is governed by training and pruning pattern of trees besides orchard topography, weather conditions, *etc.* In Kashmir, faulty pruning is one of the causes of concern in most of the orchards. Skilled manpower is required for carrying out field operations for ensuring removal of crowded branches to avoid high humidity condition through skilful pruning and proper pesticide coverage for effective chemical control. In addition, ensure periodical destruction of weeds beyond 60 cm around tree trunks after fruit-let stage.

Table 16.1: Plant Protection Schedule for the Management of Insect Pests and Diseases of Apple (2017)

Spray	Tree Stage	Fungicide per 100 lit of Water
I	Green Tip	Mancozeb 75 WP (300 g) or Propineb 70 WP (300 g) or Zineb 75 WP (300 g) or Captan 50 WP (300 g)* ˙ Spray at least 10 days after oil spray
II	Pink bud	*10-14 days after II spray* Mancozeb 75 WP (300 g) or Propineb 70 WP (300 g) or Captan 50 WP (300 g) orZiram 80 WP (200 g) or Dodine 65 WP (60 g)*or Carbendazim 12 per cent + Mancozeb 63 per cent 75 WP (250 g)* or Captan 70 per cent + Hexaconazole 5 per cent 75 WP (50 g)* * Give preference in case rainy weather hampers green tip spray
III	Petal fall	*12-15 days after III spray* Difenoconazole 25EC (30 ml) or Flusilazole 40 EC (20 ml) or Bitertanol 25 WP (50 g) or Penconazole 10 EC (50 ml)
IV	Fruit let (Pea size)	*10-14 days after IV spray* Mancozeb 75 WP (300 g) or Propineb 70 WP (300 g) or Zineb 75 WP (300 g) or Captan 50 WP (300 g) or Ziram 80 WP (200 g) or Captan 70 per cent + Hexaconazole 5 per cent 75 WP (50 g) or Trifloxystrobin 25 per cent + Tebuconazole 50 per cent 75 WG (40 g)
V	Fruit development-I	*12-15 days after V spray* Dodine 65 WP (60 g) or Kresoxim-methyl 44.3 SC (40 ml) Metiram 55 per cent + Pyraclostrobin 5 per cent 60 WG (100 g)
VI	Fruit development-II	*12-15 days after VI spray* Flusilazole 40 EC (20 ml) or Captan 70 per cent + Hexaconazole 5 per cent 75 WP (50 g) or Myclobutanil10WP (50 g) or Penconazole10EC(50 ml)
VII	Fruit development-III	*12-15 days after VII spray* Mancozeb 75 WP (300 g) or Ziram 27 SC (600 ml) or Zineb 75 WP (300 g) or Ziram 80 WP (200 g) or Chlorothalonil 75 WP (150 g)
		In case rainy weather hampers VII spray Difenoconazole 25 EC (30 ml) or Trifloxystrobin 25 per cent + Tebuconazole 50 per cent 75 WG (40 g)

Spray	Tree Stage	Fungicide per 100 lit of Water
VIII	Fruit development-IV	**(*Need based*) : a) when leaf spotting incidence is more than 20 per cent *12-18 days after VIII spray***
		Metiram 55 per cent + Pyraclostrobin 5 per cent 60 WG (100 g) or
		Captan 70 per cent + Hexaconazole 5 per cent 75 WP (50 g)or
		Hexaconazole 5 EC (50 ml) or
		Myclobutanill 10 WP (50 g) or Bitertanol 25 WP (50 g)
		(Need based): b) For *Marssonina*/Sooty blotch/Flyspeck
		Mancozeb 75WP (300 g) or Ziram 27 SC (600 ml) or
		Propineb 70 WP (300 g) or Ziram 80 WP (200 g)
IX	Pre-harvest	**(*Need based*): For long term Storage *25 days before harvest***
		Mancozeb 75 WP (300 g) or Ziram 27 SC (600 ml) or
		Zineb 75 WP (300 g) or Ziram 80 WP (200 g)

The Advisory

i. Maintain a gap of 3-4 days between insecticide and fungicide spray and avoid mixing of fungicides, insecticides or spray suspensions.

ii. Adjuvants/Stickers like Sandovit @ 50-75 ml/100 litre of suspension may be added for better efficacy of fungicides especially during rainy days, but not be used with Dodine.

iii. Avoid spraying during high temperatures/rainy weather. In case of heavy rains within 12 hours of spray, the spray is to be repeated immediately, particularly in absence of adjuvants/stickers.

iv. Same pesticide should not be repeated continuously in two sprays and stop spraying 3-4 weeks before harvesting of fruit.

v. For the safety of pollinators, spraying should be done in the morning and evening hours and bee colonies should be kept preferably away from spraying area, besides avoiding spray during full bloom period.

vi. The management strategies for cankers include removal of girdled limbs and papery bark, and scrapping of cankers with sharp knife, followed by application of wound dresser like Bordeaux paste or chaubatia paste on cut surfaces.

vii. Use balanced nutrition (N:P:K + FYM) and ensure that no water stress develops in the orchard. However, under drought like conditions, conduct 2-3 sprays of calcium chloride (dehydrated) @ 0.3-0.4 per cent after pea size stage along with boric acid @ 0.15 per cent at an interval of 10-15 days after the first fungicide spray.

References

Anonymous. 2013. *District wise area and production of major horticultural crops in Jammu and Kashmir state for the year 2012-2013.* Department of Horticulture (Kashmir), Jammu and Kashmir. Pp.1-2.

Harada, Y., Sawamura, K. and Konno, K. 1974. Diplocarponmali, sp. nov., the perfect state of apple blotch fungus Marssonina coronaria. *Annals of Phytopathological Society of Japan* 40:412-418.

Hardwick, N.V. 2006. *Disease forecasting.* In: The Epidemiology of Plant Diseases. Second edition. (Eds. B.M. Cooke and B. Kaye). Springer, pp. 239-268.

Jones, A.L. and Sutton. T.B. 2001. *Diseases of tree fruits of the East.* Michigan State University Extension 19-20: 57-60.

Lee, D.H., Back, C. G., Win, N.K.K., Choi, K.H., Kim, K. M., Kang, I.K., Choi, C., Yoon, T.M., Uhm, J.Y. and Jung, H.Y. 2011. Biological characterization of Marssonina coronaria associated with apple blotch disease. *Mycobiology* 39: 200-205.

MacHardy, W.E. 1996. *Apple Scab: Biology, Epidemiology and Management.* American Phytopathological Society, St. Paul, Minnesota, USA. 545 pp.

Marine, S.C., Yoder, K.S. and Baudoin, A. 2010. *Powdery mildew of apple.* The Plant Health Instructor. DOI: 10.1094/PHI-I-2010-1021-01.

Qazi, N.A. 2015. *Forecasting of plant diseases of temperate fruits.* In: Temperate Fruits and Nuts- A way Forward for Enhancing Productivity and Quality. (Eds. K.L. Chadha, Nazeer Ahmad, S.K. Singh and P. Kalia). Daya Publishing House, A Division of Astral International Pvt. Ltd., New Delhi. pp. 373-384.

Qazi, N.A., Khursheed, A., Beig, M. A., Munshi, N. A. and Khan, N. A. 2008. Impact of inoculum distance on initiation and progress of apple scab [Venturiainaequalis (Cke.) Wint.] Infection. *Applied Biological Research* 10: 40-43.

Sharma, J.N., Sharma, A., Sharma, P. 2004. Out-break of Marssonina blotch in warmer climates causing premature leaf fall problem of apple and its management. *Acta Horticulturae* 662:405-409.

Venkatasubbaiah, P., Sutton, T.B. and Chilton, W.S. 1991. Effect of phytotoxins produced by Botryosphaeriaobtusa, the cause of black rot of apple fruit and frog eye leaf spot. *Phytopathology* 81:243-247.

Verma, L.R. and Sharma, R.C. 1999. *Diseases of Horticultural Crops- Fruits.* Indus Publishing Co, New Delhi. pp. 724.

Zaki, F. A., Qazi, N.A., Mantoo, M.A. and Munazah Yaqoob. 2010. *Pests and diseases of temperate fruit crops and their management.* (pp.194-211). In: Sustainable Crop Protection Strategies, Vols. 1 and 2. (Ed. H.R. Sardana, O.M. Bambawale and D. Prasad), Daya Publishing House, Delhi. pp. 819.

Chapter 17

Insect-Pest Complex and Integrated Pest Management on Apple in Jammu and Kashmir, India

Barkat Hussain[1], Abdul Ahad Buhroo[2],
Abdul Rasheed War[3] and Asma Sherwani[1]

[1]Division of Entomology, Sher-e-Kashmir University of Agricultural
Sciences and Technology of Kashmir, Srinagar
[2]Department of zoology, University of Kashmir, Srinagar – 190 006, J&K
[3]International Crops Research Institute for the Semi-Arid Tropics (ICRISAT),
Hyderabad – 502 324, Telangana
E-mail: bhatbari@rediffmail.com

1. Introduction

Jammu and Kashmir is the largest producer of apple in the country because of its best temperate climate suited for its production as compared to rest of the Indian Union. The history of apple cultivation in Kashmir is traced to the old literature. With the introduction of exotic varieties of apple in Kashmir valley, new insect pests emerged, which include San Jose scale, European red mite, Codling moth and Woolly apple aphid. The establishment of these pests and subsequent losses made by these pests are alarming to the fruit growers. These pests cause losses both directly and indirectly. The direct losses are the reduction in fruit damage, quality and quantity of apple fruits and indirect losses are the costs incurred for their management. Due to the lack of well-organized and a precise quarantine system for insect pests of apple in India, the exotic pests are of major concern. Although thrips, leaf rollers and tussock moth also infest apple in Jammu and Kashmir, the losses by these insects are non-significant. As such it is necessary to depict the nature of infestation, identification and detection of different stages of apple pests and their distribution in India. The insect pests which have prime importance on apple in Jammu and Kashmir and other states of Indian union are given in Table 17.1.

Table 17.1: Important Insect-Pests and their Status on Apple in Kashmir Valley

Sl.No.	English Name	Scientific Name	Damaging Stages	Status of the Pest
1.	San Jose scale	*Quadraspidiotus perniciosus*	Both Nymphs and adults	Key pest
2.	European Red Mite	*Panonychus ulmi*	Both Nymphs and adults	Key pest
3.	Codling moth	*Cydia pomonella*	Caterpillar	Key pest in Ladakh
4.	Indian Gypsy moth	*Lymantria obfuscate*	Caterpillar	Minor pest
5	Woolly apple aphid	*Eriosoma lanigerum*	Both nymphs and adults	Major pest
6.	Tent Caterpillar	*Malacosoma indicum*	Caterpillar	Sporadic pest
7.	Apple stem borer	*Aeolesthes sarta*	Grubs	Major pest
8.	Bark beetle	*Scolytus nitidus*	Grubs	Minor pest
9.	Apple leaf miner	*Lyonetia clerkella*	Larvae	Minor pest

2. Important Insect-Pests

(a) San Jose Scale

Scientific Name: *Quadraspidiotus perniciosus* (Comstock)

Family: Diaspididae

Order: Hemiptera

(i) Distribution

This pest has been introduced in Kashmir valley and Himachal Pradesh during the first decade of twentieth century.

Detection and Identification

Nymphs: Ovoviviparous females give birth directly to the young ones that emerge out, under the scale coverings. One female is capable of giving birth to 200-400 nymphs. These tiny yellow crawlers are mobile free living for 12-24 hours and wander in a random fashion. Upon settling in the preferred places on the host tree, the crawlers insert their mouthparts and suck the sap from tender shoots/ branches of the host tree. While feeding, they secrete a white waxy material and this stage is known as a white cap stage. The waxy material latter changes to black colour and this stage is named as the black cap stage. The cover of the scale turns into various shades, grey and black.

Adult: After the first molt, male and female (immature) scales are easily distinguishable. The female covering is circular and that of the male is elongated. Males are winged and females are non-winged. The colour of the adult male is tiny yellow. Males mate with the non-winged females, which are covered by grey scales. The yellow lemon coloured female is visible when the covering is lifted. Female scales are very prolific over a 6-week period and can produce approximately 400 young nymphs. It takes 25 days for males to mature and 31 days for females. In Kashmir, two complete generations and third partial generation occurs in a year.

(ii) Nature of Infestation

San Jose scale nymphs and females attack the above ground parts. The spots on the fruits show characteristic symptoms of purplish red colour. While the infested branches show ash grey symptoms, sucking of the tree leads to reduced growth and eventually death of the tree.

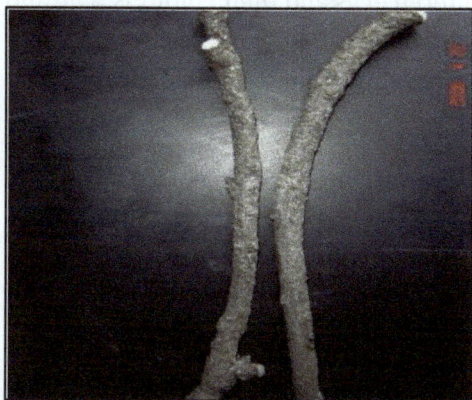

San Jose Scale Infestation on Apple Twigs and on Apple Fruit.

(iii) Monitoring and Surveillance

San Jose scale infestation can be easily visualized on the infested branches and fruits as they turn grey or black and ultimately, ash coloured. The sticky tape should be pasted on small twigs or limbs around the infested areas to determine when crawlers are active. Use degree day model to observe the emergence of San Jose scale male adults and crawler emergence. Daily monitoring is necessary to set a biofix for the San Jose scale adults (UC IPM, 2005). The lower developmental threshold for San Jose scale is 10 °C. Application of dormant season spray oils is effective for controlling the San Jose scale population (Sofi and Hussain, 2008) and the second insecticidal treatment is advised in the first week of May, at the time of crawler emergence (unpublished data). Monitor pruned twigs when winter pruning is done especially on twigs from tree tops to estimate the over wintered population.

(b) European Red Mite

Scientific Name: *Panonychus ulmi* (Koch.)

Family: Tetranchidae

Order: Trombidifoemes

(i) Distribution

In India, It is widely distributed throughout Jammu and Kashmir, Himachal Pradesh, Utter Pradesh and Meghalaya.

Detection and Identification

Egg: Eggs are laid during winter and summer. The eggs laid during winter are known as overwintered eggs. Eggs are laid on fruit spurs, twigs, near the base of buds and tree crevices. The colour of the winter eggs is brick red with a stalk. The overwintered eggs hatch during spring. The summer eggs are laid on the foliage, however, if heavy infestation of mite is noticed, eggs could be found on fruits also. The summer eggs are smaller than overwintered eggs and their colour is pale and translucent. The average hatching of eggs varies from 6.7 to 14.4 days.

Nymphs: European red mite possess three stages namely larva, protonymph and deutonymph. The larva is fairly larger than the egg, is orange-red in colour and can be distinguished from other stages by having only three pairs of legs. The protonymph and deutonymph are comparatively larger and posses four pairs of legs.

Adult: The adult female is brick red in colour and oval in shape with strong white bristles on the back of the abdomen. The bristles are with white bases, which appear as white spots on the back. The male is yellowish red in colour and

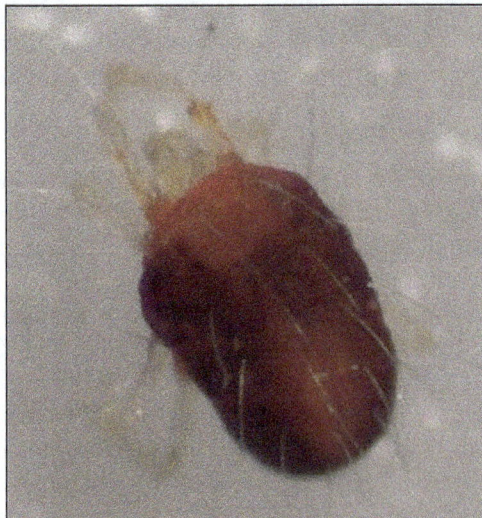

Adult Mite, *Panonychus ulmi.*

fairly smaller than female (UC IPM, 2008) with a pointed abdomen. Male is more cylindrical than female and has a tapered abdomen.

(ii) Nature of Infestation

The mites feed into leaf cells and suck out the contents and chlorophyll by their mouth parts. The infested leaves show bronzing appearance, develop large necrotic areas, and are shed prematurely. The photosynthesis rate is also reduced.

(iii) Monitoring and Surveillance

Regular monitoring is important for any survey and surveillance programmes. For red mite, monitoring should be done as soon as the overwintered eggs start to hatch. Mostly overwintered eggs of red mite hatch during spring season, which varies from place to place. Ten trees should be selected from one block of apple orchard and from each tree at least ten leaves should be examined using an insect hand lens. Record both the number of mites and number of leaves infested by red mite to estimate the mite density and the mite infested leaves. Application of delayed dormant spray oil is effective in controlling over wintering eggs.

(c) Codling Moth

Scientific Name: *Cydia pomonella* (Linnaeus)

Family: Tortricidae

Order: Lepidoptera

(i) Distribution

In India, its distribution is restricted to Ladakh region of Jammu and Kashmir State and is thought to have entered Ladakh from the North West Frontier province of Pakistan, where it is a serious pest of deciduous fruits.

Detection and Identification

Egg: Flat and pin head size eggs are transparent in colour. Eggs are laid singly by a female on the leaves and near to apple fruit. The eggs turn dark as the hatching starts. Egg period is about 6 to 14 days depending on environmental conditions.

Larva: After egg hatching, the newly emerged larva bore the fruit and feed on seeds. The larvae have a pink body and a black head. During winter, codling moth larva overwinters as full-grown larvae within thick, silken white cocoons under loose bark or fallen fruits and in the soil as well as in debris around the base of the tree (Zaki, 1999). Full grown larvae are pinkish or creamy white in colour with brown head.

Pupa: Fully developed larvae pupate in the soil, fallen fruits, on fallen leaves and mostly under the loose bark around the stem. The pupal period ranges from 10- 20 days.

Adult: *Codling* moth adults are about one inch long. Forewings are greyish dark with waxy lines and coppery areas at the tip of the wings. The body is molted greyish brown in colour. The overwintered larvae pupate in the spring and emerge out as adults during mid-May and June. The adult emergence is more often synchronized with the fruitlet stage.

(ii) Nature of Infestation

Codling moth larvae bore inside the fruit and eat towards the central core of the fruit. The larvae feed on the seeds inside the fruit and create exit holes. The waste material is pushed out through the exit holes, however, the waste material may remain associated, or in the fruit that accelerates internal rotting. Codling moth larvae directly feed on fruits and the losses are more than 80 per cent of apple fruit. Under severe infestation, fruits fall prematurely.

(iii) Monitoring and Surveillance

Fruit damage can be estimated by observing forty fruits from an apple tree. In Ladakh, apple trees are not grown scientifically, as trees are un-pruned and scattered. Apple plantations could be found on the slopes near houses. To assess the fruit damage, at least 120 fruits are examined from the three apple trees randomly from three different sites at each location. To examine fruit injury and fruit damage, 360 fruits should be selected from each hamlet. The time for estimating the fruit damage can be done from July to October.

Fruit Injury by Codling Moth.   **Boring on Fruit by the Larva.**

As the fruit growers do not have gadgets for the application of insecticides and other IPM methods, because of socio-religious constraints in Ladakh. Use pheromone-baited traps for codling moth, these traps should be installed very close to the infested branches (Hussain *et al.,* 2015). To set the biofix for codling moth in Ladakh, trap the adult population from the first week of May. Delta traps baited with codlemone lure are effective to set the biofix and to monitor the adult population. The trees should be banded with gunny bags or cardboard sheets to monitor the first generation larvae, which scan for shelter and also monitor the overwintering larvae around these bands.

(d) Indian Gypsy Moth

Scientific Name: *Lymantria obfuscate* (Walker)

Family: Lymantriidae

Order: Lepidoptera

(i) Distribution

In India, it is distributed in Himachal Pradesh, Punjab and Jammu and Kashmir. It is now considered as a minor pest on apple in Jammu and Kashmir.

Detection and Identification

Egg: A single female lays about eggs in batches. The egg batch can be found on loose bark, tree stems of host trees, old branches, and in outdoor objects. These eggs are covered with yellowish brown covering. The eggs hatch from the last week of March to first week of April.

Larva: Newly hatched larvae are small and buff-coloured, but turn black within a few hours. The latter instars are hairy and blackish, with several light orange dots

A: Male   B: Female.

Males trapped in delta traps.

on the last half of the back. The total larval period lasts for 50- 66 days. The larvae remain active from the last week of March to end of May in Kashmir conditions.

Pupa: Pupae are dark brown or mahogany-coloured and are found in sheltered areas. The pupal period lasts for 4 -12 days. The pupal stage can be observed from the first week of June to mid-June.

Adult: Males are small and dark brown with blackish markings and are strong fliers. Females have atrophied wings and are unable to fly. Each female lays about 500 to 1000 eggs. These eggs are laid in batches and are covered with a tuft of hairs. The adults are active from mid-June to mid-July. The Indian gypsy moth has only one generation in Kashmir. The longevity of adults last for 3-30 days.

(ii) Nature of Infestation

Caterpillars are voracious feeders and feed gregariously on leaves during the night time. Severe infestation by the caterpillar results in complete defoliation of leaves and poor quality or failure of fruit formation.

(iii) Monitoring and Surveillance

Monitoring of the adult population should be done when adults have started to emerge. The best time for monitoring the adult population of gypsy moth should be done by using delta traps baited with a dispar lure (Hussain *et al.,* 2015). The best time for installing the pheromone-baited traps should be started in the first week of June.

The larvae are nocturnal feeders and shelter during the day on limbs or branches of the infested tree. They cause complete leaf defoliation. To estimate the overwintered population, gypsy moths overwinter as in the egg stage, therefore, egg monitoring and survey should be carried from July to March.

(e) Wooly Apple Aphid

Scientific Name: *Eriosoma lanigerum* (Hausmann)

Family: Aphididae

Order: Hemiptera

(i) Distribution

In India, it is found in Punjab, Assam, Jammu and Kashmir, Karnatka, Megalya, Skim, Tamil Nadu, Utter Pradesh and West Bengal

Detection and Identification

Eggs: Aphids are viviparous and reproduce both asexually, parthenogenetically and sexually after mating. Sexual forms mate in winter and lay single, long, oval egg in the crevices of bark.

Nymphs: Eggs hatch in the spring into wingless, parthenogenetic viviparous females. Nymphs hibernate underground on the roots of the tree. Winter is considered as a non-reproductive period. During the onset of March, a female produces 30-116 nymphs parthenogenetically, which could be winged or wingless. Winged forms are present throughout the year and non-winged forms occur from July to October. Within 24 hours the nymphs start secreting waxy white cottony filament, hence named as a woolly apple aphid. Nymph period lasts for 11 days in summer and 93 days in winter (Thakur and Dogra, 1980)

Adult: Nymphs and adults are reddish-brown in colour with waxy white filaments on the whole body. They feed on trunks, tender branches, and twigs of apple trees. In autumn, winged aphids develop from both the aerial and root colonies. Both sexes are wingless. Nymphs on the trees migrate downwards towards root zone for hibernation. Multiple generations of woolly aphid occur throughout the year.

(ii) Nature of infestation

Both nymphs and adults suck cell sap from trunk, branches, stems, twigs, leaf petioles bark of twigs and on roots. Knots on roots and twigs become swollen due to the feeding by this aphid. Severe infestation by this aphid lead to yellowish foliage and short fibrous root system.

Wooly Apple Aphid Infestation on Branches.

(iii) Monitoring and Surveillance

Visual inspection of infested trees is the best method for assessing the damage and monitoring the population numbers. Mostly the damage is more done by the immature insects as compared to adults (winged). Use of sticky traps is commonly used for monitoring the winged aphids. As this pest is known for upward and downward migration, it is better to wrap the sticky traps around the stem just above the ground level to observe the migration cycle of this pest.

(f) Tent Caterpillar

Scientific Name: *Malacosoma indicum* (Walker)

Family: Lasiocampidae

Order: Lepidoptera

(i) Distribution

It is distributed in north-western India and is the most serious pest in Himachal Pradesh and Jammu and Kashmir (Malik *et al.,* 1972)

Detection and Identification

Eggs: Eggs encircle the small branches and twigs in the form of rings with an adhesive substance secreted by the female. This adhesive substance is known as spumaline and forms a protective covering around the egg bands. Each egg band contains about 200-400 eggs. The egg stage lasts for about 9-10 months.

Larva: Caterpillar emergence from eggs starts in early spring at the time of bud break and more precisely when the leaves start emerging out. They start feeding on new leaves by forming small webs or tents, hence named as "Tent Caterpillar". The tents often found on the crotches of limbs act as a refuge for the larvae during the night. Caterpillars move out from the tents and feed on leaves. They feed in groups and cause heavy defoliation. Shredded skin of different larval instars can be found inside the tents and these webs/tents can be easily seen.

Pupa: The pupal stage passes inside the cocoons. These cocoons are constructed loosely and are silky white in colour and oval in shape (Hill, 2008). The cocoons are present in the web and in dead tree material on the ground, or mostly inside of a rolled leaf.

Adult: The adults are brown and yellowish moths with two diagonal markings on the front wings. Only one generation is completed in a year.

Adult Tent Caterpillar.

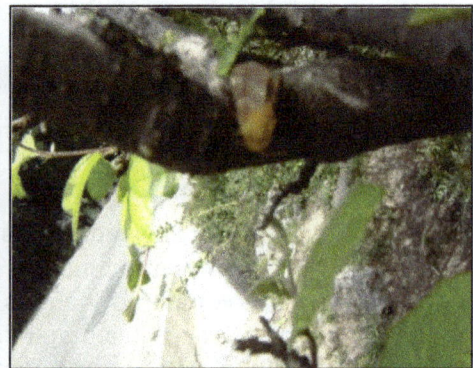

Tents made by the Tent Caterpillar.

(ii) Nature of Infestation

Caterpillars rest at night in tents and feed during the day on leaves. The leaves are skeletonised leaving behind the midrib and veins.

(iii) Monitoring and Surveillance

The real estimate of overwintering population is to monitor the egg mass counts, as this insect overwinters in the egg stage. During early spring, the number of tents made by this pest is important to estimate the pest density as the caterpillar is the damaging stage. For estimating the adult population, pheromone-baited traps should be tied to the infested branches of the trees. The light traps are also effective in sudden outbreaks in the region, where tent caterpillar occurrence is most prevalent.

(g) Apple Stem Borer

Scientific Name: *Aeolesthes sarta* (Solsky)

Family: Cerambycidae

Order: Coleoptera

(i) Distribution

In India, it is widely distributed in Kashmir and Himachal Pradesh.

Detection and Identification

Egg: Eggs are laid on the dry woody portions of the host trees singly in the cuts and silts of the bark made by the female. A single female can lay about 100 eggs, which are creamy white and are elliptical in shape. Egg hatching lasts for 7-14 days.

Larva: Grubs are dirty white with a reddish-brown head. They start feeding by boring inside the stem. Grubs remain inactive during winter, start feeding in March and remain active for two years. They feed by boring the woody portion of stems and branches.

Zig-zag Galleries made by Stem Borer.

Pupa: Pupae are small, yellow-brown and sometimes pupal cases are observed in the trunk or on the infested branches. Pupation takes place inside a tunnel made in the woody tissue by the infested grubs. The pupal stage lasts for 40 to 100 days.

Adults: Adults are dark brown in colour with mottled yellowish pubescence on the elytra. Antennae of the male are 1.5 times longer than their body length, while females are of the same length.

(ii) Nature of Infestation

The newly emerged grubs feed on the bark and make zig-zag galleries. They bore and feed on sap wood (Beeson, 1941), throwing frass out from the exit hole. Sap flow is blocked or restricted by such feeding. The vitality of the tree is reduced leads to death of the plant. The infestation can be detected on the branches and main stem by visible exit holes with a diameter of 1 to 2 cm. Sap oozes out from these exit holes. When the grubs penetrate deep into the stem or branches, coarse saw dust comes out from these exit holes, which can be seen on the ground.

(iii) Monitoring and Surveillance

Monitoring the holes made by this pest is a sound method for estimating the intensity of damage. As no pheromone has been isolated from this species, visual inspection of trees is the only method for survey and surveillance programme. Larva bore inside the trunk and branches result into the excavation of dust coming out of the exit holes. In severe case, huge infestation shows the symptoms of rotting bark. Dying and dead limbs, and yellowing of leaves is a peculiar symptom, when the attack is severe.

(h) The Bark Beetle

Scientific Name: *Scolytus nitidus* (Schedl)

Family: Scolytidae

Order: Coleoptera

(i) Distribution

It is distributed in Himachal Pradesh, Jammu and Kashmir and Uttar Pradesh. Dry and hot weather conditions increase its infestation level. It is a most serious pest in un-irrigated slopes of Kashmir Valley especially in apple orchards.

Detection and identification

Egg: The egg is slightly oval and pale white in colour. The eggs are covered by the boring dust. The female lays on an average 60 eggs, which hatch in 5 to 7 days.

Grub: Newly emerged grubs look like minute immobile white dots. Immediately after feeding, the grub becomes light creamy in colour, curved and legless. The larval development ranges from 38-50 days consisting of five larval instars.

Pupa: The pupation takes place in pupal cells at the ends of larval galleries. The pupa is soft and white. The pupal stage last for 6-18 days.

Adult: The cylindrical adult is 4.00 mm long and 1.68 mm wide (Buhroo and Lakatos, 2007). It has shining black pronotum and dark red brown elytra with declivous abdomen. The adults live for 45-60 days. This species undergoes two complete and a 3rd partial generation per year in Kashmir (Buhroo and Lakatos, 2007)

(ii) Nature of Infestation

The adult females of this pest cause damage by girdling a shot-hole in the inner bark (the phloem-cambial region) on twigs, branches or trunks of apple trees. This activity often results in falling of frass on the surface of soil. Small emergence holes in the bark are a good indication of the presence of bark beetles. Removal of the bark with the emergence holes often reveals dead and degraded inner bark. Galleries are found under the bark due to which the blocking of food and water is severely hindered. During the early part of the attack, the tree does not show symptoms, but growth is arrested. Infested trees show a reduction in foliage and fruit yield.

Emergence of Bark Beetle and Life Stages of *Scolytus nitidus*.
A: Egg stage, B: Larval stage (I and V instars), C: Pupal stage, D: Adult stage.

(iii) Monitoring and Surveillance

For adult monitoring, bark beetle should be monitored through pheromone baited traps. As the bark beetle infestation is not prevalent to all apple trees, because of the aggression pheromone. It has been seen that bark beetles are ethanol hungry and ethanol baited traps should be installed to monitor their population. Yellow sticky traps also work well for the monitoring of shot hole borers, when directly hung on the infested tree. Mostly healthy plants avoid attack by shot hole borers by the flow of sap at great pace from the wound sites. So weak trees should be monitored as these plants suspect infestation by shot hole borer and the most prominent damage could be seen on the tip regions of the branches of the host trees.

(i) Apple Leaf Miner

Scientific Name: *Lyonetia clerkella* Linn.

Family: Lyonetiidae

Order: Lepidoptera

(i) Distribution

In India, it has been reported in Jammu and Kashmir and the losses by this pest are prominent.

Detection and Identification

Egg: The egg looks like a small scar and eggs are laid inside the cuticle of the leaf.

Larva: The larva possess chewing mouthparts and a prominent head capsule. The larva also contains six thoracic and abdominal legs. The larvae are green in colour and live and feed inside the leaves. The larva leaves the mine prior to pupation through an exit in the upper epidermis of the leaf.

Pupa: Pupa forms a silken cocoon attached with a leaf like that of a "hammock" passion on the upper side of a leaf. Pupa hangs on the leaf as a hammock, sometimes it folds the leaf.

Adult: The adult of the leaf miner is very small with shiny white wings and brownish markings on the tips of the wings. The invasion of this pest in Jammu and Kashmir has been studied by Rather and Buhroo (2015).

(ii) Nature of Infestation

Damage done by the leaf miner is only present on the leaves. The larva damages the leaf tissue and feeds between upper and lower epidermis. The whitish or brown long, narrow and sinus tunnel is mined by the larvae on the upper surface of leaf. The management guidelines are not available in Jammu and Kashmir, because the attack by this pest on apple orchards is not so serious.

Leaf Miner Infestation (*Source*: www.rhs.org.uk).

Integrated Pest Management (IPM) for Insect-Pests in Kashmir Valley

Sl.No.	Name of the Pest	IPM Strategy
1.	San Jose Scale	Pruning, collection and burning of infested branches and twigs to ensure complete orchard sanitation. These twigs harbor San Jose scales during winter.
		Application of dormant spray oils @ 2 per cent before leaf emergence or at late dormant stage of apple trees.
		Spraying of Chlorpyriphos 20 EC @ 1ml/l of Water
		Spraying of Dimethoate 30 EC or Quinalphos 25 EC @ 1ml/liter of water when emergence of crawlers has been noticed.
		Encarsia perniciosi and *Aphytis diaspidis* (Parasitoids of San Jose scale) are mass produced and mass released to check the San Jose Scale population.
		Mass production and mass release of predators (*Chilocorus* Sp.) to check the San Jose scale infestation on fruits and twigs.
2.	European Red Mite (ERM)	Collection and burning of pruned twigs to reduce overwintered population as the egg laying is being noticed on these twigs by ERM.
		Application of winter spray oil @ 2 per cent at dormant stage of apple trees or summer spray oils @0.75 per cent during summer to check both overwintering and summer eggs of ERM, respectively.
		Conservation, mass production and mass release of predatory mites (*Amblyseius fallacis*) and lady bird beetles should be done to check the population of ERM.
		Spraying of Fenzaquin 10 EC or Hexythiazox 5.45 EC @ 4 ml/10 liters of water.
		Spraying of Fenpyroximate 5 SC and Propargite 57 EC @ 10 ml/liter of water
3.	Codling Moth	Collection and destruction of fallen fruits to reduce the infestation/population of codling moth in next season.
		Pruning is not practiced in Ladakh as the insecticidal film does not cover the apple tree completely.
		Burlapping of apple stems with gunny bags or cardboard material to trap the overwintered larvae of codling moth.
		Mass trapping of codling moth by use of pheromone baited traps reduce the infestation of codling moth.
		Application of insecticides like Chlorpyriphos 20 EC, Dimethoate 30 EC @ 10 ml/10 liter of water before the larvae penetrate into the fruit significantly reduces the fruit damage.
		Releasing of egg parasitoids (*Trichogramma embryophagum* and *T. cacoeciaepallidum*) target the eggs of codling moth and significantly reduce the fruit infestation. The synchronization of egg laying by codling moth should coincide with the release of egg parasitoids for a very good control.
		The gadgets used for various IPM strategies should be made easily available in the Ladakh region.

Sl.No.	Name of the Pest	IPM Strategy
4.	**Indian Gypsy Moth**	Survey and scouting for the removal and destruction of egg batches is enough for the management of this pest in well managed apple orchards. Application of Nuclear Polyhedrosis Virus (NPV) and spraying of Dimethoate 30 EC @ 1ml/liter of water if damage has been noticed in apple orchards. Monitoring the adult population by use of pheromone baited traps is very important for any pest management strategy.
5.	**Woolly Apple Aphid**	Spraying Dimethoate 30 EC and Ethion 50 EC @ 10 ml/10 liters of water. Some MM series rootstocks has shown resistance against Wooly apple aphid. Conservation, mass production followed by mass release of exotic parasite, *Aphelinus mali* is very effective to check the population of woolly apple aphid. Conservation of predators should be ensured by minimizing the pesticide application to maintain the pest defender ratio balance.
6.	**Tent caterpillar**	As this pest is reported from few pockets of Kashmir valley, the proper management strategy for this pest is not proper. It is a minor pest of apple. Go for scouting the egg bands of Tent caterpillar when regular monitoring of apple orchards is to be carried out. Application and use of NPV reduce the caterpillar damage on tree foliage. Use Dimethoate 30 EC @ 10 ml/10 liters of water when insect outbreak is observed
7.	**Apple stem Borer**	Monitor the attacked twigs and branches infested by stem borers, burn them in winter and scout the dry wood inside the orchards as the female beetles love to lay eggs on dry wood. Spraying of Chlorpyriphos 20 EC and Dimethoate 30 EC @ 10 ml/10 litres of water at the time of egg laying. Apply fumigant granules inside the live holes made by the grubs and then seal these holes in stems and branches of apple trees with adhesive tapes or mud or any material that blocks the passage of air. Pheromone trap could be a big achievement to trap adult population of apple stem borer. Mixing of paint with insecticides and then applying on tree trunks reduces the insect attack and infestation on tree trunks. The heavily infested limbs and trees are to be uprooted and burnt to slow/prevent the spread of the stem borer infestation and outbreak in un-infested orchards.
8.	**Bark beetle**	Scouting and monitoring the infested trees and branches as the bark beetle attack in patches. Application of insecticides before the adults penetrate and lay eggs on the bark by using Chlorpyriphos 25 EC and Dimethoate 35 EC @ 10 ml/10 liters of water.

Sl.No.	Name of the Pest	IPM Strategy
		Heavily infested limbs and branches attacked by bark beetles should be removed and burnt.
		The trees with some sort of stress due to some unfavourable conditions for their proper growth are to be monitored more carefully as compared to healthy ones.
		Use of pheromone traps is effective way to trap adult population and timing of insecticide applications for the management of this pest.
9.	**Apple leaf miner**	Monitoring and trapping of leaf miner by the use of pheromone baited traps.
		Application of systemic insecticides for the management of this pest when this pest attains an epidemic status.

Source: Spray Schedule. Directorate of Research, SKUAST-Kashmir.

References

Beeson, C. F. C. 1941. *The ecology and control of forest insects of India and their neighbouring countries*, Vasant press: Dehradun: 785.

Buhroo, A. A. and Lakatos, F. 2007. On the biology of the Bark Beetle *Scolytus nitidus* Schedl (Coleoptera: Scolytidae) attacking apple orchards. *Acta Silvatica et Lignaria Hungarica.* 3: 65-74.

Hill, D.S. 2008. Pests of crops in warmer climates and their control. *Springer* Science *and Business Media.*

Hussain, B; Ahmad, B and Bilal S. 2015. Monitoring and mass trapping of the Codling Moth, Cydia pomonella, by the Use of Pheromone Baited Traps in Kargil, Ladakh, India. *International Journal of Fruit Science* 2, 15(1):1-9.

Hussain, B., War, A.R., Ganie, S.A. and Bilal, S. 2015. Monitoring and testing different doses of disparlure for Indian gypsy moth, Lymantria obfuscata, in a temperate region of India (Kashmir Valley). *Acta Phytopathologica et Entomologica Hungarica 50*(1): 85-92.

Malik, R. A., Punjabi, A. A. and Bhat, A. A. 1972. Survey study of insect and non insect pests in Kashmir. *Horticulturist (J&K)* 13(3): 29-44.

Rather, S. and Buhroo, A. A. 2015. Arrival sequence, abundance and host plant preference of the Apple Leaf Miner *Lyonetia clerkella* Linn. (Lepidoptera: Lyonetiidae) in Kashmir. *Nature Science* 13(9):25-31

Sofi, M.A. and Hussain, B. 2008. Pest management strategy against black cap stage of San Jose Scale in apple orchards of Kashmir valley. *Indian Journal of Entomology* 70(4): 398-399.

Thakur, J.R. and Dogra, G.S. 1980. Woolly apple aphid, Eriosoma lanigerum, research in India. *International Journal of Pest Management* 26(1):8-12.

UC IPM, 2005. Insects and Mites (www.ipm.ucdavis.edu).*Pest Management Guidelines*: Prune. UC ANR publication No. 3464.

Zaki, F.A. 1999. Incidence and biology of codling moth, *Cydia pomonella* L., in Ladakh (Jammu and Kashmir). *Applied Biological Research* 1: 75-78.

Chapter 18

Minimizing Pesticide Residues in Apple

Malik Mukhtar and Asma Sherwani

Research Centre for Pesticides Residue and Quality Analysis,
Division of Entomology,
Sher-e-Kashmir University of Agricultural Sciences and
Technology of Kashmir, Srinagar
E-mail: drmalikmukhtar@yahoo.com

1. Introduction

Insect pests and diseases of fruit and field crops are one of the limiting factors for optimizing their yield. To mitigate the losses caused by insect pests and plant diseases, pesticides are one of the important inputs in modern agriculture. The losses in the developed and developing countries are reported to be in the range of 10-30 per cent and 40-75 per cent respectively due to plant diseases and insect pests (Roy, 2002). It is estimated that 70 per cent of pesticides is used in developed world and rest 30 per cent in developing world (Pimentel, 1987). After application, there is slow dissipation of pesticides on treated surfaces. Pesticide treated food contain an appreciable amount of residues; therefore decontamination of edible part is of paramount importance. There are myriad of decontamination techniques that minimize the amount of pesticides residues below maximum residue limit (MRL). Home processing of fruits is considered one of key element to minimize the surface residues to a great extent.

Physicochemical properties of the pesticides determine the extent of retention time of residues on food. Peel surface is the area where most of the pesticide residues are retained (Awasthi, 1993), and are therefore easily dislodged by washing, peeling or treatment with various chemical solutions (Gupta, 2006). There are different production phases and phenological stages during which a pesticide may enter into the fruit, like blooming, growth and harvesting. So there are different locations in the fruit where residues of pesticides could be present (Trewavas and Stewart, 2003). Whereas, negligible amount of systematic pesticides might get into the flesh (Lewis *et al.,* 1998). Another factor that minimize the residue deposition on fruits is the nozzle type used for spraying. Many studies have shown initial deposit of pesticides are dependent on the canopy structure. In many situations the sprayer technology like nozzle type, sprayer setting and sprayer type determine the initial deposit (Xu *et al.,* 2006; Rawn *et al.,* 2007).

2. Pesticide Residues in Apples Grown in Kashmir

Orchardists in Kashmir apply diverse pesticides belonging to several groups to contain diseases and insect pests. A study was carried out from 2013-2014 to determine the pesticide residues on apple grown in Kashmir. The study revealed residues of nine pesticides were detected in farm gate samples from various districts of Kashmir (Sheikh *et al.,* 2015). The pesticides residues detected on apple from different districts include, Clothianidin, Myclobutanil, Bitertanol, Fenazaquin, Chlorpyrifos, Phosalone, Fenpyroximate, Difenaconazole and Quinalphos (Table 18.1). The finding of the study makes it clear that various steps (pre and post harvesting of fruit) are needed to minimize the pesticide residues so there is safe consumption of apple by the consumers.

Table 18.1: Status of Pesticide Residues in Harvested Apples from Various Districts of Kashmir

Pesticide	Amount of Pesticide Residues in different Districts (µg/g)							
	Bara-mulla	Sopore	Srinagar/ Ganderbal	Anant-nag	Shopian	Kulgam	Pul-wama	MRL
Quinalphos	X	X	0.247	0.01	X	0.10	X	0.50
Phosalone	X	0.179	X	0.650	X	3.054	0.124	2.00
Fenazaquin	0.860	0.140	X	0.260	X	X	0.417	0.10
Chlorpyrifos	0.560	1.498	0.360	3.080	3.365	0.350	0.501	1.50
Fenpyroximate	X	X	0.455	X	X	0.721	X	1.00
Difenaconazole	X	0.083	X	1.721	0.149	X	0.124	1.00
Myclobutanil	0.301	1.230	X	X	3.948	X	0.570	0.50
Bitertanol	0.386	0.105	0.289	X	X	0.256	0.599	2.00
Clothianidin	0.317	0.229	X	X	0.412	X	X	0.50

(*Source*: Sheikh *et al.*, 2015).   X (Residues not detected), MRL (Maximum residue limit).

(a) Minimizing Pesticides Residues in Apple by Home Processing

(i) Washing

The preliminary step in household process is the washing and the effect of washing on the surface residues has been well studied (Kaushik and Naik, 2009; Zabik *et al.*, 2000). However, these studies suggest that the rinsability of pesticides from fruit surface is not always correlated with its water solubility and different washing procedures may be adopted for different pesticides (Angioni *et al.*, 2004; Cabras *et al.*, 1998). Washing with water can dislodge surface residues, but residues of systemic insecticides can be less affected. Pesticide residue with washing may not only be performed through water rinsing but with some chemical solutions also. Penetration of pesticide residues into the pulp of the fruit is the most vital process that might decide the fate of pesticide residue during washing. Systemic insecticides like Dimethoate and Quinalphos with water solubility equal to 23300 mg/L and 18 mg/L respectively are not reduced during washing (Cabras *et al.*, 1997). Almost 50 per cent of fenitrothion residue was removed by washing of apple with tap water when treated with 0.15 per cent of fenitrothion (Lipowska *et al.*, 1998).

Water solubility of the pesticide is not only related to washing but octanol/water partition coefficient (Kow) strengthens the prospect that partition coefficients between water washing and cuticle correlate with Kow of pesticides (Baur *et al.*, 1997). The use of appropriate chemical solution has the likelihood to solubilize waxes present on the fruit's epicuticle that may finally dissipate the residues (Angioni *et al.*, 2004). Residues still present following washing could be attributed as penetrated residue as they have passed the cuticle barrier. Care should be taken not to use high levels of sodium hypochlorite, hydrogen peroxide and potassium permanganate in washing water as these could form oxons from the organophosphorus pesticides by chemical oxidation (Ou-Yang *et al.*, 2004; Pugliese *et al.*, 2004).

(ii) Peeling

A vast number of insecticides and fungicides that are directly applied to crops have restricted penetration into the pulp of the fruit; hence residues of these pesticides are confined to the peel from where they are easily dislodgeable by peeling. Peeling of fresh apples nearly removes residues of pesticides from fruit. Residues in the pulp were 0.5 to 1 per cent of those in the peel.

(b) Removal of Pesticide Residues by Various Chemical Solutions

(i) Acidic Solution

Dipping in acidic solutions like ascorbic acid, acetic acid and citric acid at a concentration of 5 and 10 per cent for 10 minutes result in appreciable reduction of pesticide residues. Acidic solutions perform better in removing surface residues as compared to neutral or alkaline solutions. Acidic solutions of 5 and 10 per cent decontaminate 100 per cent surface residues, whereas, 80 per cent pesticide residues are eliminated with 5 and 10 per cent solutions of citric and ascorbic acid (Wheeler, 2002).

(ii) Neutral Solutions

Solutions of sodium chloride (NaCl) are widely used to remove pesticide residues from various fruits and vegetables. Salt-water washings has been found to remove pesticides satisfactorily from different fruits and vegetables (Zohair, 2001). There is a reduction of 28 to 93 per cent in organochlorine and 100 per cent in organophosphate pesticides residues when a saline solution of 5 and 10 per cent are used with 15 minutes fruit dipping (Wheeler, 2002). Reduction of pesticide residues increases with increase in concentration of NaCl solution (Ismail *et al.*, 1993).

(iii) Alkaline Solution

Sodium hydroxide (NaOH) and potassium dichromate are also used for decontamination of pesticide residues. Surface residues of pyrethroids are removed upto 50 to 60 per cent by dipping fruits in NaOH solution (Awasthi and Lalitha, 1983).

(iv) Removal of Pesticide Residues using Ozone

Ozone (O_3) is a natural component of the earth's atmosphere and perhaps one of the potent sanitizers (Khadre *et al.*, 2001). O_3 generation is achieved by passing air or oxygen through an electric discharge or by irradiating with ultraviolet light (Mahapatra *et al.*, 2005). O_3 does not harm the flavour of fruits as it changes to oxygen through autolysis (Li and Tsuge, 2006). That is the reason O_3 has the potential to remove residues of pesticides from fruits and vegetables (Gabler *et al.*, 2010). A threshold concentration of $0.075\mu l/L$ of O_3 has been suggested for continuous human exposure (US Environmental Agency, 2008). Several authors have reported the decomposition of pesticide residues by O_3 (Daidai *et al.*, 2007; Hwang *et al.*, 2001; Hwang *et al.*, 2002). Fenitrothion @ 140 ppm was completely decomposed within 40 minutes in 13 per cent ozonated solution (Tanaka *et al.*, 1992). The residues of captan, azinphos-methyl and formetanate @ 2-ppm on the surface of apples were completely reduced when immersed in 0.25 ppm solution of O_3 for 30 minutes (Ong *et al.*, 1996). A brassicaceous vegetable treated with 0.1 ppm of diazinon, parathion, methyl parathion and cypermethrin removed 53, 55, 47 and 61 per cent residues of these pesticides respectively when treated with solution containing 2 ppm dissolved O_3 for 30 minutes (Wu *et al.*, 2007).

(c) Effect of Spray Nozzle on Reduction of Pesticide Residues on Fruits

Initial deposits of pesticide residues on fruits are dependent on three processes (i) growth dilution (ii) spray deposit and (iii) dissipation due to weather factors. The deposits of pesticide residues vary within the fruits of the same tree as top and outside regions of the tree are likely to receive more spray than those inside the canopy. In many situations the initial deposit is determined by sprayer setting, nozzle and sprayer type. In a study, large variation in residue levels was observed when sprayed either with coarse or fine droplet nozzles (Rawn *et al.,* 2007). This depicts that less initial deposit and hence fewer residues remain on fruits with fine spraying as compared to coarse spraying.

(d) Other Measures that will Help to Reduce Pesticide Residues on Fruits

☆ Always follow recommended doses of pesticides as higher than recommended dose will lead to high levels of pesticide residues.

☆ Waiting period of each pesticide should be observed before harvest.

☆ Alternative methods of pest control like mechanical, cultural, physical or biological should be practiced in the field.

☆ Maintenance and calibration of sprayers should be done at proper intervals as uncalibrated sprayers lead to high initial deposit resulting in high level of residue.

☆ Spray only when the pests are above the economic injury level.

☆ Misbranded or spurious pesticides should not be used for spraying.

References

Angioni, A., Schirra, M., Garau, V.L., Melis, M., Tuberoso, C.I.G., Cabras, P. 2004. Residues of azoxystrobin, fenhexamid and pyrimethanil in strawberry following field treatments and the effect of domestic washing. *Food Additives and Contaminants* 2: 1065-1070.

Awasthi, M. D. and A. Lalitha. 1983. Comparative persistence of synthetic pyrethroids on cauliflower. *Journal of the Entomological Research* 7: 139-144.

Awasthi, M.D. 1993. Decontamination of insecticide residues on mango by washing and peeling. *Journal of Food Science and Technology* 30:132- 133.

Baur, P., Buchholz, A. and Schönherr, J. 1997. Diffusion in plant cuticles as affected by temperature and size of organic solutes: similarity and diversity among species. *Plant, Cell and Environment* 20: 982-994.

Cabras, P., Angioni, A., Garau, V.L., Melis, M., Pirisi, F.M., Cabitza, F. and Cubeddu, M. 1998. Pesticide residues on field-sprayed apricots and in apricot drying processes. *Journal of Agricultural and Food Chemistry* 46: 2306-2308.

Cabras, P., Angioni, A., Vicenzo L., Garau, V.L., Minelli, E.V., Melis, M. and Pirisi, F.M. 1997. Pesticides in the distilled spirits of wine and its by products. *Journal of Agricultural and Food Chemistry* 45: 2248-2251.

Daidai, M., Kobayashi, F., Mtsui, G. and Nakamura, Y. 2007. Degradation of 2, 4-dichlorophenoxyacetic acid (2,4-D) by ozonation and TiO2/UV treatment. *Journal Chemical Engineering Japan* 40(9): 378-384.

Gabler, F. M., Smilanick, J. L., Mansour, M. F. and Karaca, H. 2010. Influence of fumigation with high concentrations of ozone gas on post harvest gray mold and fungicide residues on table grapes. *Post harvest Biology and Technology* 55 (2): 85-90.

Gupta, A, 2006. *Pesticide residue in food commodities*. Agrobios India.

Holland, P. T. D., Hamilton, B. Ohlin and M. W. Skidmore. 1994. Effect of storage and processing on pesticides residues in plant products. *Pure and Applied Chemistry* 66(2): 335-356.

Hwang, E.S., Cash, J. N. and Zabik, M. J. 2001. Post harvest treatments for the reduction of Mancozeb in fresh apples. *Journal of Agricultural and Food Chemistry* 49(6): 3127-3132.

Hwang, E.S., Cash, J. N. and Zabik, M. J. 2002. Degradation of Mancozeb and Ethylenethiourea in apples due to post harvest treatments and processing. *Journal of Food Science* 67(9): 3295-3300.

Ismail, S. M. M., A. Ali and R. A. Habiba. 1993. GC-ECD and GC-MS analysis of profenofos residues and its biochemical effects in tomatoes and tomatoes products. *Journal of Agricultural and Food Chemistry* 41:610-615.

Kaushik, G., Satya, S. and Naik, S.N. 2009. Food processing a tool to pesticide residue dissipation - A review. *Food Research International* 42, (26-40): 0963-969.

Khadre, M. A., Yousef, A. E. and Kim, J. G. 2001. Microbial aspects of ozone applications in food: a review. *Journal of Food Science* 66(9): 1242-1252.

Lewis, D. J., Thorpe, S.A., Wilkinson, K. and Reynolds, S. L. 1998. The carry through of residues of maleic hydrazide from treated potatoes, following manufacture into potato crisps and jacket potato crisps. *Food Additives and Contaminants* 15:506-509.

Li, P. and Tsuge, H. 2006. Ozone transfer in a new gas-induced contactor with microbubbles. *Journal of Chemical Engineering of Japan* 39(11): 1213-1220.

Lipowska, T., Szymczyk, K., Danielewska, B. and Szteke, B. 1998. Influence of technological process on fenitrotion [fenitrothion?] residues during production of concentrated apple juice - short report. *Polish Journal of Food and Nutrition Sciences* 7(48): 293-297.

Mahapatra, A. K. Muthukumarappan, K. and Julson, J. L. 2005. Applications of ozone, bacteriocins, and irradiation in food processing: a review. *Critical Reviews in Food Science and Nutrition* 45(6): 447-461.

Ong, K. C., Cash, J. N., Zabik, M. J., Siddiq, M. and Jones, A. L. 1996. Chlorine and ozone washes for pesticide removal from apples and processed apple sauce. *Food Chemistry* 55(2): 153-160.

Ou-Yang, X.K., Liu, S.M. and Ying, M. 2004. Study on the mechanism of ozone reaction with parathion-methyl. *Safety and Environmental Engineering* 11:38-41.

Pimentel, D. 1987. Pesticides: energy use in chemical agriculture. In: Marini-Battelo GB (eds) *Towards a second green revolution* Elsvier, Amsterdam, pp. 157-176.

Pugliese, P., Moltó, J.C., Damiani, P., Marín, R., Cossignani, L. and Manes, J. 2004. Gas chromatographic evaluation of pesticide residue contents in nectarines after nontoxic washing treatments. *Journal of Chromatography A* 1050: 185-191.

Rawn, D.F.K., Quade, S.C., Shields, J.B., Conca, G.C., Sun, W.F., Lacroix, G.M.A., Smith, M., Fouqout, A. and Belanger, A. 2007. Variability in captan residues in apples from a Canadian orchard. *Food Additives and Contaminants* 24 (2): 149 - 155.

Roy, N.K. 2002. *Pesticide residues and their environmental implications.* In: Chemistry of pesticides (Ed. N. K. Roy) CBS, New Delhi, pp 265-279.

Sheikh, B. A., Malik, M., Ashraf, A., and Somia, B. 2015. *Pesticide Residues: Kashmir Apples- Findings of Valley Wide Survey in 2013-2014.* Series/DR-SKUAST-K/2015/06, SKUAST-Kashmir

Tanaka, K., Abe, K., Sheng, C. Y. and Hisanaga, T. 1992. Photocatalytic wastewater treatment combined with ozone pre-treatment, *Environmental Science and Technology* 26(12): 2534-2536.

Trewavas A and Stewart, D. 2003. Paradoxical effects of chemicals in the diet on health. *Current Opinion Plant Biology* 6:185-190.

US Environmental Protection Agency 2008. National ambient air quality standards for ozone. *Federal Register* 73(60): 16436-16514.

Wheeler, W. 2002. Role of research and regulation in 50 years of pest management in agriculture. *Journal of Agricultural and Food Chemistry* 50: 4151-4155.

Wu, J. G., Luan, T. G., Lan, C. Y., Lo, W. H. and Chan, G. Y. S. 2007. Efficacy evaluation of low-concentration of ozonated water in removal of residual diazinon, parathion, methyl-parathion and cypermethrin on vegetable. *Journal of Food Engineering* 79(3): 803-809.

Xu, X., Wu, P., Thorbeck, P. and Hyder, K. 2006. Variability in initial spray deposit in apple trees in space and time. *Pest Management Science* 62: 947 - 956.

Zabik, M.J., El-Hadidi, M.F.A. J. Cash, N., Zabik, M.E. and Jones, A.L. 2000. Reduction of azinphos-methyl, chlorpyrifos, esfenvalerate, and methomyl residues in processed apples. *Journal of Agricultural and Food Chemistry* 48: 4199-4203.

Zohair, A. 2001. Behavior of some organophosphorus and organochlorine pesticides in potatoes during soaking in different solutions. *Food and Chemical Toxicology* 39: 751-755.

Chapter 19

Viral Diseases of Apple and Other Temperate Fruits in Kashmir: An Overview

Bilal A. Padder, M D Shah, Mushtaq Ahmad,
Iretfa Mohmmad, Aadil Ayaz Mir,
M. S. Dar and Asha Nabi

Plant Virology and Molecular Plant Pathology Laboratory,
Division of Plant Pathology,
Sher-e-Kashmir University of Agricultural Sciences and
Technology of Kashmir, Srinagar
E-mail: bapadder@rediffmail.com

1. Introduction

Temperate fruits belong to family Rosaceae and include pome (apple, quince and pear) and stone fruits (apricot, peach, plum, almond and cherry). Consumption of temperate fruit on daily basis may reduce the risk of cardiovascular and certain cancer diseases. The saying "An apple a day keeps a doctor away" fits best as eating a fruit keeps our body metabolically and physiologically active against intruders. Nature has bestowed Kashmir valley with temperate climatic that is suitable for cultivating pome and stone fruits. Farming community soon realized the importance of pome and stone fruits and started cultivating it over a large area. Among the various temperate fruits, apple cultivation comprises about 80 per cent of area, the remaining is under stone fruits, and the fruit industry is backbone of state's economy.

Various fungal and viral diseases that cause huge loss to small framers of the state both qualitatively and quantitatively infect temperate fruits of Kashmir. Various fungal diseases are kept under control by judicious application of fungicides from pink bud until harvest. However, non-availability of antivirals to manage virus infected plant besides remains infected until death further aggregates the problem. For many years, viral diseases of pome and stone fruits were not considered dangerous in Kashmir valley. However, with climate change and advancement in detection tools helped in identification of viruses in Kashmir valley. The viruses cause a wide range of symptoms ranging from symptomless (latent) to general decline in vigour and productivity. Leaf symptoms include distortion or twisting, mottling, rolling, necrotic spots, shot holes, and unusual colour patterns. Fruits may show reductions in size and quality, distortions in shape, and alterations such as ring spots, mottling, and line patterns. Viruses cause huge economic loss to all sectors of production chain. For example, *Plum pox virus* (PPV) cause severe damages and has enormous economic and social impacts. Tens of millions of euros and dollars have been spent for controlling this pathogen without success. The major viruses identified on pome and stone fruits in the Kashmir valley belong to several families and genera (Table 19.1). These viruses were identified based on serology and molecular detection tools.

Table 19.1: Apple and other Temperate Fruit Viruses Prevalent in Kashmir Valley

Virus	Host
Apple chlorotic leaf spot virus (ACLSV)	Apple, pear, peach apricot, cherry
Apple stem grooving virus (ASGV)	Apple, Pear, Cherry
Apple stem pitting virus (ASPV)	Apple, Pear, Cherry
Prunus necrotic ringspot virus (PNSRV)	Most of the stone fruits
Apple mosaic virus (ApMV)	All pome and stone fruits
Prune dwarf virus (PDV)	Stone fruits
Plum pox virus (PPV)	Stone fruits

Virus infecting pome and stone fruits are present in all fruit growing areas of world and the losses incurred are well known. The farming community in developed countries manage the dreadful infectious agents by planting virus free certified plant material in their orchard. However, lack of information about the presence of viruses in the valley and planting material without proper certification will have adverse effects on farming community of Kashmir in future. Keeping in view the significance of viruses to the farming community, the present chapter address these issues and describe the presence and detection of viruses infecting pome and stone fruits in Kashmir. We will also highlight various management practices that are essential for managing them.

2. Viruses

(a) Symptoms

(i) Apple Chorotic Leaf Spot Virus (ACLSV)

The virus was detected first time from apple trees in the USA (Mink and Shay, 1959). The virus is generally latent. However, if apple cultivar is grafted on Marubakaido rootstocks the infected trees chlorotic leaf spots. Most of the stone fruits infected with ACLSV are symptomless but few susceptible cultivars show bark splitting, severe fruit deformation, graft incompatibility and necrosis. In apricot and plum, virus cause severe deformation of fruit, often confused with the "sharka" disease due to plum pox virus. Because of these symptoms, it is named as "pseudopox". Virus is graft transmitted and no vector is currently known. ACLSV is not known to be transmitted by seed or pollen (Yoshikawa, 2001).

(ii) Apple Stem Grooving Virus (ASGV)

ASGV was first reported from Virginia, USA in 1960s. The virus also infects kiwifruit (Massart *et al.*, 2011). The virus is latent on apple cultivars but cause symptom grooving, brown line and graft union abnormalities when infected cultivar is grafted on the sensitive rootstock like *Malus pumila* (Virginia crab). In nurseries, virus infected plants are less vigorous. ASGV is transmitted by infected propagative material and no vector is currently known.

(iii) Apple Stem Pitting Virus (ASPV)

Apple stem-pitting disease was first described as incompatibility between certain apple cultivars and the rootstock *M. sylvestris* in the USA in the 1940s. ASPV is mostly latent in the commercially grown cultivars. The virus produces characteristic symptoms on the indicator woody hosts like leaf chlorosis, xylem pitting and top working. Most of the pear cultivars are susceptible and show cholortic leaf spots during first few years of growth. As the age of plant increases symptoms become latent (Jelkmann and Paunovic, 2011). Virus is transmitted by grafting and through infected propagative material (Jelkmann and Paunovic, 2011).

(iv) Apple Mosaic Virus (ApMV)

The virus was described for the first time in apple (Bradford and Joly, 1933) and Padder *et al.*, 2011, first time reported from Kashmir. The severity of symptoms in apple depends on cultivar susceptibility. Fruits do not develop diagnostic symptoms. Apple trees infected with the virus show pale yellow to bright cream irregular spots or bands along major veins on spring leaves. The spots may become necrotic on severely affected leaves after exposure to summer sun and heat. The symptomatic leaves drop prematurely. The distribution of symptomatic leaves may be erratic throughout the tree or limited to a single limb (Figure 19.1). In pear, Apple mosaic virus (ApMV) infection is usually symptomless (Petrzik and Lenz, 2011). In stone fruits, it causes typical yellow line pattern, bright yellow blotches, rings, bright yellow vein clearing, and/or oak-leaf pattern (Diekmann and Putter, 1996; Nemeth,

Figure 19.1

1986). Symptoms generally appear at the beginning of summer and, in some cases, are present only on a limited number of leaves randomly distributed on the plants (Paunovic *et al.,* 2011).

ApMV infects a large number of woody hosts and is frequently found in mixed infections with PNRSV, and PDV on their common stone fruit hosts. In apple, it often occurs together with ACLSV, ASPV, ASGV, and other apple-infecting viruses. ApMV is only transmitted by vegetative propagation and by grafting. In addition, slow natural spread in nurseries via root grafting occurs (Dhingra, 1972; Hunter *et al.,* 1958). No insect vector is known for the virus. ApMV is not pollen-transmissible (Barba *et al.,* 1986; Digiaro *et al.,* 1992; Sweet, 1980).

(v) Prunus Necrotic Ring Spot Virus (PNSRV)

It causes serious disease in nurseries. Symptom severity is determined by the virus isolate and host cultivar (Nyland *et al.,* 1976), and synergistic interactions of Prunus necrotic ringspot virus (PNRSV) with other viruses, for example Prune dwarf virus (PDV), result in more severe disease symptoms. PNRSV is easily transmitted by routine plant propagation methods (Cole *et al.,* 1982; George and Davidson, 1963; Mink, 1992; Nyland *et al.,* 1976) and by root grafting in orchards. PNRSV is also carried on and in pollen grains and is readily transmitted through seeds (Amari *et al.,* 2009; Aparicio *et al.,* 1999). In addition, it is transmitted by different thrips species carrying infected pollen to healthy plants.

(vi) Prune Dwarf Virus (PDV)

In plum, PDV causes stunting and leaf malformation and shortened internodes. In Italian prune, it decreases the length of shoots, their diameter, number of leaves, and the photosynthetic total area (Hadidi and Barba, 2011). In cherry, PDV may cause leaf chlorotic spots, rings and diffuse mottling, and possibly stem pitting and flat limb. Fruits can be malformed and their production is reduced. In some apricot cultivars, PDV has been reported to induce gummosis on the trunk. In most peach cultivars, PDV induces mild stunting while leaves become dark green and more erect than those of non-infected trees, but infection by severe isolates can cause important yield reduction and poor quality of fruits. Peach's infected with both PDV and PNRSV (peach stunt disease, PSD) display bark splitting, increased sucker production and yield is reduced by up to 60 per cent. The virus causes economic losses on stone fruit trees, especially in sour and sweet cherry, almond, and peach (Nolasco *et al.*, 1991; Rampitsch *et al.*, 1995; Uyemoto and Scott, 1992). PDV is transmitted by grafting (buds, scions), pollen and seed. Pollen transmission depends on many factors such as fruit tree species and circumstances affecting pollination. Pollen transmission in sweet and sour cherry shows the highest transmission rates (George and Davidson, 1964; Gilmer, 1965). Seed transmission occurs in sweet cherry, sour cherry, mahaleb, and myrobalan, but infection rates vary with the species (Caglayan *et al.*, 2011).

(vii) Plum Pox Virus (PPV)

Sharka disease was detected in the early 1900s in Bulgaria and it was first described by Atanasoff, (1932). PPV induces sharka disease, the most devastating disease of stone fruit trees worldwide. Its presence in a country may create trade restrictions at the international, national, and local levels. Most of the susceptible plum, apricot, peach, and cherry cultivars show leaf symptoms that appear as pale or yellowish green rings, spots, or leaf mottling. Some peach cultivars may also show discoloration on flowers. Affected plum fruits are deformed and show rings, irregular lines, and poxes on the surface. Colored rings and bands appear on the skin of apricot fruits and apricot stones show pale rings or spots. Peach fruit symptoms are mostly restricted to the skin where pale rings and diffuse bands appear before maturation (Barba *et al.*, 2011). The virus is transmitted by grafting and other vegetative propagation techniques and by aphids such as *Aphis spiraecola* and *Myzus persicae* in a non-persistent manner. There is no evidence for either pollen or seed transmission (Glasa *et al.*, 1999; Myrta *et al.*, 1998; Pasquini *et al.*, 2000; Pasquini and Barba, 2006). The costs associated with the disease involve not only direct losses in fruit production, commercialization, eradication, compensatory measures, and lost revenue but also indirect costs including those from preventive measures such as quarantine, surveys, inspections, control of nurseries, diagnostics, and the impact on foreign and domestic trade. The loss from sharka during the last few decades, at the global level, is estimated at about 18 million metric tons of apricots and about 45 million metric tons of European plums with an estimated value of 3600 and 5400 million Euros, respectively (Cambra *et al.*, 2006).

(b) Detection

Virus disease management depends on the accurate detection and is the most important part in controlling the pome and stone fruit viruses. Early virus detection in fruit trees, propagating material, mother plants and rootstocks is a prerequisite for the control of diseases induced by these pathogens and to guarantee a sustainable agriculture. Temperate fruit virus detection falls into two major categories *viz.,* biological and molecular. Former is based on the interaction of virus with its hosts and is characterized by the symptom inspection on the host. In case of latent infection, mechanical inoculation to susceptible indicator plants (herbaceous indicator plants) or grafting on woody indicator plants will assist in virus detection. Detection based on biological indexing is time consuming and require herbaceous/woody indicators. Most of temperate fruit viruses produce characteristic local lesions and symptoms on mechanical transmission (Figure 19.2).

Figure 19.2

Molecular detection of viruses is easy, accurate, and sensitive and takes only few hours to detect a particular virus. Serology and PCR based amplification of coat protein (CP) gene are two molecular virus identification methods. For most of important viruses of pome and stone fruits, commercial antibody kits are available that are used to detect viruses through ELISA (Figure 19.3). ELISA can also be used for the identification of virus serotypes by using specific monoclonal antibodies.

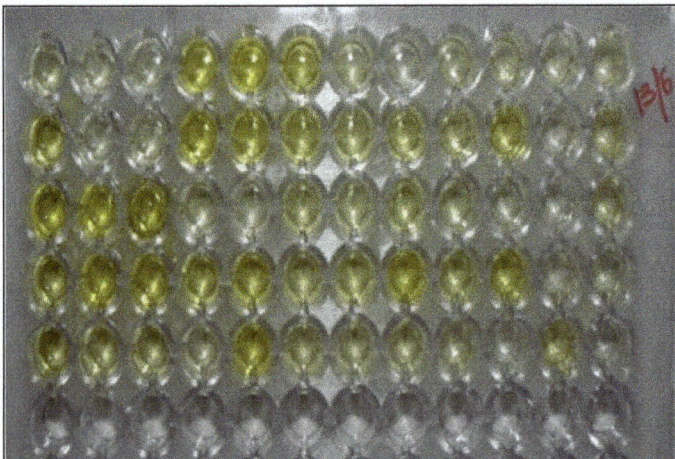

Figure 19.3

Molecular detection based on amplification of CP gene is robust, quick and easy detection tool for the viruses. As most of the viruses infecting pome and stone fruits are RNA viruses, it is necessary to use RT-PCR for amplification of CP. Over the years, RT-PCR technique has been optimized and there are many primers for an individual virus. With the development of multiplexing, it is easy to identify multiple viruses in a single sample (Figure 19.4).

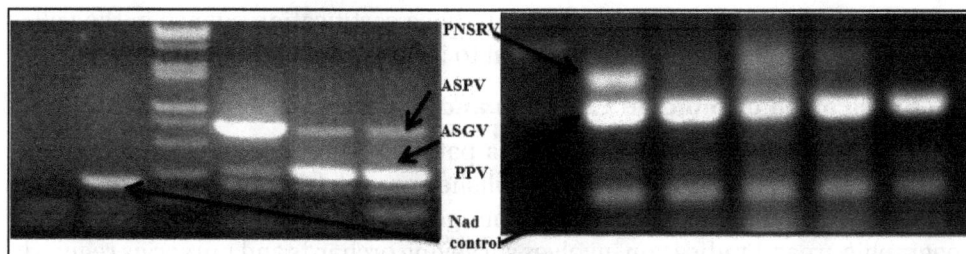

Figure 19.4: Molecular Detection of Viruses Infecting Pome and Stone Fruits in Kashmir.

(c) Management

Temperate fruit viruses are not only responsible for important economic losses but also cause direct or indirect damage to host plants like reduction in vegetative growth, market quality of fruit *etc*. Because of their significant effect on fruit quality, shape and tree health has resulted in the imposition of control measures both nationally and internationally. Virus diseases are difficult to manage owing to unavailability of antiviral chemicals, so need a different management capsule to tackle the diseases. Various management strategies for pome and stone fruit viruses are listed below.

1. Exclusion of viruses by crop quarantine
2. Exclusion by crop certification
3. Eradication of infected cultivars and rootstocks
4. Control of virus vectors
5. Selection of tolerant or resistant cultivars

(i) Exclusion of Viruses by Crop Quarantine

Trade route between states and countries provides a pathway whereby plants and their associated pests are easily and rapidly transported. Hence, quarantine regulations are set to eliminate the movement of pathogens into geographic area where they were not previously known to occur. These phytosanitary measures are seen by the government based organisations. In Kashmir valley, pome and stone fruits has been transported from various European countries and there is possibility of pest introduction. It is necessary to have strict vigilance and carry out detection to avoid economic loss to the farming community of valley.

(ii) Exclusion by Crop Certification

Propagative material such as rootstocks, buds, shoots are the main source of temperate fruit virus spread. To reduce risks, developed countries have put in place many regulations that guarantee the production of high-quality planting material. The most important is certification of planting material and represents an effective way to assure about trueness of plant cultivar type and sanitary status. In order to reduce the virus spread, it is necessary to have certification agency in the valley that will assure quality planting material to its growers.

(iii) Eradication of Infected Cultivars and Rootstocks

This principle aims at eliminating a pathogen after it is introduced into an area but before it has become well established or widely spread. It can be applied to individual plants, fields, or regions, but generally, it is not effective over large geographic areas. Eradication involves surveying orchards and nurseries regularly, and immediately removing infected trees before the virus spreads. For example, many attempts were made to eradicate new foci of PPV in countries, regions or geographical areas, but it has been difficult to eliminate the pathogen. Eradication is effective only if it is done timely.

(iv) Controlling Viral Insect Vectors

A small number of viruses infecting temperate fruit trees are transmissible by vectors. The most important example is represented by PPV, transmitted by different species of aphids in a non-persistent manner. The control of insect vectors plays an important role in the management of systemic diseases, but it must be used together with other control measures such as eradication of infected plants and use of certified propagation material.

The use of insecticide treatments results only in reducing the populations of potential vectors, without preventing the transmission of non-persistently transmitted viruses such as PPV. This is because these viruses are prevalently spread by transient species. However, oil application has been proposed to reduce the spread of PPV in nurseries.

(v) Selection of Tolerant and/or Resistant Crop Cultivars

Apart from the control of the virus vector and the use of virus-free material, the development of virus-resistant varieties appears to be the most effective approach to achieve control of plant viruses, especially for perennial crops that can become infected during their long life span. The use of resistant or tolerant cultivars and/ or rootstocks could be the most important aspect of virus disease management, especially in areas in which virus infections are endemic. The conventional breeding for virus-tolerant or resistant fruit tree cultivars using available germplasm is a long-term strategy, and the development and production of these cultivars may take decades, if successful. In particular, the selection is slow and difficult due to the transfer of undesirable characteristics and other constraints typical of fruit trees such as long biological cycle with extended juvenile phase and high level of heterozygosis. Many studies have been performed on fruit tree cultivars for evaluation of their susceptibility to different viruses, based on field observations under natural infection pressure or involving models obtained by grafting or chip budding or by viruliferous vector transmission in the field or under greenhouse conditions. Obtained data in most cases were not comparable as they were influenced by virus isolates, virus inoculation methods, classification of plant response to viruses and types of diagnostic tests with different sensitivity used for evaluation of germplasm infection.

References

Amari, K., Burgos, L., Palla's, V. and Sa'nchez-Pina, M. A. 2009. Vertical transmission of Prunus necrotic ringspot virus: Hitch-hiking from gametes to seedling. *Journal of Virological Methods* 90: 1767–1774.

Aparicio, F., Myrta, A., Di Terlizzi, B., and Palla's, V. 1999. Molecular variability among isolates of Prunus necrotic ringspot virus from different Prunus spp. *Phytopathology* 89: 991–999.

Atanasoff, D. (1932). Jahrbuch Universita"t Sofia. *Agronomische Fakulta"t*, 11, 49.

Barba, M., Hadidi, A., Candresse, T., and Cambra, M. 2011. Plum pox virus. In A. Hadidi, M. Barba, T. Candresse, and W. Jelkmann (Eds.), *Virus and virus-like diseases of pome and stone fruits* (pp. 185–198). St. Paul, MN: APS Press.

Barba, M., Pasquini, G., and Quacquarelli, A. 1986. Role of seeds in the epidemiology of two almond viruses. *Acta Horticulturae*193: 127–130.

Bradford, F. C. and Joly, L. 1933. Infectious variegation in the apple. *Journal of Agricultural Research* 46: 901–908.

Caglayan, K., UlubasSerce, C., Gazel, M. and Varveri, C. 2011. Prune dwarf virus. In A. Hadidi, M. Barba, T. Candresse, and W. Jelkmann (Eds.), *Virus and virus-like diseases of pome and stone fruits* (pp. 199–206). St. Paul, MN: APS Press.

Cambra, M., Capote, N., Myrta, A. and Lla'cer, G. 2006. *Plum pox virus and the estimated costs associated with sharka disease.* OEPP/EPPO Bulletin, 36, 202–204.

Cole, A., Mink, G. I. and Regev, S. 1982. Location of Prunus necrotic ring spot virus on pollen grains from infected almond and cherry trees. *Phytopathology* 72: 1542–1545.

Dhingra, K. L. 1972. Transmission of apple mosaic by natural root grafting. *The Indian Journal of Horticulture* 29: 348–350.

Diekmann, M. and Putter, C. A. J. 1996. *FAO/IPGRI technical guidelines for the safe movement of germplasm.* (pp. 1–110). Rome: No. 16. Stone Fruit. Food and Agriculture Organization of the United Nations, Rome/International Plant genetic Resources Institute.

Digiaro, M., Savino, V. and Di Terlizzi, B. 1992. Ilarvirus in apricot and plum pollen. *Acta Horticulturae* 309: 93-98.

George, J. A. and Davidson, T. R. 1964. Further evidence of pollen transmission of necrotic ringspot and sour cherry yellows viruses in sour cherry. *Canadian Journal of Plant Science* 44: 383-384.

George, J. and Davidson, T. R. (1963). Pollen transmission of necrotic ringspot virus and sour cherry yellows viruses from tree to tree. *Canadian Journal of Plant Science* 43: 276-288.

Gilmer, R. M. 1965. Additional evidence of tree-to-tree transmission of sour cherry yellows virus by pollen. *Phytopathology* 55: 482-483.

Glasa, M., Hricovsky, I. and Kudela, O. 1999. Evidence for non-transmission of Plum pox virus by seed in infected plum and myrabolan. *Biologia Bratislava* 54: 481-484.

Hadidi, A. and Barba, M. 2011. Economic impact of pome and stone fruit viruses and viroids. In A. Hadidi, M. Barba, T. Candresse, and W. Jelkmann (Eds.), *Virus and virus like diseases of pome and stone fruits* (pp. 1-8). St. Paul, MN: APS Press.

Hunter, J. A., Chamberlain, E. E. and Atkinson, J. D. 1958. Note on the transmission of apple mosaic by natural root grafting. *New Zealand Journal of Agricultural Research* 1: 80-82.

Jelkmann, W. and Paunovic, S. 2011. Apple stem pitting virus. In A. Hadidi, M. Barba, T. Candresse, and W. Jelkmann (Eds.), *Virus and virus-like diseases of pome and stone fruits* (pp. 35-40). St. Paul, MN: APS Press.

Massart, S., Jijakli, M. H. and Kummert, J. 2011. Apple stem grooving virus. In A. Hadidi, M. Barba, T. Candresse, and W. Jelkmann (Eds.), *Virus and virus-like diseases of pome and stone fruits* (pp. 85-90). St. Paul, MN: APS Press.

Mink, G. I. 1992. Prunus necrotic ringspot virus. In J. Kumar, H. S. Chaube, U. S. Singh, and A. N. Mukhopadhyay (Eds.), Plant diseases of international importance: 3. *Diseases of fruit crops* (pp. 335-356). Englewood Cliffs, NJ: Prentice Hall.

Mink, G. I. and Shay, J. R. 1959. A survey for stem pitting in Indiana Apple varieties. *Plant Dis. Reptr. Suppl 254*: 18-21.

Myrta, A., Di Terlizzi, B. and Savino, V. 1998. Study on the transmission of plum pox potyvirus through seeds. *Phytopathologia Mediterranea* 37: 41-44.

Nemeth, M. 1986. Virus, mycoplasmas and rickettsia diseases of fruit trees. Budapest: AkademiaiKiado. Posnette, A. F., and Ellenberger, E. C. (1957). The line pattern virus disease of plum. *The Annals of Applied Biology* 45: 74-80. Poul, F. and Dunez, J. (1990). Use of monoclonal

Nolasco, G., Neves, M. A. and Faria, E. A. 1991. Distribuic,a˜o no Algarve de vý′rus do grupoIlarvirusemamendoeira e suasconseque^nciasnaproduc,a˜o: 1aaproximac,a˜o. *Revista de la Facultad de CienciasAgrarias* 15: 33-37.

Nyland, G., Gilmer, R. M. and Moore, J. D. 1976. Prunus ringspot virus group. In U.S. Dept. of Agr. Handbook: 437. *Virus diseases and noninfectious disorders of stone fruits in North America* (pp. 104–132). Washington, DC: U. S. Government Printing Office.

Padder, B. A., Shah, M. D., Ahmad, M., Hamid, A., Sofi, T. A., Ahanger, F. A. and Saleem, S. 2011. Status of apple mosaic virus in Kashmir Valley. *Applied Biological Research 13*(2): 117-120.

Pasquini, G. and Barba, M. 2006. *The question of seed transmissibility of Plum pox virus*. OEPP/EPPO Bulletin 36: 287–292.

Pasquini, G., Simeone, A. M., Conte, L. and Barba, M. 2000. RT-PCR evidence of the non-transmission through seed of Plum pox virus strains D and M. *Journal of Plant Pathology* 82: 221–226.

Paunovic, S., Pasquini, G., and Barba, M. (2011). Apple mosaic virus in stone fruits. In A. Hadidi, M. Barba, T. Candresse, W. Jelkmann (Eds.), *Virus and virus-like diseases of pome and stone fruits* (pp. 91–96). St. Paul, MN: APS Press.

Petrzik, K. and Lenz, O. 2011. Apple mosaic virus in pome fruit. In A. Hadidi, M. Barba, T. Candresse, and W. Jelkmann (Eds.), *Virus and virus-like diseases of pome and stone fruits* (pp. 25–28). St. Paul, MN: APS Press.

Rampitsch, C., Eastwell, K. C. and Hall, J. 1995. Setting confidence limits for the detection of Prune dwarf virus in Prunusavium with a monoclonal antibody-based triple antibody sandwich ELISA. *Annals of Applied Biology* 126: 485–491.

Sweet, J. B. (1980). Fruit tree virus infections of woody exotic and indigenous plants in Britain. *Acta Phytopathologica* 15: 231–238.

Uyemoto, J. K. and Scott, S. W. (1992). Important diseases of Prunus caused by viruses and other graft-transmissible pathogens in California and South Carolina. *Plant Disease* 76: 5–11.

Yoshikawa, N. (2001). *Apple chlorotic leaf spot virus*. CMI/AAB Descriptions of Plant Viruses 386 No. 30 revised

Chapter 20

Nematode Pests as Vectors of Plant Viruses in Fruit Ecosystem

G. M. Lone and F. A. Zaki

Division of Entomology,
Sher-e-Kashmir University of Agricultural Sciences and
Technology of Kashmir, Srinagar
E-mail: lonegm1555@gmail.com

1. Introduction

Indian population is growing at an annual rate of 1.9 percent and would reach 1.4 billion by 2025 requiring 380 million tons (MT) of food grains against the current production of 246.2 MT (2011-12). Ecologically, India is one of the most diverse countries where fruits as a back bone of horticulture are grown over an area of 5.50 million hectares with a production of over 55 million metric tonnes/ year and which accounts for about 10 per cent of the total world's fruit production. Horticulture sector has a significant impact on the growth of the country's economy and India is the second largest producer of fruit after China and is known as fruit basket of the world. Fruit industry as a back bone is generating employment to millions of people besides providing nutritional security. Among the major fruits grown in India are mangoes, grapes, apple, apricot, orange, banana, guava, litchi, papaya *etc.*India stands 6th in apple production in the world but ranks at 53rd position in terms of productivity. It is because the average productivity of apple is nearly 6-8 t/ha, which is lower than the average yield of advanced countries *e.g.* Belgium (46.22 t/ha), Denmark (41.87 t/ha) and the Netherlands (40.40 t/ha).The productivity of apple in Jammu and Kashmir state fluctuated between 9-13 metric tonnes or less since last sixteen years (Banday *et al.,* 2009) which is the cause of serious concern to the fruit growers, research workers and development agencies. Same is the condition in other states of the country with respect to their major respective fruit crops.In this way our total production is much below the annual requirement of present population if present level of per capita consumption is 46g of fruits per day (Chadha and Pareek, 1993) which is against the requirement of 92g (with a minimum of 85g) as prescribed by ICMR. Several factors are attributed to this trend in productivity and quality and among which incidences of new viral diseases in fruit cultivation are in fact emerging as a serious threat to all types of fruit orchards in India.

During the survey, surveillance and monitoring for pests, diseases and physiological disorders in apple ecosystem of Kashmir under Directorate of Research SKUAST-Kashmir in 2012 and 2013 it was observed that among the various above surface and under surface pests and diseases in fruit ecosystem,the apple trees and other woody perennials growing on the same land consistently for a long period provide relatively suitable underground habitat thereby offering good opportunities for harbouring the herbivorous animal fauna in abundance and these are voraciously feeding in underground fruit ecosystem and during that survey an outbreak of new unknown viral infection was detected on apple trees emerging as a new unmanaged serious threat to fruit ecosystem.

Among the various pests and diseases, the information on nematodes parasitizing on fruit trees besides acting as vectors of plant viruses and their detailed debilitating role in fruit yield is scanty because of many reasons. Nematodes are the most abundant, wide spread and diversified group comprising 80-90 per cent of all multicellular animals that occur in every environment. These are second only to insects in the number of species in animal kingdom. Among these some are animal parasites, some are plant parasites and some are bio agents *etc.*

Among the plant parasites under phylum Nematoda, the longidorid nematodes are polyphagousecto parasites of plant roots.In general, the species of *Longidorus* (needle) and of *Xiphinema* (dagger) are better adopted to feed as ectoparasitic pests on the roots of woody fruit trees and other herbaceous plants. The first report of the damage to plant roots as pest by *Longidorus* was made by Horner and Jensen (1954) and by *Xiphinema* by Schindler (1957). Later Hewitt *et al.* (1958) gave the first report of virus transmission by Longidorid nematodes.

These ectoparasites in general are large bodied with thick cuticle, with long and muscular two part oesophagus, spear attenuated with long extensions, distinct stoma and well developed gonads and all stages feed on root tips, reproduce by meiotic pathogenesis and are capable of transmitting viruses from plant to plant. Trichodorids are also having unique characters such as spicules musculature, presence of single testis, gubernaculum, prominent excretory pore and the absence of rectum. They all are more abundant in sandy, warm and course structured irrigated soils where fruit trees are heavily attacked and population is seen high at depths below 20 cm.

Most severe ectoparasite nematode problems occur in regions where good host plants like apple, mango *etc.* are grown too frequently as constant hosts for too long time on the same land. During my extensive type of surveying and surveillance programme from 2006-2011 in fruit ecosystem of Kashmir it was seen that by gregarious feeding of such ectoparasites like longidorids and trichodorids,these cause a decline in root growth by feeding on root tips, halting growth by transformation of root tips into terminal grafts and besides providing primary infection courts by their mechanical style tthrushing for secondary infection of other micro-organisms like bacteria and other fungal pathogens. The loss due to nematodes in general to Indian agro-horticulture is estimated at about Rs. 210 crores annually (Jain *et al.,* 2007) but to see the potential of these longidorids and trichodorids alone under the super family Longidoroidea in the capacity of a pest or as causal organism of any disease or as vector of viruses in fruit ecosystem no joint efforts have been made by entomologists, pathologists and nematologists under the umbrella of plant protection and with the result the annual losses incurred by the depredation of such types of parasitic nematodes is almost incalculable.

2. Nematode Virus Interaction

There is interaction between PPNs and plant viruses in two groups. 1. Specific interrelationship between certain ectoparasitic nematode species and some viruses which they transmit. This has attracted much attention since their report or discovery in 1958 and the subject has gained its importance. 2. The other aspect is the general effects of plant viruses on various plant parasitic nematodes. This type of interaction has not been investigated deeply yet.

(a) Specific Interaction

After many years of microscopic study, the first reliable proof of transmission of plant viruses by plant parasitic nematodes was given when Hewitt *et al.* (1958) succeeded in transmitting grapevine fan leaf nepo virus by *Xiphinemaindex*. This discovery incited an intensive search for nematode vectors of other soil viruses, diseases *etc.* and lead to research on many aspects of the taxonomy, biology and ecology of both the nematode vectors and the viruses. Their detailed information we can have and judge from the reviews of Lamberti *et al.* (1975), Bajaj, H.K. Jairajpuri, and S.Waseem Ahmad *etc.* Practically all the plant parasitic nematodes feeding on virus infected plants ingest the virus particles but according to previous records out of total estimated of 2600 nominal species of plant parasitic nematodes only about 30 species were virus vector nematodes. All belong to the family longidoridae and trichodoridae of the order Dorylaimida.

(b) Nematodes Transmitting Plant Virus

Of the five genera of the family longidoridae, only some members of the genera *Xiphinema* and *Longidorus* are virus vectors. Within the longidorids seven out of

172 valid species of the genius *Xiphinema* are known to be virus vectors, when these figures are subject to continuous changes due to progress in taxonomy and improvement in transmission techniques because at present 281 are known species of a single genus *Xiphinema*. Similarly seven out of 83 previous species of *longidorus* and thirteen out of 50 species of *trichodorus* and *paratrichodorus* are able to transmit the viruses from plant to plant.

Longidorid nematodes are characterized by an axial stylet nearly 200 um long consisting of two parts, one an elongated needle (spear)-odontostyle with second part of its extension-odentophore-usually half of the length of odontostyle. This stylet is connected to a typical dorylaimidoesophagus having glandular and muscular bulb connected to it by a slender food canal. Juveniles are recognized by having both functional and replacement odontostyle for the next stage, situated in the wall of the anterior oesophagus. Here their tail is short, hemispheroid or conoid, rarely elongate or filiform.

3. Distribution and Hosts

Some important available reports of virus vector nematodes from our country and outside the country are as under (Tables 20.1 and 20.2).

Table 20.1: Reports of some Nematode Vectors from Putside India

Nematology	Year	Nematodes Recorded	Host with Place of Reporting
Hewitt *et al.*	1958	X.index	Transmission ofgrape vine fan leaf mosaic virus in USA
Cadman,	1963	Species of *Xiphinema* Species of *Xiphinema*	From woody perennials,
Pitchrer,	1965		Detailed judgement through review papers and assessment of loss estimation
Cohn,	1975	Species of *Xiphinema*	
Bloombergand Sutherland	1971		
Graffin and Epstein	1964	X. americanum	-do-
Dalmasso	1970	*Xiphinema* and *Longidorus* species	Pear and apple orchards (France)
Ogiga and Estey	1973	*Xiphinema* spp.	Apple orchards at Canada.
Ferris and Mckenry	1974	*Xiphinema* spp.	On grapes in California
Sutherland	1974	X. baseri	Douglas fir nursery in California
Scognamiglio and Verma	1979	*Xiphinema*	Ornamentals, fruit trees, pulses and cash crops in 22 localities of Compania, Italy.
Coomans and Heyns	1985	X. capenseand X. Bolandium	Vine yards in South Africa
Bitterlin and Gonsalves	1987	Presence of viruliferous nematode X. Rivesi	With the Persistence of tomato ring spot virus (TmRSV)
Jaffe *et al.*	1987	X. americanum	New York apple at Pennsylvania.
Lamberti and Roca	1987	X. basiri,X.brevicolle, X. index, X. insigne and X. Intermedium	Fruits crops in Pakistan.

Nematology	Year	Nematodes Recorded	Host with Place of Reporting
Roca *et al.*	1987	Six species of *Longidorus* and nine species of *Xiphinema*	Italy
Qasim *et al.*	1988	*X. americanum, X. Index* and *X. Basiri*	From root samples of apple, almond, grape and plum from Baluchistan
Lamberti et al.	1992	Six new species of *Xiphinema*	Spain, Italy, Israel and Turkey
Brown *et al.*	1994	Eight *Longidorus*, one *Paralongidorus*, and seven *Trichodorus* species, seven *Paratrichodorus* and four *Trichodurs* species	As natural vectors of nepo viruses As natural vectors of tobra viruses
Liskova *et al.*	1992 and 1993	Nine *Longidorus*, one *Paralongidorus* and seven *Xiphinema* species	Vineyards, orchards, hedgerows and forests of Slovakia Republic
Maqbool and Nasira	1995	Framed a list of species of the genera, *Longidorus*, *Paralongidorus, Trichodorus, Paratrichodorus* and *Xiphinema*	Pakistan
Wang *et al.*	1996	16 species belonged to 4 genera as, *Longidorus, Paralongidorus. Xiphinema* and *Trichodorus*	From 165 orchards and vine yards in 22 provinces of China
Sturhan *et al.*	1997	3 *Longidorus*, 1 *Paralongidorus*, 6 *Xiphinema*, 2 *Trichodorus* and 3 *Paratrichodorus*	*X. americanum* indigenous to New Zealand when *L.elongatus, X.Diversicaudatum*and *T.primitives* were newly introduced from Europe.
Ploeg	1998	*L.africanus*	In Bermuda fields of California at depths of 60 to 90 cm
Pan *et al.*	2000	Host range and distribution of species of longidorids and trichodorids	In Xiamen, Fujian province China.
Ganguly *et al.*	2000	*X. butanense* of Blue pineand *X. bambusi*of Bamboo	East Bhutan
Andret *et al.*	2004	*X. index*	Transmission of grapevine fan leaf virus (gflv)
McNeil *et al.*	2006	*Trichodorids* and *Xiphinema*spp.	Reported as vectors of some plant viruses in New Zealand
Tzortzakakis *et al.*	2006	*X. index, X.italiae* and *X. pachtaicum*	viticulture areas of Greece from existing grapevine fields
Wasim	2007	Three new and nine known species of Dorylaimida	Singapore

4. Viruses Transmitted by Nematodes

Plant viruses transmitted by nematodes form a hererogenous collection and these can be differentiated from one another on the basis of particle morphology and on the basis of relationship with vectors.These occur in two groups –Nepo and Tobra viruses according to virus classification scheme given by Harrison *et al.* (1971).

Table 20.2: Reports of some Nematode Vectors from India

Nematology	Year	Nematodes Recorded	Host with Place of Reporting
Bajaj and Jairajpuri	1976 and 1979	*Xiphinema, Longidorus*and *Trichodorus*	Descriptions of all the Indian species of *Xiphinema* have been brought together followed by detailed discussions on the relationships, geographical distribution, economic importance etc.
Jairajpuri and Bajaj	1978	*X. basiri*	North Indian
Baghel and Bhatti	1982	*X. insigne*	Citrus (Hissar Haryana)
Waliullah *et al.*	1993	*X. basiri,X.americanum, Longidorus*and *Paralongidorus*	Strawberry (Srinagar Kashmir)
Khan and Khana	1997	*X. basiri* and *X.americanum*	Citrus (Himachal Pradesh)
Singh and Khan	1997	Five new species of *Xiphinema*	Banana, orange, *Litchi* and apricot (North and North-Eastern India.)
Waliullah and Koul	1997	*X. basiri*	Cherry (Kashmir).India
Nath *et al.*	1998	*X. radicicola*	Pine apple (Tripura) India
Baqri	1999	Listed 81 spp.in the order of Dorylaimida	Plant and soil nematodes (West Bengal)
Mukherjee *et.al.*	2000	High diversion of longidorids	Tripura
R a m a a n d Dasgupta	2000	*X. elongatum*	Coconut (West Bengal)
Siddiqui, A.U.	2001	*Xiphinemaudaipurensis*	Roses (Udaipur) Rajasthan Rajasthan
Zaki and Manto	2003	*X. species*	Apple (Kashmir)
Chaudhury *et al.*	2004	species of *Longidorus* and *Trichodorus*	Assam, India
Khan and Verma	2005	*X.basiri,L.citri, L.brevicaudatus, L.attenuatus, P.microlaimus*and *P. neoformis*etc.	Apple (Himachal Pradesh)
R a t h o r e a n d Ganguly	2005	Longidorids	Southern Gujarat, India
Adekuncle et al.	2006	*X.diversicaudatum,Longidorus* spp. *Trichodorus* spp.	Nagrota,Palampur, Sundernagar (H.P.)
Wasim	2007	Anew genus and three new and nine known species of dorylaimida	Singapore.
Lone,G.M.	2011	4 species of *Xiphinema* and 3 species of *Longidorus*	Apple (Kashmir}

There are at present 36 known nepo viruses of which 11 have a recognized nematode vector. Allnepo viruses have iso-diametric particles with icosahedral symmetry and are 23-30 nm in diameter. Their genome is bipartite, with two functional ribonucleic acids (RNA-1 and RNA-2), separately encapsulated. These nepo viruses are divided into two, three or four sub groups. Some nepo viruses are still incompletely characterized so final grouping is yet not possible.

The number of tobra viruses has increased from two to three only namely tobacco rattle, pea early browning and pepper ring spot tobra viruses. They have rigid rod shaped particles of three different lengths respectively: very short ±45nm, short 50-110 nm and long 185-200nm.

Table 20.3: *Xiphinema* Vector Nematodes and Transmitted Nepo Viruses

Nematode	Virus	Host Range of Virus
X.americanum senulato	Tomato ring spot	Fruits, vegetables, ornamentals
	Cherry rasp leaf	
X.americanum senustricto	Tobacco ring spot	Fruits, vegetables, ornamentals
	Peach rosette mosaic	
X.californicun	Tobacco ring spot	Fruits,vegetables, ornamentals
	Grapevine yellow vein strain	Grapevine
X.diversicaudatum	Arabis mosaic	Fruits,vegetables, ornamentals
	Strawberry latent ring spot	
X.index	Grapevine fan leaf	Grapevine
X.italiae	Grapevine fan leaf	Grapevine
X.rivesi	Tomato ring spot	Fruits, vegetables, ornamentals

4. Viruses Transmitted by Nematodes

Plant viruses transmitted by nematodes form a hererogenous collection and these can be differentiated from one another on the basis of particle morphology and on the basis of relationship with vectors.These occur in two groups –Nepo and Tobra viruses according to virus classification scheme given by Harrison *et al.* (1971).

There are at present 36 known nepo viruses of which 11 have a recognized nematode vector. Allnepo viruses have iso-diametric particles with icosahedral symmetry and are 23-30 nm in diameter. Their genome is bipartite, with two functional ribonucleic acids (RNA-1 and RNA-2), separately encapsulated. These nepo viruses are divided into two, three or four sub groups. Some nepo viruses are still incompletely characterized so final grouping is yet not possible.

The number of tobra viruses has increased from two to three only namely tobacco rattle, pea early browning and pepper ring spot tobra viruses. They have rigid rod shaped particles of three different lengths respectively: very short ±45nm, short 50-110 nm and long 185-200nm.

Nepo viruses are recovered from a wide range of naturally infected weeds but grapevine fan leaf occurs naturally only in grapevine.

Table 20.4: *Longidorus* Vector Nematodes and Transmitted Nepo Viruses

Nematode	Virus	Host Range of Virus
L.apulus	Artichoke Italian Latent Italian strain	Vegetables
L.attenuatus	Tomato blackring	Fruits, vegetables, ornamentals
L.diadecturus	Peach rosette mosaic	Fruits
L.elongatus	Raspberry ringspot Scottish strain Tomato blackring Beet ringspot strain Peach rosette mosaic	Fruits
L.fasciatus	Artichoke Italian Latent Greek strain	Vegetables,
L.macrosoma	Raspberry ring spot English strain	Fruits
L.martini	Mulberry ringspot	Mulberry

5. Nematode Feeding and Virus Acquisition

The root surface is exposed by the longidorid nematodes by their rapid mechanical thrusting and twisting of stylet. Before the plant cell contents are ingested, secretions from the oesophageal glands are charged through the stylet into the cell. During feeding process, the periods of pumping (ingestion) and of silent salivation changed from time to time a nematode may remain at feeding site for hours. In this condition, the odentostyle can thrust several cell layers deep into the root tissues and eventually reach the vascular bundle.

Table 20.5: *Trichodorus* and *Paratrichodorus* Vector Nematodes and Transmitted Nepo Viruses

Nematode	Virus	Host Range of Virus
T.cylindricus	Tobacco rattle	Wide
T.hooperi	Tobacco rattle	Wide
T.primitious	Tobacco rattle	Wide
T.similis	Tobacco rattle	Wide
T.viruliferus	Tobacco rattle Pea early browning	Wide Leguminosae
P.allius	Tobacco rattle	Wide
P.anemines	Tobacco rattle Pea early browning	Wide Leguminosae
P.minor	Tobacco rattle Pepper ring spot	Wide vegetables
P.nanhus	Tobacco rattle	Wide

Nematode	Virus	Host Range of Virus
P.pachydermus	Tobacco rattle	Wide
	Pea early browning	Leguminosae
P.porosus	Tobacco rattle	Wide
P.teres	Tobacco rattle	Wide
P.tunisiensis	Tobacco rattle	Wide

Trichodorid nematodes explore the root surface in similar way in search for a suitable feeding site. These feed on epidermal cells and root hairs only. Why? The cell walls are penetrated by rapid thrusts of onchiostylet.Then secretions of oesophageal glands are used to form a feeding tube through a cell wall and subsequently are injected into the cell and homogenize the cytoplasm. This is then rapidly ingested and then accordingly the nematode moves to another cell.

Both nematode groups acquire viruses when ingesting cytoplasm of virus infected plants. The time needed for successful acquisition is short *e.g. Xiphinema index* has shown to acquire grape vine fan leaf nepo viruses in less than 5 minutes. However, the number of successful transmissions *i.e.* the number of viruliferous nematodes increases with increasing access periods.

6. Virus Retention

During feeding process after ingestion of virus particles with the cytoplasm, the nematode vectors retain their corresponding viruses for some time.Generally *Xiphinema*s pp are said to retain the viruses they transmit for 10-12 months and most *Longidorus* vectors for shorter periods, usually shorter than three months. Some excellent examples are as.

1. English strain of raspberry ring spot nepo viruses was detected in its vector *Longidorusmacrosoma* after 60 months of starvation.
2. *Xiphinemarivesi* transmitted tomatoring spot nepo viruses to bait plants upto two years after storage in soil with host plants. These findings indicate that virus vector relations are not the same for all combinations. It was further shown that retention periods can vary with the temperature.
3. Mulberry ring spot nepo viruses was retained by its vector *Longidorus martini* in host free soil for more than 18 months at 0-9 °C, for 13 months at room temperature and for three months at 20-24 °C.
4. Tobacco rattle tobra virus has been shown to persist in trichodorid vectors for period's upto two years.

Such long retention periods have been observed solely in populations without access to host plants and apply mainly to adults. Nematodes lose some virus particles whenever they feed during salivation.

Nepo and tobra viruses are retained in their vectors at distinct sites. In *Longidorus* vectors, the virus particles associate with the interior surface of the lumen of the odontostyle and with the guiding sheath while in case of *Xiphinema*, the virus particles are adsorbed to the cuticular lining of the lumen of the odontostyle and of the oesophagus and in case of *Trichodorus* and *Paratrichodorus* virus particles are attached to the cuticle lining of the oesophagus and not to the onchiostylete. All surfaces serving as retention sites are shed during moulting together with the outer cuticle of the body. Consequently the adhering virus particles are lost and cannot be transferred from one developmental stage. The viruses retained do not multiply or nor involved in vector metabolism. There is no proof of having any direct influence of plant viruses on nematodes and the mechanism of retention is still incompletely known. Evidence shows that nematode transmitted viruses are selectively and specifically adsorbed at the retention sites in their associated vectors when cell contents of virus infected plants are ingested. Experiments indicate that in some *Xiphinema*s pp, the specific retention is based on a recognition process between lectin like molecules associated with the virus protein coat and carbohydrates present in the cuticular lining of the oesophagus.

The ability to transmit virus must also be considered in respect to biological differences between populations of one species; local population of a vector species are usually most efficient in transmitting local virus isolates. Thus geographical isolation can lead to a high specificity, as shown for strawberry latent ring spot nepo virus from Italy, which is transmitted by *X.diversicaudatum* from Italy (Brown, 1985).

7. Virus Transmission

Virus transmission occurs when the dissociated particles are inoculated with the stream of saliva into plant cells during feeding. For a successful infection the virus particles must be viable and the plant cell must not be seriously damaged by the nematode attack. Otherwise, the virus is not able to replicate and to pass into neighbouring cells.This is particularly true for tobra viruses and their trichodorid vectors. It takes only about 3 minutes to complete the feeding cycle on an individual cell, from the first stylet thrust to pierce the cell wall until leaving the emptied cell, of which the ingestion of the cytoplasm only takes 30 seconds. Therefore, a successful inoculation can only occur when the feeding cycle is not completed and the cytoplasm not entirely removed. In Longidorid vectors the feeding periods are much longer than in trichodorids. Individuals have been observed staying at one feeding site for hours or even days. When a longidorid nematode stops feeding at a cell, the stylet is inserted next deeper cell so that a column of cells fed upon by the nematode remains, progressing from the epidermis into deeper cell layers. Hence, there is simple time for virus particles to become established (Weischerand Wyss, 1976).

8. Geographical Distribution of Vectors and Viruses

Whatever the distribution of nematodes is available on the floor of Science of Nematology, it is the result of a natural development based on geographical events and of dissemination caused by man's activities. The natural occurrence and distribution can best be studied in undisturbed habitats or neglected habitats. In disturbed or cultivated lands nematode species may have been distributed or introduced with propagation material or soil used as shipsballast even over larger distances. Since due to lack of thorough surveys over larger areas reliable information on the geographical distribution of phytonematodes is not sufficient, therefore, historical biography is a new field in Nematology, which can give interesting information about the origin of nematodes and viruses and the evolutionary development of their association. The best example is the study of ectoparasitic nematodes of family longidoridae under the aspects of phylogeny and biogeography where of the two genera with virus transmitting species, *Xiphinema* was found cosmopolitan genus originating from Gondwana land, from where it spread to Laurasia before the breakup of Pangaea some 180 million years ago. The main speciation subsequently occurred in Africa, where the majority of the species are found.

For *Longidorus* the situation is less clear. The probable origin lies in South-East Africa and India, from where it spread to Laurasia, with the main speciation occurring in Europe. *Longidorus* species are found mainly in the Northern Hemisphere. Little is known about the historical biogeography of trichodorid nematodes. At present they are most frequent in Europe and North America. However, the increasing number of reports *e.g.* from Africa, India, Japan and New Zealand, indicate a wide spread occurrence where single species are often having a regional or local distribution.

The discovery that species of Longidoridae and Trichodoridae as vectors of plant viruses (Hewitt *et al.,* 1958; Sol *et al.,* 1960) developed great interest in the scientists under plant protection worldwide and at present even 281 species have been described under a single genus *Xiphinema* including 38 species from India alone The number of described species drastically increased from 25 Longidoridae *{Longidorus* and *Xiphinema*) and 15 *Trichodorus* species in 1960 to currently more than 400 Longidoridae *{Longidorus, Paralongidorus, Xiphinema, Xiphidorus, Para xiphidorus*) and about 90 Trichodoridae (*Trichodorus, Paratrichodorus, Monotrichodorus*).

Generally nepo viruses with wide host range have a wide distribution, whereas viruses with few host range are more restricted. An exception is grapevine fan leaf nepo virus, which is present in all major grape growing areas of the world. Although there is already a highly specialized virus associated with *Vitis vinifera* it has been widely disseminated with infected propagating material over the centuries. In similar way arabis mosaic nepo virus may have reached New Zealand and the American tomato ring spot nepo virus to Australia (Martelliand Taylor, 1989). Of the three recognized tobra viruses, tobacco rattle tobra virus has a worldwide distribution; whereas pea early browning tobra virus is restricted to Europe and pepper ring spot tobra virus to South America. Within these virus species there are serologically distinct strains having different nematode vectors and a different distribution.

9. Methodology

The research activities related to problems of nematode virus interaction increased rapidly when Hewitt *et al.* (1958) reported the successful transmission of grape vine fan leaf nepo virus by *Xiphinema index*. A great number of such associations were reported in the following years and among which a few were contradictory due to inappropriate methods applied.The main problems detected were difficulties in species determination and the lack of adequate standardized methods for sampling and transmission experiments. Many of the older specie descriptions were incomplete and do not allow exact identification. *Xiphinema americanum* Conn, 1913 is a good example. It has been considered to be one species for about 50 years. Now it is regarded a group of 38 closely related species (Lamberti and Cerone, 1991). In similar way at present *Xiphinema insigne* single specie showed much variation on the basis of tail structures. All former records of *X. americanums* p.f. as vector is being checked in order to obtain reliable information while studying nematode virus interaction. Under natural and experimental conditions certain requirements are to be fulfilled. There are certain specific techniques for sampling for "Virus vector nematodes and their viruses (Brown and Boag, 1988 and Brown *et al.*, 1990).It was shown that the distribution of nematode-transmitted viruses is highly aggregated. The optimum sampling depth is between 10 and 40 cm depending on nematode species, host plant root system, soil characteristics and climate.

To establish the vector status of a nematode species several criteria have to be met for transmission experiments and among which the most important are as follows:

1. Nematode and virus must be correctly identified.
2. The nematode being tested must be the only possible vector.
3. The virus must be avoidable to the nematode.
4. The condition must be suitable for the transmission to take place.
5. Virus contamination of the bait plant must be avoided.
6. Bait plant must be shown to contain the virus being tested.

10. How to Confirm Virus is Transmitted by Nematode

For transmission experiments with longidorid nematodes and nepo viruses, Trudgill *et al.* (1983) developed a test system that meets all the requirements (6 points mentioned earlier). In this system well identified virus free nematodes are placed around the roots of source plants with an identified virus. After an appropriate period of excess, the nematodes are extracted, counted and in small groups transferred to virus free bait plants. Some of the nematodes are examined under electron microscope for the presence of virus particles. Later the bait plants are tested for virus infection. Virus and vector must be re-identified at the end of test. Ploeg *et al.* (1989) and Brown *et al.* (1989) did the same for trichodorid nematodes. On the basis of sixth criteria, the authors made a critical review and came to the conclusion that only 14 out of 46 published results on virus transmission by longidorids and 13 out of 40 reports on trichodorids were supported by adequate evidence and therefore, were valid as shown in the Tables 20.1–2.3. The other reports were rejected for the following three reasons.

1. There was no systemic infection: the viruses found could have been contaminants on the roots.

2. The descriptions of methods used were inappropriate.

3. Vector and/or virus were not adequately identified.

Since this procedure is not applicable to study the associations between trichodorid nematodes and tobra viruses. Therefore, Brown *et al.* (1989) developed a special testing system—the use of small (0.5cm^3) plastic capsules containing a single bait and inoculated with a single nematode allows the specific identification of each individual nematode transmitting a given virus.

11. General Nematode-Virus Interaction

1. Plant viruses have no ascertainable direct influence on carrier nematodes, vectors or non-vectors. The indirect effects are based on changes in host plant mechanism caused by viruses and nematodes respectively.

2. Development and multiplication of nematodes can be enhanced or inhibited depending on the virus, nematode or plant species.Inhibitory effects on R.K. nematodes *M.javanica* were observed in cucurbits infected with water melon mosaic poly virus. Virus infection retarded the establishment of these nematodes in the roots as compared with healthy plants. All the effects favourable or determental are more pronounced when nematode inoculation was preceded for 2-3 weeks by virus infection.

3. The concomitant presence of nematode and virus in most cases results in a synergistic effect that aggravates the plant damage. Little is known about the physiological or biochemical basis of enhancing or inhibiting nematode development by a virus infection of the host.

4. Locality-wise status, horizontal distribution and seasonal population fluctuation of most prevalent virus vectors and identification of known and new nematode vectors and their transmitted diseases is must in fruit ecosystem of our diversified country.

5. To see the potential of longidorids and trichodorids under the super family Longidoroideain the capacity of a pest or as causal organism of any disease or as vector of viruses in fruit ecosystem no joint efforts have been made

by entomologists, pathologists and nematologists under the umbrella of plant protection and with the result the annual losses incurred by the depredation of such types of parasitic nematodes is almost incalculable.

12. Longirodirs in Fruit Ecosystem of Kashmir

It is found that almost 95 per cent of taxonomically identified longidorids were found as females and males were no where and so pathogenesis is assumed as rule and according to the time limits of the research programme, the species of only two genera namely *Xiphinema* and *Longidorus* were distinguished by morphological features and among which some species are considered as new reports in Kashmir valley on apple trees. The results revealed that more or less population of longidorids were found associated with apple trees in all the orchards of each village of the prominent fruit belts of Kashmir. Mixed population from the composite soil samples comprised nine populations. One group belong to subfamily Xiphinematinae, family Xiphinematidae of super family Longidoroidea (Thorne, 1935) and order Dorylaimida. This group included four species of genus *Xiphinema* viz., *X. insigne; X. index; X. americanum* and *X. diversicaudatum.* The other group belong to sub-family-Longidorinae, family Longidoridae of same super family Longidoroidea and order Dorylaimida. It included three species of genus *Longidorus* viz., *L. elongatus; L. mirus* and *L.brevicaudatus.*

☆ Among various longidorids *Xiphinema insigne; X. index* and *L. elongatus* are found well established in apple orchards of 25-35 aged trees and comparatively more in neglected apple orchards than in managed and disturbed orchards.

☆ The population build-up of all these vary greatly depending on the distance, depth and availability of fresh roots.

☆ There is not found any significant difference in energy flow of longidorids from locality to locality and from belt to belt.

☆ The root zone between horizontal distance of 90 and 135 cm away from the base of the apple tree is found most susceptible to predominant longidorids

☆ The region at the depth of 0 to 30 cm between 90 to 135 cm horizontally is more susceptible than the region vertically between 31 to 45 cm.

☆ Less population of predominant longidorids is found beyond 46 cm depth. During autumn the population of predominant longidorids in the region between 31 to 45 cm is nearly at par with the same population in the region between 0 to 30 cm depth.

☆ The possibility of high population build-up of these virus vector nematodes with the approach of global warming cannot be ruled out and the period from May to September is very crucial for their multiplication and dispersal

☆ The identification of *X. diversicaudatum, X. americanum, L. elongatus, L. mirus and L. brevicaudatus* are new records from apple trees of Kashmir, India.

References

Anonymous, 2008. *Area and production statement for the year 2007-2008.* Department of Horticulture, Jammu and Kashmir Govt., Srinagar.

Bajaj, H.K. and Jairajpuri, M.S. 1976. Two new species of *Xiphinema* from India. *Nematologia Mediterranean* 4 : 195-200.

Bajaj, H.K. and Jairajpuri, M.S. 1979. Review of the genus *Xiphinema* Cobb 1913 with the description of species from India. *Records of the Zoological Survey of India*75: 255-325.

Bloomberg, W.J. and Sutherland, J.R. 1971. Phenology and fungus-nematode relations of corky root disease of Douglas fir. *Annals of Applied Biology* 69: 265-276.

Boydston, R.A., H. Mojtahedi, J.M. Crosslin, P.E. Thomas, T. Anderson, and E. Riga. 2004. Evidence for the influence of weeds on corky ringspot persistence in alfalfa and Scotch spearmint rotations. *American Journal of Potato Research* 81:215-225

Brown, D.J.F., Robertson, W.M. and Trudgill, D.L. 1995. Transmission of viruses by plant nematodes. *Phytopathology* 33: 223-249.

Cadman, C.H. 1963. Biology of soil borne viruses. *Annual Review Phytopathology* 1: 143-172.

Chiristie, J.R. and Perry, V.G. 1951. Removing nematodes from soil. *Proceedings of Helminth. Society Washington* 18: 106-108

Cobb, N.A. 1913. New nematode genera found inhabiting fresh water and non-brackish soils. *Journal of Washington Academic Sciences* 3(16): 432-444.

Cobb, N.A.1918. Estimating the nema-population of soil with special reference to sugarbeet and root-gall nemas*Heteroderaschachtii* Schmidt *and Haterodera radicicola* (greef) Muller, and with a description of *Tylencholaimusaequalis*n. sp. USDA *Agriculture Technical Circular Bur. Pt. Ind. U. S. Department of Agriculture*1: 1-48.

Cohn, E. 1975. Relation between *Xiphinema* and *Longidorus* and their host plants. In: *Nematode vectors of plant viruses* (Eds. F. Lamberti, C.E. Taylor, and J.W. Seinhorst). Plenum Press, London and New York, pp. 365-386.

Cohn, E. and Mordechai, M. 1969. Investigations on the life cycles and host preference of some species of *Xiphinema* and *Longidorus* under controlled conditions.*Nematologica* 15, 295-302.

Khan, M.L. and Khana, A.S. 1997. Nematodes associated with Citrus crops in Himachal Pradesh. *Indian Journal of Nematology* 27(2):266-67.

Lamberti, F. and Roca, F. 1987. Present status of nematodes as vectors of plant viruses. In:*Vistas on Nematology* (Eds. J.A. Veech and D. W. Dickson). 25th Anniversary Publication Society, *Nematol*, pp. 321-328.

Lone, G.M., Zaki, F.A. and Waliullah, M.I.S.2012. Report of *Xiphinema index*, Loose, C.A.1949. Notes on free living and plant parasitic nematodes of Ceylon.*J.Zool. Soc. India*1:23-29.

Pitcher, R.S. 1965. Inter-relationships of nematodes and other pathogens of plants. *Helminthological Abstracts* (Series B) 34: 1-17.

Qasim, M., Hashmi, S. and Maqbool, M.A. 1988. Distribution of parasitic nematodes and their importance in fruit production of Baluchistan. *Pakistan Journal of Nematology* 6(1) 17-22.

Raski, D. J. 1988. Dagger and needle nematodes. pp. 56-59. In: *Compendium of Grape Diseases* (R. C. Pearson and A. C. Goheen, Eds.). American Phytopathological Society Press, St Paul, Minnesota. 93 p.

Roberts, I. M. and Brown, D. J. F. 1980. Detection of six nepo viruses in their nematode vectors by immunosorbent electron microscopy. *Annals of Applied Biology* 96:187-192.

Thorne and Allen, 1950. (Dorylaimida: Longidoroidea) Associated with neglected apple orchards of Baramulla, Kashmir, India. *Indian Journal of Nematology* 42(1):71-74.

Note: The publication is part of Ph. D. thesis submitted by senior author to Sher-e-Kashmir University of Agricultural Science and Technology, Shalimar Srinagar J&K.

Chapter 21

Physiological Problems in Apple and their Management

Farooq A. Khan[1], Sajad A. Bhat[1] and Raj Narayan[2]

[1]Division of Basic Sciences,
Sher-e-Kashmir University of Agricultural Sciences and
Technology of Kashmir, Shalimar – 190 025, Srinagar, J&K
[2]CITH Research Station, Mukteshwar, Nainital, Uttarakhand
E-mail: fakphtskuastk@rediffmail.com

1. Introduction

Apple is one of the most important fruit crops of Kashmir. Besides providing employment and income to lacs of people, it contributes more than Rs. 4000 crore towards state economy. It is the main cash crop of Jammu and Kashmir State. Chilling requirement (below 7°C) of most of the apple varieties is about 1000 hours or more for optimum growth and production of quality fruits. Temperate conditions of the Kashmir valley with chilling temperatures during winter months are most suitable for quality apple production. In 2016-17, total area under apple cultivation was 144.82 and 18.14 thousand hectares and production was 1688.41 and 38.42 thousand metric tonnes in Kashmir and Jammu province respectively (DoH, 2017). With the introduction of improved technology and methods of cultivation, both production and productivity has increased over the years. However, various unfavourable biotic and abiotic factors (Figure 21.1) come in the way of realizing the achievable yield potential and quality of apples which results in a greater loss to the fruit industry.

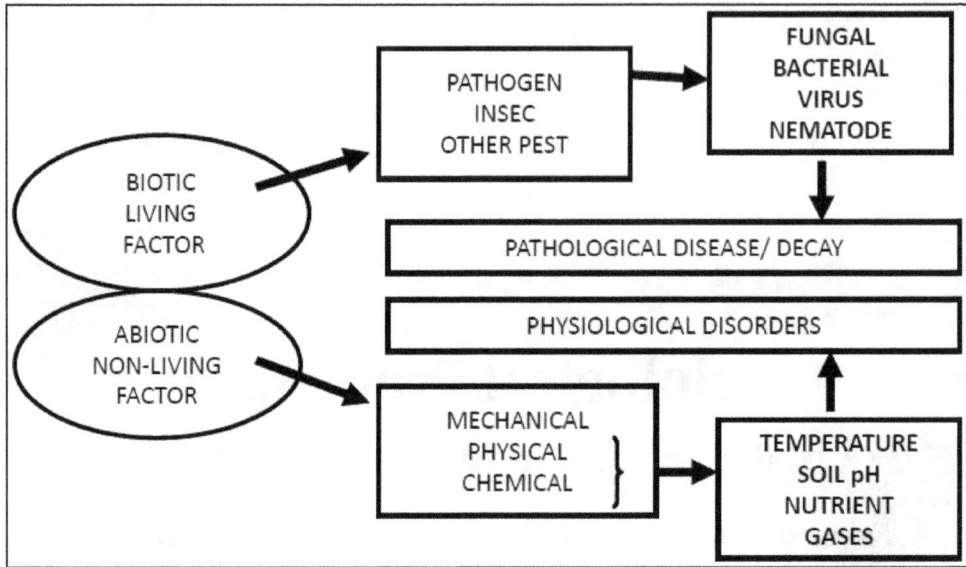

Figure 21.1: Pathological Disease and Physiological Disorders.

Physiological (abiotic) disorders are problems that are not caused by insects or diseases but rather by climatic factors (temperature, rain, humidity, sun light, *etc.*) and management practices (training and pruning, irrigation, fertilization, harvest procedure, *etc.*) that change the micro-climate endured by the plant or plant part. The action of environmental factors that are outside the optimum ranges leads to the deterioration of physiological processes during the pre- and post harvest period which in turn leads to the incidence of physiological disorders.

(a) Identification of Physiological Disorders

Some physiological disorders are easy to identify, but others may be difficult or even impossible to identify. In diagnosing plant damage, a series of detective steps have to be followed, gathering clues from the general situation down to an individual plant or plant part to determine the most probable cause of the damage (Figure 21.2). To help in identifying physiological disorders it is important to know that:

☆ Physiological disorders are often caused by the lack or excess of something that supports life or by the presence of something that interferes with life.

☆ Physiological disorders can affect plants in all stages of their lives.

☆ They occur with the absence of infectious agents therefore cannot be transmitted.

☆ Plant reactions to the same agent vary widely, from little reaction to death.

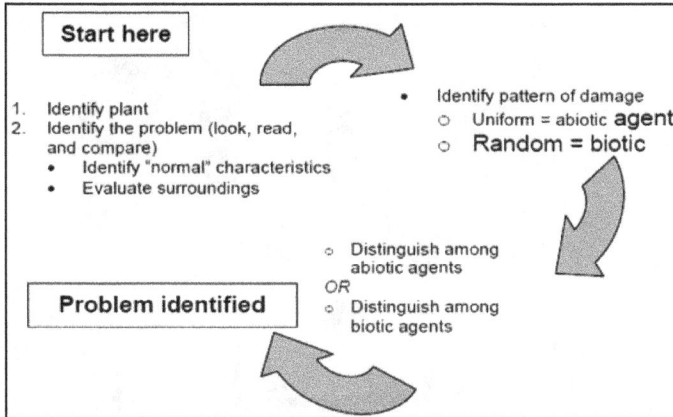

Figure 21.2: Systematic Approach for Determining the Cause of Damage.

☆ Dealing with physiological disorders often means dealing with the consequences from a past event.

☆ There is generally a clear line of demarcation from damaged and undamaged tissue.

☆ Physiological disorders are serious in themselves but often serve as the 'open door' for pathogens to enter.

(b) Types of Physiological Disorders in Apple

A number of physiological disorders developed in apples because environmental and cultural conditions during pre- and post harvest periods often are significantly different from those encountered by plants in an ideal environment. Various physiological disorders of apples are caused due to the basic reasons of nutrient deficiency, sudden fluctuation in temperature, poor soil conditions, improper moisture availability during cultivation, harvest maturity, post harvest handling techniques, storage conditions *etc.* Environmental factors implicated in the occurrence of physiological disorders include irradiance (intensity, photoperiod and spectral quality), humidity, CO_2 concentration, air temperature/movement, growing medium temperature/moisture level, and mechanical effects. Depending on the causal factors, physiological disorders may be (i) pre-determined *i.e.* induced during pre-harvest period and become visible during storage *e.g.*, bitter pit, and (ii) storage-induced *i.e.* induced and developed due to post harvest factors during storage *e.g.*, storage scald.

The major physiological disorders of apples (Figure 21.3) are described here under:

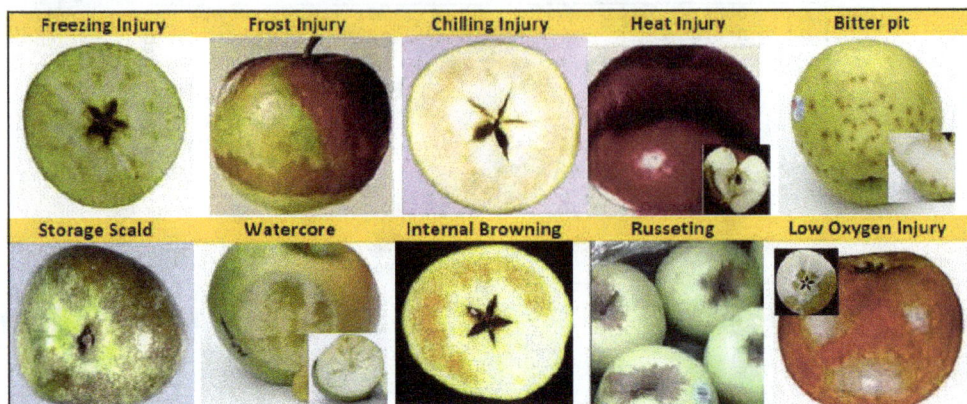

Figure 21.3: Major Physiological Disorders of Apple.

1. Freezing Injury

Freezing injury occurs when the temperatures of the tissue falls below its freezing point. Due to the presence of dissolved substances in the cell sap the freezing points of fruits are slightly below (1°C to a few °C) the freezing point of water (freezing point depression). The sweeter the fruit the lower is its freezing points.Freezing may occur before harvest during frosty weather or after harvest as a result of improper conditions of carriage or storage. Apples frozen on tree suffer less residual damage than fruit detached from the tree and exposed to the same conditions. Both temperature and duration of exposure determine the extent of damage. Fruits which have been exposed to freezing temperatures are weakened and their storage life is curtailed.

(i) Symptoms

Minor freezing injury on the surface of the fruit results in a brown discoloration of skin. Internal freezing injury results in brown discoloration of the vascular strands, which, through desiccation, may develop cavities in the damaged areas during subsequent cold storage. During storage the lesion may remain intact or become sunken, depending on the extent of injury to the sub-epidermal tissue, which losses moisture more rapidly after freezing damage. Skin injury and internal injury may occur independently or coincidently in the same fruit. Fruits affected with freezing injury may also fail to ripen. In severe freezing ice crystals may protrude from the tissue, cells are killed and release fluid on thawing, leaving the fruit wilted and often discolored. Fruits soften and become mealy.

(ii) Physiology

Freezing injury may kill vascular elements, which become prominently brown. More extensive freezing cause large ice crystals to form, which rupture cell membranes and walls and result in perfuse enzymic browning and ultimate cell death. When it is thawed, severely frozen fruit is brown, soft and wet. Ice formation in intercellular spaces gives tissue a water soaked appearance but does not cause tissue injury. Such fruit, however, is likely to have a higher rate of respiration and evaporation.

(iii) Control

Irrigation is an alternative to the use of orchard heaters. Cull out freeze-damaged fruits from the harvested crop. There being no external signs, but the damaged fruits can be separated by floatation in a specially prepared solution. Have accurate and well positioned thermostats in fruit stores. Fruit that is suspected of being frozen should be moved carefully. Frozen fruit should be marketed promptly or processed soon after the freezing was noted.

2. Chilling Injury

Chilling injury (CI) occurs at temperatures well above freezing point. Apple cultivars susceptible to low temperature injury include Cox' Orange Pippin, Starking Delicious, McIntosh, Jonathan, Sturmer Pippin and Barmley's Seedling.CI is associated with low levels of phosphorus and to a lesser extent with low levels of potassium and magnesium

(i) Symptoms

CI in apples is characterized by a diffuse browning of the outer cortex. The affected area is often well defined, usually moist rather than dry and mealy, and in the early stages is separated from the skin by an area of healthy tissue. Both normal and affected tissues may have a fermented flavor. Vascular elements become browner than the cortex tissue. The fruit is likely to be spongy when compressed.

(ii) Physiology

CI stimulates respiration and ethylene production rates of fresh fruits. Low temperature prevents normal metabolism of fruit tissue, and the complex biochemical reactions associated with respiration proceed along alternative pathways. Changes in Isoprenoid metabolism and higher levels of abscisic acid, acetic acid, and acetate esters have been implicated in LTB. External and internal browning is related to oxidation of phenolics by polypherol oxidase. Brown substances (polyphenols) accumulate in the vascular tissue within the peel.

(iii) Control

Primarily the susceptible cultivars should be stored at temperatures above the critical value. If storage at a lower temperature is required applying of oil, wax or film and conditioning of the fruit are recommended. CA storage permits the use of slightly lower temperatures or longer storage periods. Similar benefits are obtained with the use of sealed polyethylene bags. Removal of ethylene from the storage atmosphere may help to reduce CI. Increased Ca level in the fruit and pre-harvest application of GA_3 can also reduce the problem.

3. Heat Injury

Heat injury may occur either before harvest due to high temperatures and radiant energy or after harvest due to certain post harvest treatments. Sunscald is a common form of pre-harvest heat injury in apple which is especially a problem where temperatures are high and skies are clear.Granny Smith and other light skinned apple varieties are more sensitive to sunburn. Also, sensitivity may be associated with low calcium concentration in the fruit.Hot water washing or hot air drying can also result in heat injury of apple at 55 °C for 30 second.

(i) Symptoms

Initial symptoms are white, tan or yellowed patches which turn brown within a few weeks in cold storage. Injured cortex tissue is brown and firm and may become spongy and sunken. Hot water or air injury appears after several weeks of storage as a diffuse browning of the skin, similar to common scald, and as in scald, the unblushed portion are preferentially affected. Prominent calyx lobes of Delicious apples are quite susceptible because they rise in temperature more quickly than the rest of the apple when exposed to hot air or hot (38°C-40°C) water.

(iii) Physiology

Sunburn results from heat stress to the fruit leading to injury of the affected cells. The temperature of the surface of the fruit can be as much as 18°C (32°F) above air temperatures when the fruit is exposed to solar radiation and 8 to 9°C (14 to 16°F) warmer than the shaded side of the fruit. The transpiration or evaporation of water from the apple fruit helps to cool the fruit while it is attached to the tree. Upon harvest from the tree or during periods of water stress on the tree, fruit surfaces and flesh can increase too much higher temperatures resulting in greater sunburn.

(iii) Control

The best method of control is to avoid sudden exposure of fruit to intense heat and solar radiation. Proper tree training and pruning are critical. Avoid water stress and perform summer pruning very carefully to avoid excessive sunburn. Careful sorting to remove affected fruit upon packing is the only solution once the injury has occurred. Injury will be easier to spot after a few weeks of storage because of the change in color from yellow to brown.

4. Bitter Pit

Bitter pit (BP) is considered as Ca deficiency disorder and is influenced by climate and orchard cultural practices. Granny Smith, Yellow Newtown, Jonathan, Golden Delicious, Red Delicious, Cox's Orange Pippin, Grimes Golden, Merton, Worcester, Star Krimson, Marigold, Northern Spy, and York Imperial are highly susceptible cultivars. Older trees with larger crop load are less susceptible while young trees that are just coming into bearing are the most susceptible. Immature fruit are more susceptible to bitter pit than fruit harvested at the proper harvest maturity.

(i) Symptoms

Brown lesions of 2-10 mm in diameter develop in the flesh of the fruit. The tissue below the skin becomes dark and corky. At harvest or after a period of cold storage the skin develops depressed spots on the surface. These most often start to appear as water soaked spots on the skin near the calyx. These spots generally turn darker and become more sunken than the surrounding skin and are fully developed after one to two months in storage.

(ii) Physiology

Initiation of symptoms may begin four to six weeks after petal fall when affected tissues have a higher rate of respiration and ethylene production. This is a period of greater protein and pectin synthesis with greater migration of organic ions into the affected areas. A mineral imbalance in the apple flesh develops with low levels of Ca and relatively high levels of K and Mg. Low levels of Ca impair the selective permeability of cell membranes leading to cell injury and necrosis. Other explanations for the cause of BP include the dissolution of the middle lamellae by oxalic and succinic acid, and changes in proton secretion and potassium permeability.

(iii) Control

Summer sprays of calcium chloride ($CaCl_2$), calcium nitrate (Ca $(NO_3)_2$ and/ or a post harvest dip in a calcium solution are recommended. At least three sprays should be applied at one month intervals beginning in mid-June. Also select large fruit from upright limbs of light cropped, vigorous trees 2-weeks before harvest. Dip the fruit in a solution of 2,000 ppm ethephon in water to hasten the ripening process. Hold the fruit for two weeks at room temperature. If BP develops, delay the harvest as long as possible. The delay will allow the bitter pit to fully develop. The affected fruit is then removed during the packing process.

5. Storage Scald

Storage scald (also called common scald or superficial scald) has been a serious concern for apple growers for as long as apples have been stored and marketed commercially. Granny Smith, Rome Beauty, Delicious, Winesap and Yellow Newtown are very susceptible, Gala and Fuji are moderately susceptible. Incidence and severity of storage scald is favored by hot, dry weather before harvest, immature fruit at harvest, high nitrogen and low calcium concentrations in the fruit, and inadequate ventilation in storage rooms or in packages.

(i) Symptoms

Irregular brown patches of dead skin develop within 3 to 7 days upon warming of the fruit following cold storage which can become rough when severe. Symptoms may be visible in cold storage when injury is severe. In this case, the symptoms intensify upon warming the fruit. Scald is usually not evident until after 3 months of storage. Storage scald can involve only the skin or extend 6 mm (¼-inch) into the flesh depending on the variety. It can be more severe on the greener side of the fruit.

(ii) Physiology

It is a type of chilling injury. The alpha-farnesene, a naturally occurring volatile terpene in the apple fruit, is oxidized to a variety of products (conjugated trienes). These oxidation products result in injury to the cell membranes which eventually result in cell death in the outermost cell layers of the fruit. Ethylene promotes the formation of alpha-farnesene and oxygen is required to oxidize alpha-farnesene to conjugated trienes. The differences in the concentration of naturally occurring antioxidants hold the key to the cause of storage scald.

(iii) Control

The most common method used to control scald is application of an antioxidant immediately after harvest. Antioxidants like Diphenylamine (DPA) should be applied within one week of harvest for maximum control. Wait 16 hours after DPA application before cooling of fruit. Low oxygen CA storage can provide a non-chemical control method in some cases. Oxygen concentrations between 0.5 and 1.0 per cent can significantly delay the development of scald, perhaps up to 10 months.

6. Watercore

Watercore (WC) is a serious physiological disorder of apples that occurs on the tree. It increases fruit susceptibility to internal breakdown (IB) caused by low oxygen (O_2) and/or high carbon dioxide (CO_2) concentrations and reduced potential storage life of apples. In addition, WC apples are reported to be more susceptible to fungal rots. Susceptible varieties include Jonathan, Delicious, Stayman, Winesap, Granny Smith and Fuji. Watercore is associated with high maturity fruit and low night temperatures. Another type of watercore which is unrelated to maturity may occur during unusually hot weather. This watercore is found more on the exposed side of the apple and may be associated with sunscald.

(i) Symptoms

The problem is characterized with water-soaked regions in the flesh, hard and glassy in appearance, only visible externally when very severe. Water-soaked areas are found near the core and around the primary vascular bundles but may occur in any part of the apple or involve the entire apple. Affected fruit may be particularly sweet and attractive at harvest but will deteriorate rapidly in storage. Fruit with WC have a greater density (90-95 per cent of the water density) than fruit without watercore (70-75 per cent of the density of water).

(ii) Physiology

The water soaked appearance of WC affected fruit results from the accumulation of sorbitol-rich solutions in the intercellular spaces. Sorbitol is the carbohydrate source translocated into the fruit from the tree which must be converted to fructose by the apple fruit. Sorbitol may be translocated to the fruit faster than it can be assimilated perhaps due to an inability of the apple tissue to convert sorbitol to fructose. The browning and breakdown that results from severe watercore is likely due to reduced gas diffusion in the affected tissue and may involve an accumulation of ethanol and acetaldehyde.

(iii) Control

The most effective way to reduce the incidence of WC is to avoid delayed harvests. Fruit should be harvested before water core develops extensively. Lots with moderate to severe water core should not be placed in CA storage but should be marketed quickly.

7. Internal Browning

Internal browning (IB) also called brown heart, thought to result from CO_2 injury, has been reported in Fuji, Cox's Orange Pippin, Braeburn and Jonathan apples which cause considerable losses of CA-stored fruit. The problem is associated with later harvested, large, and over-mature fruit and with higher CO_2 concentrations in storage. Because IB is not detectable externally, except in very severe cases, affected fruit can be discovered by buyers or consumers there by affecting future confidence in the product.

(i) Symptoms

Brown discoloration in the flesh, firm but moist, usually originating in or near the core. Brown areas have well defined margins and may include dry cavities resulting from desiccation. Symptoms can range from a small spot of brown flesh to nearly the entire flesh being affected in severe cases. When the entire apple is affected, a margin of healthy, white flesh usually remains just below the skin. Symptoms develop early in storage and may increase in severity with extended storage time.

(ii) Physiology

Evidences suggest that IB occurs as a result of CO_2 injury to the apple. Injury incidence and severity increase with increasing concentrations of CO_2 in the storage. Variability in susceptibility of apple varieties and in apples of different maturities may be due to anatomical differences (cell size, size of intercellular spaces) rather than biochemical differences. Diffusivity of CO_2 decreases as fruit mature and ripens which results in higher internal CO_2 concentrations and thus more chance of injury. The accumulation of sorbitol in the intercellular spaces also reduces gas diffusion in the affected tissue develop IB symptoms.

(iii) Control

Harvest at the optimum maturity, especially for CA storage and maintain CO_2 concentrations below 1 per cent. Assure good air circulation in storage rooms to prevent pockets with higher CO_2 concentrations. Avoid heavy wax coatings and thoroughly and rapid cool fruit after waxing and packaging.

8. Russeting

Russet is a common problem of apple which greatly reduce the market value of the fruit. High temperatures at night and high humidity are conducive to russet initiation. Mechanical injury from frost, hail, or wind also predisposes the fruit to russet. Russet is often reported in Cox's Orange Pippin, Discovery, Dunn's Seedling, Golden Delicious, Newtown, Rome Beauty, Stark, and Stayman's Winesap. However, some strains of Golden Delicious are russet-free.

(i) Symptoms

Russet is a condition in which phellogen (cork) tissue appears on the surface of an apple. The disorder may be of a slight nature or one in which the entire surface is covered by a network of corky tissue. Russet is more prevalent on lateral than on terminal fruit, on exposed than on shaded fruit, and on fruit from lower-altitude growing areas.

(ii) Physiology

The disorder is initiated 11- 30 days after the petal fall. Thin cuticles, irregular cell division, and fruit cracking are factors involved in the development of russet. Russeted fruit has higher levels of sucrose, glucose, total sugars, and acids, but phenolic concentration is lower than in non-russeted fruit. Leachates of the flowers of apples cause russet, and it is thought that rainfall leaches these substances from flowers and deposits them on the fruit surface. Some fungicidal and chemical sprays (*e.g.*, $ZnSO_4$), as well as spray surfactants, can be phytotoxic and cause russeting of fruit. A severe case of powdery mildew can also cause russet on apples. The powdery mildew fungus produces obvious leaf symptoms that point to its probable culpability as the principal cause of russeted fruit.

(iii) Control

Preventive procedures include the use of dimethoate, wettable sulfur, silicone preparations, and fungicides as foliar sprays. The preferred treatment is GA_{4+7}, applied at 5 to 10-day intervals during full bloom. Gibberellic acid encourages elongation of epidermal cells, which are better able to withstand internal pressures within the fruitlet.

9. Low Oxygen Injury

It is also a type of storage disorder. Critical oxygen levels differ among apple cultivars; Delicious are tolerant of atmospheres below 1 per cent O_2, whereas McIntosh in some growing areas are injured in atmospheres below 2 per cent O_2. McIntosh is highly susceptible to low-oxygen injury, whereas, Northern Spy, Empire, and Spartan rarely show symptoms. Low-oxygen injury varies with cultivar, oxygen concentration, length of exposure, and temperature. Low-temperature breakdown, core flush, and core browning can be accelerated in low-oxygen atmospheres.

(i) Symptoms

Injured apples develop dark brown water--soaked lesions in the skin, which may extend into sub-epidermal tissue. The lesions have definite margins, resemble soft scald, and are present on the blushed and non-blushed sides of the apple. As the severity of the injury increases, varying amounts of cortex and core tissue may become brown, moist, and water-soaked. McIntosh apples stored in low oxygen sometimes develop water-soaked bruise-like lesions, and the fruit may have a purplish cast in the red areas of the skin. Other symptoms of low-oxygen injury include purpling and the development of brown, corky tissue below the skin or through the 'cortex. The corking disorder is exacerbated by advanced maturity and low storage temperatures. Brown corky tissue can also develop in late -stored Cox's Orange Pippin apples. The lesions occur in the inner and outer cortical regions, mostly in the stem half of the fruit.

(ii) Physiology

Accumulation of ethanol or a product of ethanol metabolism may be the basic cause of low-oxygen injury. Reduced O_2 levels render pome fruits more sensitive to CO_2, and therefore narrow limits may have to be implemented for CO_2 to avoid CO_2-related problems. Apples exposed to short periods of reduced oxygen supply can recover with aeration if the alcohol concentration in the tissue does not exceed 110 mg/l00 g of fresh weight. Anaerobic conditions (<1 per cent O_2) for 4 days or more with Cox's Orange Pippin apples result in irreparably damaged fruit. Red Delicious apples, on the other hand, require 8-10 weeks of storage in 0 per cent O_2 before injury is evident.

(iii) Control

Concentration of O_2 should be monitored properly in the storage chamber. The effects of low-oxygen injury are mitigated by high humidity in the storage atmosphere.

References

Beattie, B.B., McGlasson, W.B and Wade, NL. 1989. *Post harvest diseases of horticultural produce* Vol. 1: Temperate Fruit.NSW Agriculture and Fisheries. CSIRO. AUSTRALIA. 84 pp. illus.

Bhat, S.A and Khan, F.A 2010. *Physiological disorders of apple and pear*: In a training Compendium on "Advanced Technologies for Post harvest Management of Food Crops", MY Ghani, FA Khan and SA Bhat eds.). Pp. 88-96. Sponsored by SAMETI and conducted by SKUAST-K at Shalimar.

Bishnu,P.K., Eckhard, G and Moritz K. 2013. *Russeting in apple and pear*: a plastic periderm replaces a stiff cuticle, AoB Plant, 5: 48. Published online 2012 Dec 17. doi: 10.1093/aobpla/pls048

DoH (Directorate of Horticulture, Kashmir). 2017. www.Hortikashmir.gov.in. *Area and production 2016-17.*

Eksteen, G.J and Combrink, J.C. 1987. *Manual for the identification of post harvest disorders of pome and stone fruit.* Fruit and Fruit Technology Research Institute, Stellenbosch, South Africa. 42 pp. illus.

Elgar, J.H., Burmeister, D.M., Warkin, C.B 1998. Storage and handling effects on a CO_2- related internal browning disorder of Braeburn apples. *Hort Science*, 33, 719-722.

Giraud, M., Westercamp P. Coureau C, Chapon JF and Berrie A. 2001. *Recognizing Post harvest diseases of apples and Pears.* CTIFL. Paris, FRANCE. 100 pp. illus.

Jones, A.L. and Aldwinckle H.S. 1990. *Compendium of apple and pear diseases.* APS Press. St.Paul, MN. USA.

Khan, F.A. and Shahid, M. 2013. *Physiological disorders of fruits and their management*: In Eco-friendly Management of Plant Diseases (Shahid Ahmad and Udit Narayan, eds.), Daya Publishing House. Pp. 131-155.

Knoche, M. and Khanal, B.P. 2011. Russeting and microcracking of 'Golden Delicious' apple fruit concomitantly decline due to gibberellin A_{4+7} application. *J. Amer. Soc. Hort. Sci.* 136(3): 159–164.

Lurie, S. 1998. Post harvest heat treatments – review. *Post harvest Biology and Technology* 14: 257-269.

Marlow, G.C. and Loescher, WH. 1985. Sorbitol metabolism, the climacteric and watercore in apples. *J. Amer. Soc. Hort. Sci.* 110:676-680.

Merritt, R.H., Stiles, W.C., Havens, A.V., and Mitterling, L.A. 1961. Effects of preharvest air temperatures on storage scald of Stayman apples. *Proc. Amer. Soc. Hort. Sci.* 78:24-34.

Pierson, V.F., Ceponis, M.J.And McColloch, L.P. 1971. *Market diseases of apples, pears and quince.* Agricultural Handbook No. 376. USDA/ARS. Washington, D.C. USA. 131 pp. illus.

Snowdon, A. L. 1990. *A color atlas of post harvest diseases and disorders of fruits and vegetables.* Vol. 1: General Introduction and Fruits. CRC Press, Inc. Boca Raton, FL. USA. 302 pp. illus.

Chapter 22

Mechanization Options for Apple Based Production System in India

Junaid N. Khan, J. Dixit and Rohitashw Kumar

Division of Agricultural Engineering,
Sher-e-Kashmir University of Agricultural Sciences and
Technology of Kashmir, Srinagar
E-mail: junaidk1974@gmail.com

1. Introduction

The state of mechanization in the hilly regions of India are still not well established. In the hilly and mountainous region traditional tools and implements evaluated and developed in isolation by small group of farmers had remained as the only mechanical gadgets available for cultivation practices. *Desi* plough, Shalimar plough, plain sickle, *tangroo,* shovel, sowing by *kera* and broadcasting *etc.*are still used by the farmers in these regions. Animal power continued to dominate as the most frequently used source of power. The drudgery and low efficiency associated with the use of these tools and implements was a long standing problem which demands attention, as early as possible. With both manually operated as well as mechanically operated implements, the use of mechanical power like power tillers, reapers and to develop suitable matching implements for these sources of power are essentially required for improving the productivity of crops.

In agriculture, field operations are time-bound. Timeliness of operations is crucial to success in farming. Any delay in sowing of rainy season crops results in drastic reduction in productivity. Under irrigated conditions, delayed sowing could be compensated by increasing the seeding rate and fertilizer use. Moisture is not a limiting factor under those circumstances. In dryland agriculture soil moisture dictates the priorities with regard to field preparation and sowing. Field preparations are swift and limited.

Sowing has to be expeditious to take advantage of limited soil moisture. Swiftness in sowing should not be at the cost of optimal plant stand. In order to obtain the desired plant stand, three things are important, adequate soil moisture at seeding depth, placement of seed and fertilizer at the desired depth in the moist zone and of course, the optimal seeding rate. This requires precision with regard to rate and depth of seeding which is possible only through seed-cum-fertilizers drill, which regulates seeding at calibrated depth and distance. Since seeding is the foundation of good plant stand and subsequent growth and development of plants, it is here that mechanization is called for. With *desi*-plough, with or without seeding funnel, it is not possible to ensure the desired plant stand. Under dryland situations, plant stand are often sub-optimal and consequent result in low productivity. That means delayed sowing and low productivity. In order to circumvent this, farmers hire tractors for field preparation and sowing operation. Often times direct seeding of pearl millet is resorted to so as to get full advantage of adequate soil moisture. Mechanization thus offers its potential comparative advantage in terms of time utilization efficiency. It is now evident that mechanical power and high capacity implements are important components in improving timeliness of operations coupled with the desired precision.

Reduction in cost of cultivation is of paramount importance in agriculture. Higher labour inputs and prolonged field operations increase the cost of cultivation. With the use of improved machines not only the cost of cultivation gets reduced but the profitability is also improved. There are machines, which perform two-three operations at a time with precision. Another advantage of using improved machines lies in increasing the input use efficiency, which is crucial in agriculture. Inputs like seeds, fertilizers *etc.* are not only costly but also scarce. With precision placement of seeds and fertilizers in the right quantity and depth by improved machinery saving of inputs is ensured. Much of the labour and inputs is wasted when indigenous implements are used for field preparation and sowing. For increasing the input use efficiency, adoption of improved implements is essential.

Besides tillage, sowing and fertilizer application, inter-culture operations require attention with regard to timeliness. Weed management has to be done early, 20 to 30 days after sowing, as any delay thereafter causes irreversible damage. It has been estimated that about one ton dry matter produced by weeds per hectare causes a loss of 20 to 30 Kg of N and 8 to 10 cm of water, resulting in 20-30 per cent yield loss. Improved tools like weeder, bullock drawn two row sweep and the tractor drawn weeders are far more efficient in timely and effective weed management, besides reducing the cost of operation and man-hours requirements. In fact, weed management is a drudgery operation requiring larger man-hours, which increase the cost of operation. Availability of labour is a great problem for carrying out timely weed management. Harvesting and post harvest processing and other operations, which also require mechanization inputs, face a lot of loss and less productivity. The harvesting of major horticultural crops of the state is still in infant stages.

2. Status of Fruit Crops

Jammu and Kashmir state is well known for its horticultural produce both in India and abroad. The state offers good scope for cultivation of horticultural crops, covering a variety of temperate fruits like apple, pear, peach, plum, apricot, almond and cherry. Fruit crops have witnessed most significant increase in area contributing to the shifts at a very high pace with bright prospects towards horticulture sector. Important reason being handsome returns to the farmers. Horticulture, in the state has developed as an industry and more and more land is apportioned to this sector each year. Horticulture is gaining momentum in the state and its contribution to GSDP remains around 7-8 percent over the past few years.Presently, a total of 347,223 hectares of land are under fruit cultivation in Jammu and Kashmir with 230,187 hectares in Kashmir and 117,036 in Jammu region. Of this 1.76 lakh hectares are under apple cultivation alone.

The availability and quantity of sufficient labour for hand harvesting is a major concern to many growers (Ibanez *et al.,* 1997). Fruit harvest is a very seasonal activity and the uncertainty of stable, skilled harvest labour at the right time is a major problem. The shortage, stability and cost of labour for hand harvesting, declining fruit prices and increasing competition from lower cost world producers. This is driving force behind development in mechanized harvesting. The 'shake and catch' bulk-harvesting systems developed in the 1980's and 1990's were unsuitable for fresh market fruits because of excessive fruit damage and lack of uniformity in fruit maturity on the trees. Nearly 75 per cent of the country's temperate fruits, mainly apples, are grown in the state.

3. Constraints in Mechanization

It is recognized that selective mechanization hold the key towards a second green revolution in rainfed lands. There are quite a few constraints in this regard. Some of these are mentioned below:

- ☆ Un-organized farm sector
- ☆ Poor infrastructure set up
- ☆ Small and fragmented holdings
- ☆ Poor resource base of farmers
- ☆ Costly inputs – diesel, seeds, fertilizers, pesticides etc
- ☆ Non-availability of loans from cooperative and commercial banks for purchase and hiring of improved implements.
- ☆ Non-existing of small units for manufacturing prototypes and repairs of farm machinery
- ☆ Poor outlet resources for farm produce

There could be other location specific constraints. It is suggested that a survey with regard to constraints is important locations to be carried out and documented. It would be of great help to policy makers in chalking out concrete plans to introduce selective mechanization in these areas. The curable measures are definitely going to impact the productivity and efficiency of horticultural crops like apple in the state.

Way Forward

Based on the specific constraints, concrete steps could be taken to introduce selective mechanization in these areas, keeping socio-economic framework in view. No constraint can be a hurdle, where there is a will to implement reforms towards improving the lot of rainfed farmers. Promoting custom hiring services will open up opportunities for mechanization. Let it be a small beginning, but a beginning in selected areas has to be made. Towards achieving second green revolution, besides timeliness of operations and precision agriculture through selective mechanization, other important factors like genetic improvement, efficient management of natural resources like soil and water are equally important. Increasing input use efficiency, improving water use efficiency, prevention of post harvest losses, value addition, market access, lowering production cost, effective transfer of technology, development of industry linkages are some of the potential means to increasing productivity and profitability in rainfed areas. For intensification of agriculture farm power availability should be increased from the present level of about 0.60 kW/ha to about 2.0 kW/ha in Jammu and Kashmir, by 2020 as well as other states of Western Himalaya.For timely and efficient plant protection, aero blast sprayers, orchard sprayers and electro-static spraying equipment are required to be introduced.

Custom hiring has not been given a fair trial yet. This job cannot be done by the government. It has to be a private enterprise with a good network of depots at strategic points. Tractor hiring for tillage and sowing including seed-cum-fertilizer drill, efficient tools for intercultural operations and crop specific harvesting, machines and combines are required. These depots could also serve as inputs supply points and farm produce outlets. If processing and storage facilities could also be provided at the custom hiring depots, it would be a much desirable feature. A good network of roads will facilitate timely market accessibility.

4. Mechanization in Horticultural Crops

The following different aspects should be consider for mechanization in horticulture crop production:

☆ Fruit crop mechanization equipment for pit making, transplanting of saplings, pruning, spraying in tall crops, harvesting of fruits *etc.* need to be identified/adopted/developed and popularized.

☆ Crop mechanization equipment for seed-bed preparation, planting, transplanting of seedlings, interculture, irrigation, spraying harvesting *etc.*, need to be identified/designed and introduced.

☆ Modern manual and power operated garden tools and equipment will have to be introduced and popularized for landscape and horticulture crops.

☆ Post harvest equipment and technology are needed for grading, drying, cooling, evaporative cooling, storage, cold storage and handling of farm produce to improve their quality and shelf-life. Cold chains for transport of perishable products of fruits with minimum losses.

☆ Agro-processing activities should be promoted in the production catchments to reduce losses.

☆ Expansion and upgrading of horticultural activities requires appropriate short and medium term storage facility.

☆ Improved harvesting equipment like serrated sickles, fruit pluckers and fruit harvester should be introduced.

☆ Power operated trenchers, angle dozers, drudgers, buck scrapers and other earth moving machinery will be required to be introduced in the region for making farm ponds, bunds and terraces, irrigation channels and waterways.

(a) Mechanical Equipment's/Technology for different Crops

(i) Fruit Picker/Harvester

The farmers normally harvest the fruits like apple, pear, peach *etc.* manually by making use of orchard ladders or indigenous methods of harvesting which are very crude and result in lot of drudgery to the workers and a great damage to fruits. Keeping in view the hardships faced by the farmers, a pull and cut type harvester (Figure 22.1) was fabricated by Division of Agricultural Engineering especially for the Delicious varieties of the apple. The variety has a longer stalk unlike the American variety. The delicious variety has a longer stalk which results in easy cutting action of the fruit. The blades of the fruit harvester can be mounted at different angles. The blades are activated by means of clutch mechanism for performing cutting action. A net conveyor over a steel ring below the blades helps in collecting the fruit after cutting. Thus, avoiding damage due to free fall. The height of pull and cut harvester can be altered by an adjustable tapered pipe and can reach a maximum length of 2.5m. The harvester can also be effectively used by womenfolk, who otherwise find it very difficult to go for harvesting operations. The smaller size of the steel ring makes it reach any part of the tree without getting entangled in the branches of the tree.The harvesters fabricated work on the principle of hold and twist type device is based on the hand picking mechanism where individual fruit is held between jaws of the harvester and then twisted to shear off the stock. Taking this factor into account a number of manually operated harvesters were fabricated in the Division with an objective of reducing drudgery associated with manual method of harvesting.

Figure 22.1: Apple Harvester.

(ii) Power Tiller Operated Leaf Collector

Power tiller operated leaf collector was fabricated for reducing disease spread in apple. The platform for the power tiller was also fabricated for securing appropriated tension in the power transmission belt. Two mild steep angle iron of size 38 x 38 mm were welded and connected with nuts and bolts along with two round shafts of 20 mm diameter for making platform to install leaf collector. The pitch diameter of pulley collector shaft was taken as 10.18 mm according to maximum speed available on the shaft of the power tiller. The centre distance of

Figure 22.2: Power Tiller Operated Leaf Collector.

63.5 – 68.5 cm was calculated and a pitch length of belt was calculated as 152.4 cm. The leaf collector was installed on the fabricated platform of power tiller (Figure 22.2). A plastic suction pipe of 10 cm diameter and 10 meter length was attached to the suction end of the leaf collector. The other end of the suction pipe was connected to a movable rectangular end. The leaf collector was tested at different variable speeds of the power tiller.

(iii) Mechanical and Robotic Harvesting Technologies

New mechanical and robotic harvesting technologies now a days are used to increase fruit production efficiency (Bourely *et al.*, 1990; Peterson *et al.*, *1998;* Robinson *et al*, 1990). The demand for fresh fruit is increasing globally as the standard of living increases. Harvesting is the most labour intensive operation in fruit production accounting for up to 60 per cent of the total labour requirements.The latest developments and availability of mechanical and robotic harvesting technology, has identified potential opportunities and benefits for the use of this technology used for deciduous fruit industries. The technology for harvesting options for the tree of fruit crops such as walnut, apricot and berries and new developments in sensor technology and for non-destructive fruit quality assessment.The key features of technology are:

☆ New developments in mechanical harvesting aids provide the most immediate opportunities to improve harvest efficiency in quality fresh fruit production. The Dutch designed Pluk-O-Trak was particularly impressive because of its high harvest rates, extremely low fruit bruising and damage, versatility for different orchards and operations, low running costs, quietness of operation, ease of maintenance and suitability to pickers.

☆ Robotic bulk harvesting is a medium term proposition which needs further improvement for use on apples but has immediate application for cherries and possibly pears and plums. The main limitations for fresh apples are the detachment of stalks in some varieties, non-uniformity of fruit maturity on the tree and the potential for bruising in delicate varieties. The system is suited to narrow inclined canopies.

☆ Robotic selective harvesting of individual fruits is and exciting longer term proposition. The advances in orchard production systems, mechanical, artificial intelligence and sensor technology since the first prototypes were developed in France in the 1990's now make robotic harvesters an achievable and commercially viable proposition.

There were significant improvements in fruit removal rates and harvest efficiency (Bourley, pers. Com., 2001). In Japan researchers at Iwate University are conducting work on the robotic harvesting of Fuji apples (Kataoka*et al*, 1999). They have developed a robotic harvesting hand for apples. The Pluk-O-Trak system was developed in Holland for apples and is now used on a range of other fruit including stone fruits, pear and oranges (Sanders, pers. com., 2001). Peterson's bulk harvesting system has made major improvements in three main areas.

★ The tree training system

★ The control system and-Vision

★ Fruit selection

(iv) Pluk-O-Trak

The most impressive harvesting aid in Europe was the '*Pluk-O-Trak*'. The *Pluk-O-Trak* was manufactured in the Netherlands by Munckhof. Several hundred machines are now being used throughout Europe and other parts of the world. The unique feature of the *Pluk-O-Trak* is the 6 to 8 conveyor chutes, which can be positioned close to the picker and fruit in canopy for ease of picking. The picking conveyors can be adjusted in all directions to minimize the distance to the picker and the crop. Harvest rates are high, 250kg/hr/person. The machine can reach tops of trees from 3.25 to 4.25 m high and can be accommodated in very narrow rows from 3 to 6 m wide. Many of the mechanical adjustments can be converted to hydraulic control. The *Pluk-O-Trak* is also very versatile and can be used for pruning and hand thinning fruit operations. *Pluk-O-Trak'* harvesting Morganduft apples for processing is shown in Figure 22.3.

Figure 22.3: 'Pluk-O-Trak' Harvesting Golden Delicious Apple in France.

Technology, particularly in sensors for use in non-destructive fruit quality assessment and fruit grading is advancing rapidly around the world. The instrumented 'Glove' (Photograph courtesy of CEMAGREF) (Figure 22.4) with

Figure 22.4: Instrumented Glove.   **Figure 22.5: Ladder for Fruit Plucking.**

the help of lightweight aluminum ladder can be utilized for harvesting of the fruit (Figure 22.5).

5. Policy Issues

☆ Design and development of ergonomically suitable tools and equipments for hill agriculture.

☆ Need to develop women friendly tools and equipments for hills.

☆ Need of ergonomically suitable light weight power tiller for high hills.

☆ Introduction of appropriate and suitable horticultural tools

☆ Strategies for hill farm mechanization

☆ Testing facilities for agricultural machinery and agro-products for quality control should be created in the region

☆ For creating awareness amongst the farmers, extension workers and entrepreneurs of the region and display centres of improved agricultural machinery.

☆ To encourage agricultural mechanization on sound footing there should be a state policy for agricultural mechanization.

☆ For efficient use of stationary farm power units and equipment, the electricity to the farmers, should be provided at subsidized rates.

Acknowledgements

Authors are highly thankful to the Division of Agricultural Engineering for providing necessary information.

References

Bourely, A., Rabatel, G., Grand d'Esnon, A. and Sivila, F. 1990. *Fruit harvest robotization: 10 years of CEMAGREF experience on apple, grape and orange.* In Proc. AGENG '90 Conference, 178-179, Berlin, Germany.

Brown, G.K. 2001. *Challenges and benefits for mechanical harvesting of fruit crops.* 6th International Symposium on Fruit, Nut, and Vegetable production engineering 11-14 September 2001, Potsdam Germany. Abstracts: 23.

Ibanez, M., Hetz, E. and Vevegas, A. 1997. Evaluation of two walnut and apple harvesting systems. *Agro-Ciencia* 13(3): 325-329.

Kataoka, T., BulAnon, D.M., Hiroma, T., Ota, Y. 1999. *Performance of a robotic hand for apple harvesting.* Transactions of the ASAE Paper No. 993003. St Joseph, Mich.: ASAE.

Peterson, D.L., Miller, S.S. and Kornecki, T.S. 1998. *Over-the row harvester for apples.* Transactions of the ASAE 28(5):1393-1397.

Robinson, T.L., Millier, W.F., Throop, J.A., Carpenter, S.G., Lakso, A.N. 1990. Mechanical harvest ability of Y-shaped and pyramid-shaped 'Empire' and 'Delicious' apple trees. *Journal of the American Society for Horticultural Science 1.*

Chapter 23

Advances in Pre- and Post harvest 1-Methylcyclopropene Technology to Improve Shelf Life of Apple

Xi Chen[1], Kit Yam[1], Han Zhang[1], Nazir Mir[1], Robert Oakes[1], A. H. Pandith[2], Haroon Naik[2] and Khuram Mir[3]

[1]*Rutgers University, New Brunswick, NJ USA and Decco Postharvest, Philadelphia, PA USA*
[2]*Sher-e-Kashmir University of Agricultural Sciences and Technology of Kashmir, Srinagar, India*
[3]*Harshna Naturals, ICG Lasssipora-Pulwama, Kashmir, India*
E-mail: chenxince4@gmail.com

1. Introduction

Global apple production reached 76 million tonnes in 2014, with India as the sixth largest producer (1.9 million tonnes; US$ 1.4B) and Jammu and Kashmir contributes major chunk in India (1.4 million tonnes; US$ 780MM). Jammu and Kashmir (J&K) is proudly known as fruit bowl of India and its produce drives major revenue for the region's economy. Among all fruits grown in J&K, apple is the mainstay.

J&K is blessed to have favorable weather and geographical conditions for apple growth, and as a result Kashmiri apples have distinguished color, shape, and unique taste compared to others.Today, 700k Kashmiri families are dependent

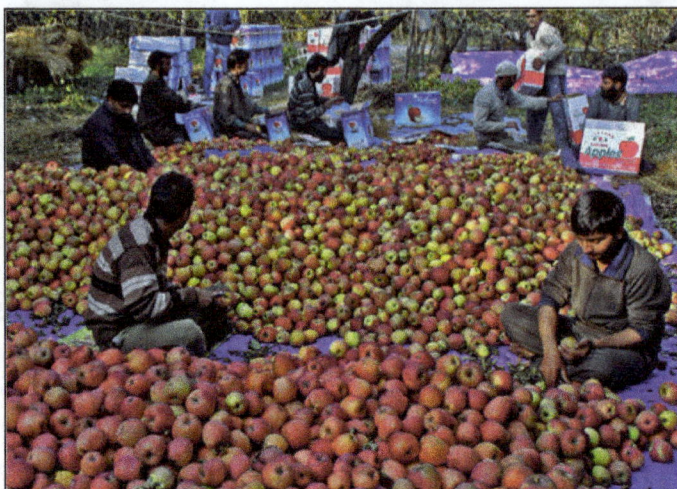

upon agriculture. However, much of the value of the Kashmiri apple, as little as 25 per cent of this fruit is actually of high quality as compared to 80 per cent in Italy. Several factors have led to this loss in quality and value.

Pre-and post harvest losses are the major issues limiting shelf life, and growth of local industry and economy in J&K. Pre- and post harvest losses occur when apples are harvested over several days and stored on the ground prior to market, and thus quality of fruit declines between harvest and the moving retail channel. Currently no treatment is available to J&K apple growers to protect quality and value of their produce. Following harvest (postharvest), only 3.5 per cent of apples are stored in Controlled Atmosphere (CA) storage facilities. During shipment to fruit aggregators and during transit to distant markets, no cold storage and refrigerated trucks are available to ship to market. J&K apples arrive in the market after the Himachal Pradesh fruit which drops the market value of J&K fruit.

Several important post harvest tools are available to extend the freshness, quality and nutrition of fruits and vegetables which have been implemented in developed and developing countries. These include refrigeration, modified atmosphere packaging, controlled atmosphere (CA) storage, ultra-low oxygen (ULO) storage and dynamic controlled atmosphere (DCA) storage. Yet, due to investment and infrastructure constraints, many of these technologies are not readily available.

This chapter introduces an advanced technology designed for pre- and post harvest application to extend shelf life of apples. This technology uses innovative approach to formulatea sprayable water-based 1-methylcycopropene (1-MCP) liquid formulation, which is able to manage the undesirable effects of ethylene when the fruits are still attached to the tree and during post harvest storage. Ongoing research has been conducted by Sher-e-Kashmir University of Agricultural Sciences, Kashmir, with technical support from Dr. Nazir Mir (Rutgers University, US) to demonstrate the bio-efficacy of this technology.

2. 1-methylcycopropene (1-MCP) (Protecting Apple Quality from Farm to Table)

(a) Science behind 1-MCP

Ethylene, a phytohormone produced by all higher plants, plays a critical role in regulating several physiological and developmental processes in plants. The responses include fruit abscission, fruit degreening, fruit ripening, color development, release of dormancy, promotion of fruit maturity, leaf/flower senescence, and various plant responses to abiotic and biotic stresses. In various fruit including apples, controlling both endogenous (produced within plants) and exogenous (environmental ethylene/ethylene application) ethylene are of importance in order to extend shelf-life and maintain fruit quality (Davies, 2010; Davies, 2013).

After many years of research, a simple, small gas molecule, 1-methylcyclopropene (1-MCP) was identified and later patented by Drs. Blankenship and Sisler at North Carolina State University in 1996 as an effective agent to bind to ethylene receptors to which 1-MCP has 10 times higher affinity than ethylene (Serek, *et al.*, 1967; Serek, *et al.*, 1994; Blankenship and Dole 2003). Since its discovery, research has been conducted to investigate its mechanism of action, effective application level, delivery mode, and different stabilizers and carriers/encapsulants (Daly and Kourelis, 2000; Baritelle, *et al.*, 2001; Mir, *et al.*, 2001; Song, *et al.*, 2006; Zhang, *et al.*, 2011; Xin, 2012; Yoo, 2012; Sarker and Liu, 2015; Mir, 2016). It is now well known that 1-MCP is effective in reducing pre-harvest drop, extending the firmness, taste and quality of stored fruit (Mir and Beaudry, 2000; Yuan and Carbaugh, 2007). The combination of 1-MCP and controlled atmosphere storage is known to extend post harvest shelf life of apples for more than 8 months. 1-MCP is effective at use rates of 0.250 to 1.0-parts per million (ppm). Given the volume of apple production, it is a very cost-effective technology. In addition, pre-harvest application of 1-MCP reduces fruit drop, which expands the harvest window enabling grower to both maintain fruit quality on the tree and at the same time more effectively manage the labor force involved in fruit harvest. 1-MCP is also shown to be effective in reducing superficial scald (Fan, *et al.*, 1999; Rupasinghe, *et al.*, 2000), which opens new avenues for apple industry as DPA, a traditional chemical used for superficial scald, was found to be carcinogenic, and its use is limited in many countries.

Because 1-MCP is in gaseous form and flammable at high concentration, it is challenging for storage and application. Since its discovery, different techniques have been investigated and developed to encapsulate, adsorbor stabilize 1-MCP. Various forms of 1-MCP complex reported in literature include α-cyclodextrin(Daly and Kourelis, 2000), cucurbituril(Zhang *et al.*, 2011), sodium salt of 1-MCP(Song *et al.*, 2006), metal coordination polymer network(Mir 2016, NatureChemistry-Editorial 2016, URQUHART 2016), 1-MCP emulsion(Xin, 2012), 1-MCP in situ generation system(Yoo, 2012), and boron/1-MCP complex(Sarker and Liu, 2015).

However, due to various reasons such as regulation and consistency of product performance, few systems are commercially viable on the market.

Also of importance is the safety of 1-MCP which has been registered in over 45-countries and is classified by the US EPA as a bio-pesticide and is exempt from a tolerance, *i.e.* assumed to have no measureable residues on treated produce. 1-MCP has been approved in the US for numerous crops including apple, apricot, Asian pear, avocado, banana, broccoli, kiwi fruit, melon, nectarine, peach, pear, persimmon, plum, plumcot and tomato. 1-MCP is registered and used in nearly all countries which produce apples to extend freshness and quality (Table 23.1).

Table 23.1: Global Registration of 1-MCP

Europe, Middle East and Africa			Asia		North America	Latin America
Austria	Greece	Portugal	Australia	New Zealand	Canada	Brazil
Belgium	Hungary	Slovenia	China	Philippines	Mexico	Chile
Croatia	Ireland	South Africa	India	Taiwan	USA	Colombia
Czech Republic	Israel	Spain	Indonesia	Thailand		Costa Rica
Denmark	Italy	Switzerland	Japan	Viet Nam		Ecuador
France	Kenya	Turkey	South Korea	Malaysia		Guatemala
Germany	Netherlands	UK				
Poland						

(b) Opportunities and Challenges of 1-MCP Technology

Post harvest application of 1-MCP is well established in apple industries especially for developed countries. For example, it is estimated that 80 per cent of apples are treated with 1-MCP in the US. However, it is limited in most developing countries, because, controlled atmosphere storage and refrigeration are not always available. This situation provides more opportunities for pre-harvest spraying application as the application is more flexible and can be combined with other spraying chemicals or fertilizers. The most challenging issue for pre-harvest 1-MCP application is the facile loss of 1-MCP in solution form since 1-MCP is a gas and most current 1-MCP gas carriers/encapsulants are water soluble. Upon mixing in water, the complete release of 1-MCP may take place in less than 1 hour which makes the biological performance almost ineffective under commercially required spray time of at least 3 to 4 hours. It not only reduces its effectiveness, but also creates safety issue in the spraying tank since 1-MCP is explosive above certain concentration.

There are attempts to use non-aqueous solution for 1-MCP pre-harvest application. For example, US Patent Application Number 2013/0065764 (Jacobson and Zhen, 2013) discloses a formulation which comprises suspended 1-MCP encapsulated materials into non-aqueous organic and synthetic fluids and then bringing the formulation into contact with plant and plant parts. The authors' report that the 1-MCP complex in solution remains in the solid form, minimizing the contact between the 1-MCP compound complex and water, leading to the retention of MCP in the solution for a longer time. The authors do not show any 1-MCP release kinetics data. Moreover, the composition of the disclosed formulations may be inapplicable to ripe or near ripe fruit or plant parts due to the potential for undesirably long residual life of some of the synthetic or organic components of the formulation post application. There are also products that use 1-MCP precursor salt for pre-harvest application, but they are not commercially viable due to the use of harsh organic chemicals in the formulation.

US Patent Application Number 2012/0264606 A1 (Kostansek, 2012) describes an oil medium for suspending encapsulated 1-MCP particles. The authors then process the suspension in the media mill to produce particles of less than 2 micrometers. When the 1-MCP solution was made from these oil based formulations and then passed through the nozzle of a sprayer, the 1-MCP retention in the spray solution was much better. Authors do not report 1-MCP release kinetics from these oil based formulations. As oils are not miscible with water, producing a homogenous solution to cause consistently a desired effect may be a challenge

Our research group has conducted extensive studies on delivery systems of 1-MCP since 2000. In this chapter, we present our selected successful research data and scientific method to reduce the loss of 1-MCP during mixing in water and demonstrate the commercial potential of liquid form sprayable 1-MCP. In addition, the formulations presented do not limit to pre-harvest application, they can also be used as liquid dip or in storage room for post harvest application.

(c) Concepts for Developing a 1-MCP Liquid Formulation

The concept for developing 1-MCP liquid formulation involves modifying 1-MCP encapsulant and/or create physical entrapment (Mir, 2014a; Mir, 2014b; Mir, 2015). The data presented in this chapter is obtained from various formulations comprising of water soluble, environmentally safe and as far as possible, food use approved ingredients that significantly reduce the loss of the 1-MCP from aqueous solutions leading to sufficient efficacy of liquid formulation or spray solution required to cause a desirable biological effect.The concept can also be applied to develop packaging and aerosol systems (Mir, 2014c).

These formulations are mainly combinations of polyols and hydrocolloids, which suspend solid form of 1-MCP/carrier complex. Upon mixing in water, the formulation would entrap water to form colloidal solutions which provides a tortuous path for 1-MCP to diffuse, thereby controlling its release. The hydrocolloidal solutions are preferably to be shear thinning, *i.e.*, viscosity of the solution reduces when shear is applied, thus it is easy to spray using a commercial sprayer.

(d) Selected Data

(i) 1-MCP Quantification

Quantification of 1-MCP was done using gas chromatography (GC) based on the method described in the literature (Mir *et al.,* 2001). 1-MCP gas samples are taken periodically from the enclosed chamber and the percentage released is plotted over time. Cyclodextrin based material was used as 1-MCP carrier/encapsulant.

(ii) Effect of Polyols alone on Controlling 1-MCP Release

Dispersion and containment of 1-MCP in polyols is evaluated using glycerol (99.9 per cent pure, Sigma Aldrich Co., St. Louis, MO). Encapsulated 1-MCP (25 mg) was dispersed in 1 mL of glycerol. The mixture is then placed in an airtight chamber to quantify the amount of 1-MCP released over time. Other useful polyols for the formulation include D-sorbitol, di, tri, tetrols and other sugar alcohols, and/or mixtures of them. Experiments were conducted at room temperature (~22 °C). Results (Figure 23.1) showed that1-MCP gas is released rapidly when encapsulated 1-MCP powder is dissolved in water (control). Complete 1-MCP release was achieved within 1 hour. By comparison, when the encapsulated 1-MCP powder was dispersed in glycerol, the release rate of 1-MCP was drastically reduced, allowing only 7.53 per cent to be released over a 16 hours holding period, thereby, allowing the liquid formulation to be biologically effective for longer time compared to control.

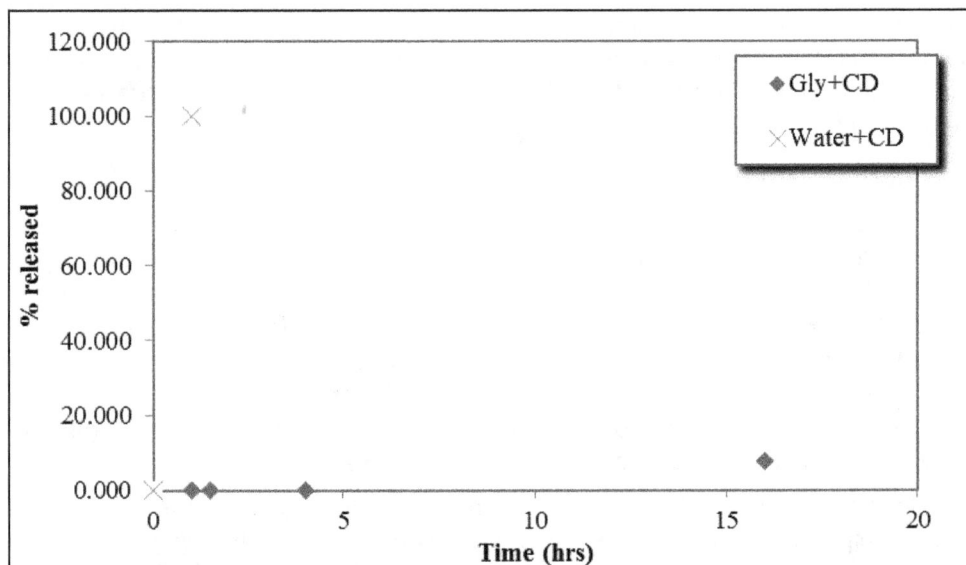

Figure 23.1: Effect of Polyol (Glycerol) on Release of 1-MCP.

(iii) Effect of Polyol and Hydrocolloid on Controlling 1-MCP Release

Colloidal gels are made by hydrating hydrocolloids with water as a dispersion medium to form a gel. Xanthan gum (CP Kelko, Atlanta, GA) was evaluated because it is a shear thinning hydrocolloid which possesses high viscosity when stagnant and low viscosity when being sprayed (high shear). It allows the formulation to be more stable before spraying and can cling on the fruits after it is sprayed. Three concentrations of xanthan gum were evaluated: 0.5 per cent, 0.05 per cent and 0.005 per cent, and 12.5 mg of 1-MCP was dispersed in 37.5 mLxanthan gum solution without any further agitation. The mixture was then placed in an airtight chamber to quantify the amount of 1-MCP released over time. Results (Figure 23.2) show that xanthan gum was effective to retain 1-MCP. After 1 hour, only around 6 per cent of 1-MCP released, which is considerably less than water. Different concentrations, from 0.005 per cent to 0.5 per cent, did not make much difference, indicating that a low concentration of xanthan gum is sufficient to retain 1-MCP.

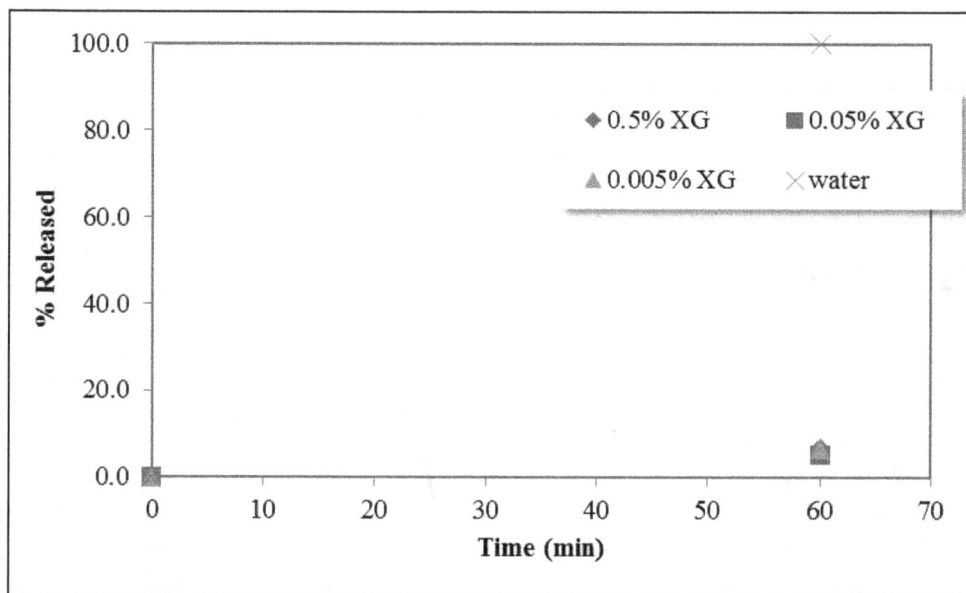

Figure 23.2: Effect of Hydrocolloid (Xanthan gum) on Release of 1-MCP.

(IV) Effect of Combination of Polyol and Hydrocolloid

Two types of polyol, glycerol and 70 per cent D-sorbitol (98 per cent pure, Sigma Chemical Co., St. Louis, MO), were evaluated. Different concentrations of xanthan gum dispersed in water, ranging from 0.005 per cent to 1 per cent were evaluated for their efficacy in controlling release of encapsulated 1-MCP dispersed in glycerol or D-sorbitol and also compared to the effect of water alone (Control) on controlling release of 1-MCP.The results (Figure 23.3) show that increasing xanthan gum concentration in water from 0.005 to 1 per cent slows down the release of 1-MCP dispersed in glycerol or D-sorbitol from 26.64 per cent to 0 per cent and 23.24 per cent to 0 per cent in 1 hour, respectively. Replacing xanthan gum

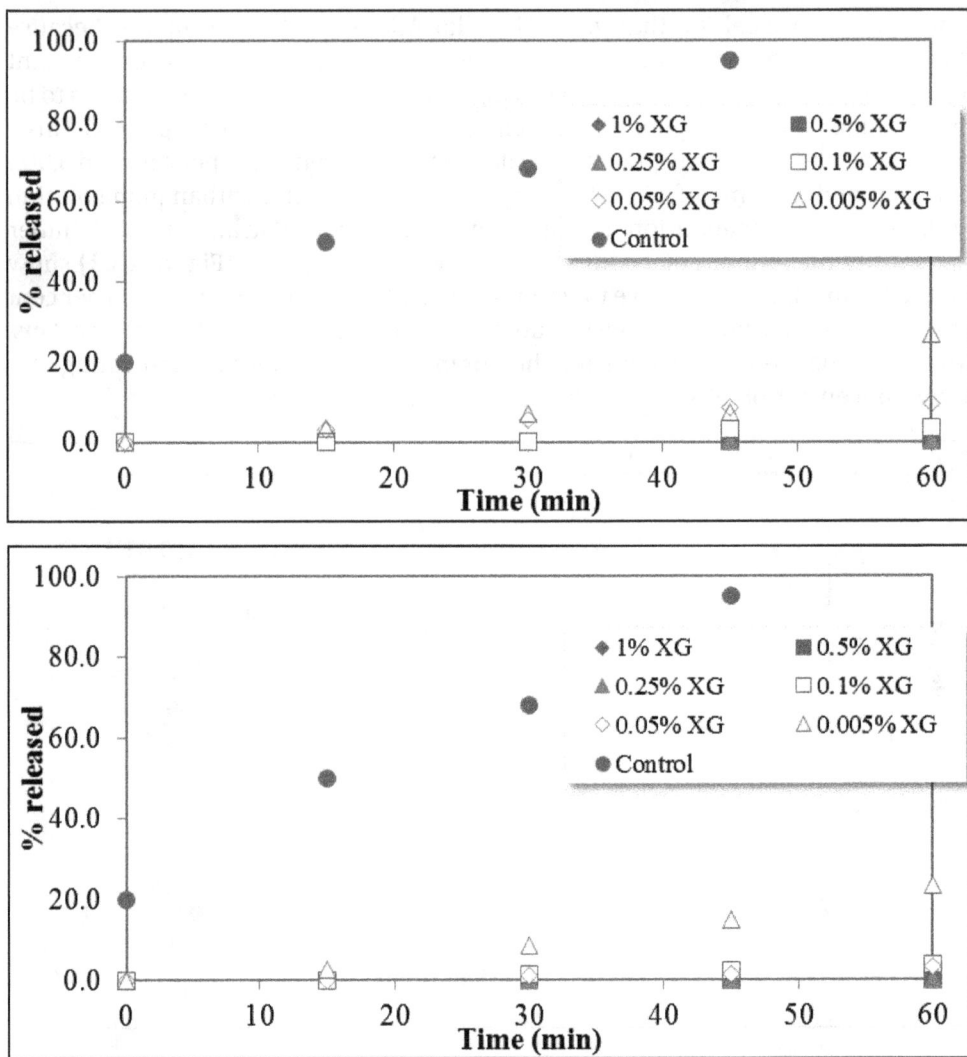

Figure 23.3: Effect of Combination of Polyol (Top: Glycerol; Bottom: D-sorbitol) and Hydrocolloid (Xanthan gum) on Release of 1-MCP.

with 100 per cent water (Control) results in 100 per cent release of encapsulated 1-MCP in 1 hour.

(v) Addition of Clay in the Formulation

Clay was evaluated because its intermolecular electrostatic forces can modify the structure of hydrocolloid, which in turn affect the release of 1-MCP. In this study, release of 1-MCP from the concentrated blend of polyol/hydrocolloid/clay without the addition of hydrated colloidal solution was carried out in a 500 mL jar and the amount of 1-MCP released was quantified using GC for 40 hours. The following formulations were evaluated and all contained 25 mg of encapsulated 1-MCP dispersed in the preferred formulation. The formulations were (1) 0.5 mg of hydroxyl propyl cellulose (HPC) in 10 mL glycerol; (2) 0.25 g of HPC and 0.25 g of laponite in 10 mL of glycerol; (3) 0.5 g of HPC and 0.5 g of laponite in 10 mL glycerol (4) 0.5 g of hydroxyl propyl cellulose (HPC) in 9 mL glycerol and 1 mL of polysorbate (surfactant); (5) 0.25 g of HPC and 0.25 g of laponite in 9 mL glycerol and 1 mL polysorbate; (6) 0.5 g of HPC and 0.5 g of laponite in 9 mL glycerol and 1 mL polysorbate.Results (Figure 23.4) showed that 1-MCP released was less than 30 per cent in a period of 40 hours in all formulations. Addition of laponite slightly improved the retention of 1-MCP in the formulation.

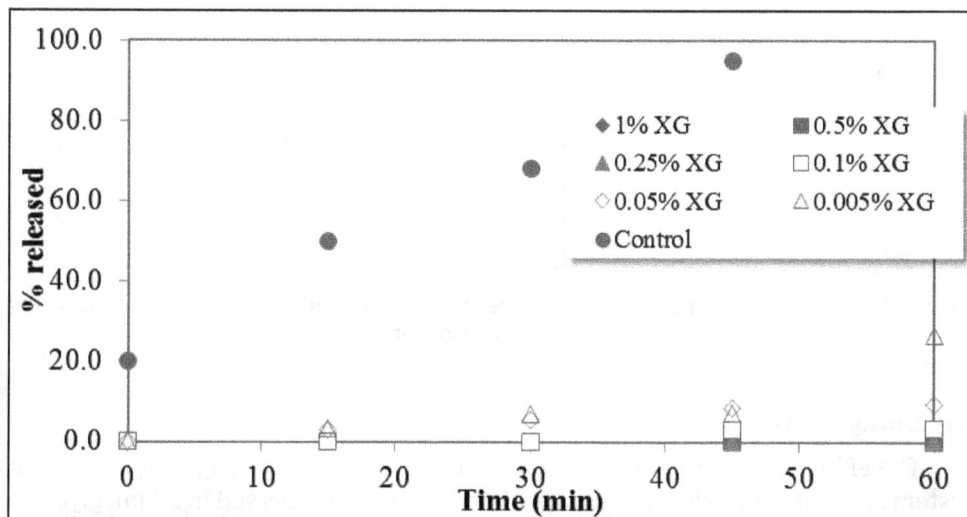

Figure 23.4: Effect of Combination of Glycerol/Hydrocolloid or Glycerol/Hydrocolloid/ Clay on Release of 1-MCP.

Another study was also conducted for the following four formulations with hydrocolloid solution (a) 0.5 g of hydroxyl propyl cellulose (HPC) in 9 mL of glycerol and 1 mL polysorbate,(b) 0.25 g of HPC and 0.25 g of laponite in 9 mL of glycerol and 1 mL polysorbate, (c) 0.5 g of HPC and 0.5 g of laponite in 9 mL glycerol and 1 mL polysorbate, and (d) 0.5 g of HPC, 0.5 g of laponite and 0.5 g of tetra sodium pyrophosphate in 9 mL glycerol and 1 mL polysorbate. Upon dispersion, 25 mg of encapsulated 1-MCP is mixed to either formulation. To make the formulation, the contents were stirred well in a 500 mL glass jar to ensure that 1-MCP is uniformly dispersed. Xanthan gum (0.05 per cent) was mixed into the encapsulated 1-MCP dispersed formulation. The solution was then placed inside an airtight chamber to quantify the amount of 1-MCP released over time. Results (Figure 23.5) showed that 1-MCP release is controlled to about 13-16 per cent in 2 hours when the encapsulated 1-MCP is blended to formulations (a), (b), (c), and (d) and hydrated xanthan gum solution is added to the blend. Results showed almost identical retention of 1-MCP in all formulations.

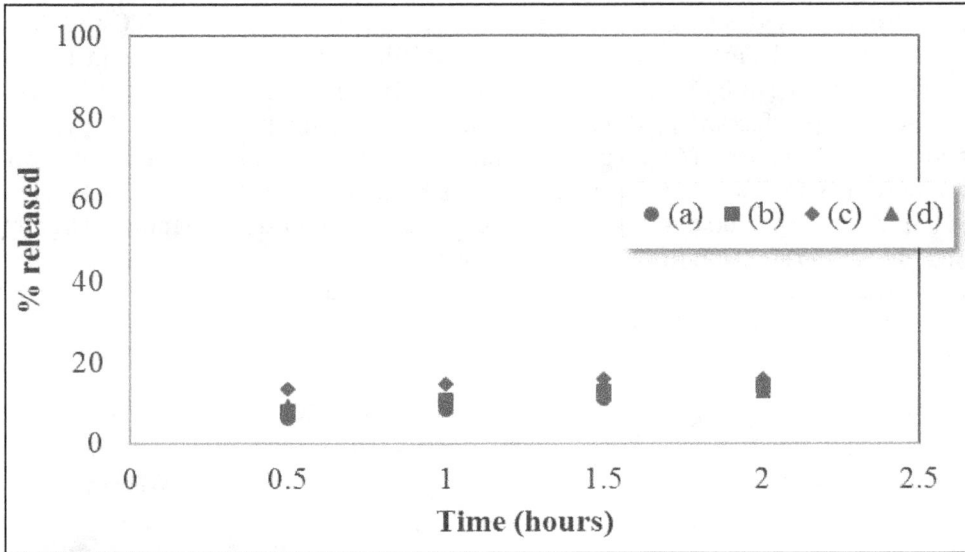

Figure 23.5: Effect of Combination of Glycerol/Hydrocolloid or Glycerol/Hydrocolloid/ Clay with Hydrated Xanthan Gum on Release of 1-MCP.

(vi) Spraying Test

The effect of delivering the preferred solution using spraying over a stagnant system is carried out. 25 mg of encapsulated 1-MCP is dispersed in 23 mL glycerol or D-sorbitol. The contents are stirred well in a 500 mL mason jar to ensure that 1-MCP is uniformly dispersed. 0.05 per cent xanthan gum is hydrated in water and 77 mL of the colloidal solution is mixed with the above 1-MCP/polyol mixture. The percentage 1-MCP released over time is evaluated by (a) placing the mixture in an airtight chamber (stagnant) (b) spraying the mixture in an airtight chamber. Results (Figure 23.6) showed that for both glycerol and sorbitol, spraying caused about 3 to 4 times increase in release of 1-MCP compared to stagnant in about 60 min, indicating the effect of decreasing the droplet size of the solution by spraying under pressure vs. not spraying or stagnant on the release of 1-MCP from the formulation.

(vii) Other Pre-harvest Ethylene Modification Methods

Other plant growth regulators such as Naphthaleneacetic acid (NAA) and Aminoethoxyvinylglycine (AVG) have been used unsuccessfully to reduce pre-harvest drop. Fruit softening occurs with NAA and delayed fruit color development has been reported for AVG (commercial name ReTain®). In addition, AVG which inhibits ethylene biosynthesis requires application before the onset of ethylene climacteric which is difficult to accurately predict under orchard conditions.

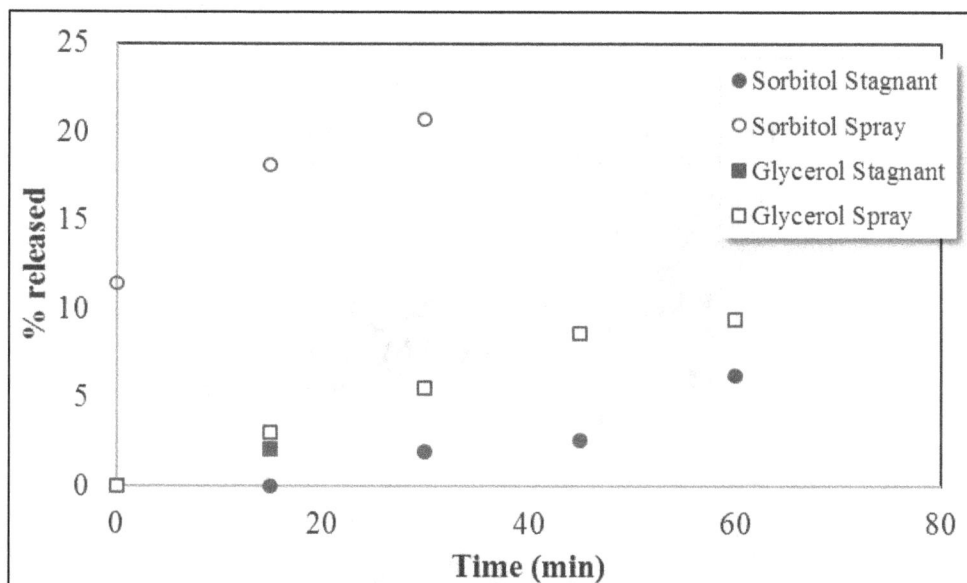

Figure 23.6: Release Profile of 1-MCP after Spraying.

3. Effort in J&K for Reducing Apple Waste

The Late Jenab Mufti Mohammad Sayeed directed Sher-e-Kashmir University of Agricultural Sciences, Kashmir (SKUAST-K) to "carry out trial experiments after working out modalities to see if our fruit growers can use this new technological invention." This research targets growth of the apple industry from Rs 3,000 crore (US$ 450MM) to Rs 15,000 crore (US$2.25B) within 5-years via high density orchards.

(a) SKUAST-K Research Program (2016–2017)

In coordination with Dr. Nazir Mir (Rutgers University, New Brunswick, NJ), SKUAST-K researchers and Khuram Mir, and in spite of recent difficult political conditions in Kashmir have initiated the first 1-MCP pre-harvest and post harvest trials. Orchard application of pre-harvest 1-MCP formulation were completed. Post harvest research trials were also conducted in 4-m^3 plastic tents and at Khuram's commercial Controlled Atmosphere facility which will demonstrate the effective protection by 1-MCP of fruit from the negative effects of ethylene (*e.g.* fruit softening and loss of flavor/quality) during storage and through the entire supply chain from Kashmir to distant markets in India and nearby countries. Collaboration between Dr. Mir's Rutgers Lab, SKUAST-K and Khuram Mir (Harsha Naturals) has been highly effective in shipping materials for the R and D trials, evaluation of the trial results and plans for commercialization in the near future.

Figure 23.7: SKUAST-K Research Team.

Figure 23.8: Commercial Trial with 1-MCP on Kashmiri Apples.

Applications of 1-MCP in large controlled atmosphere cold storage rooms will be followed by commercial introduction next season to growers and packing houses.

Figure 23.9: Commercial Apple Storage CA Rooms in Kashmir.

References

Baritelle, A. L., G. M. Hyde, J. K. Fellman and J. Varith 2001. Using 1-MCP to inhibit the influence of ripening on impact properties of pear and apple tissue. *Post harvest Biology and Technology 23(2): 153-160.*

Blankenship, S. M. and J. M. Dole 2003. 1-Methylcyclopropene: a review. *Post harvest Biology and Technology* 28(1): 1-25.

Daly, J. and B. Kourelis (2000). Synthesis methods, complexes and delivery methods for the safe and convenient storage, transport and application of compounds for inhibiting the ethylene response in plants, Google Patents.

Davies, P. 2013. *Plant hormones: physiology, biochemistry and molecular biology*, Springer Science and Business Media.

Davies, P. J. 2010. *The plant hormones: their nature, occurrence, and functions*, Springer.

Fan, X., J. P. Mattheis and S. Blankenship 1999. Development of apple superficial scald, soft scald, core flush, and greasiness is reduced by MCP. *Journal of Agricultural and Food Chemistry* 47(8): 3063-3068.

Jacobson, R. M. and Y. Zhen 2013. Oil formulations comprising cylcopropene compounds. *US 20130065764 A1.*

Kostansek, E. C. 2012. Oil formulations. *US 2012/0264606 A1*

Mir, N. 2014a. Situ mixing and application of hydrocolloid systems for pre- and post harvest use on agricultural crops. *US 8802140.*

Mir, N. 2014b. Hydrocolloid systems for reducing loss of volatile active compounds from their liquid formulations for pre- and post harvest use on agricultural crops. *US 8822382*

Mir, N. 2014c. Active compound formulation package and its subsequent release for use on plant and plant parts. US. *US20140326620 A1.*

Mir, N. 2015. In situ mixing and application of hydrocolloid systems for pre- and post harvest use on agricultural crops. *US 9005657*

Mir, N. 2016. Complexes of 1-methylcyclopropene with metal coordination polymer networks. *US 9394216 B2.*

Mir, N. A. and R. M. Beaudry 2000. *Use of 1-MCP to reduce the requirement for refrigeration in the storage of apple fruit.* IV International Conference on Post harvest Science 553.

Mir, N. A., E. Curell., N. Khan., M. Whitaker and R. M. Beaudry 2001. Harvest maturity, storage temperature, and 1-MCP application frequency alter firmness retention and chlorophyll fluorescence of Redchief Delicious' apples. *Journal of the American Society for Horticultural Science* 126(5): 618-624.

Nature Chemistry-Editorial 2016. Frameworks for commercial success. *Nat Chem.* 8(11): 987-987.

Rupasinghe, H., D. Murr, G. Paliyath and L. Skog 2000. Inhibitory effect of 1-MCP on ripening and superficial scald development in 'McIntosh'and 'Delicious' apples. *The Journal of Horticultural Science and Biotechnology* 75(3): 271-276.

Sarker, M. I. and L. S. Liu 2015. Boron complexes with gradual 1- Methylcyclopropene releasing capability. *US20150366212 A1.*

Serek, M., E. Sisler and M. Reid 1994. 1-Methylcyclopropene, a novel gaseous inhibitor of ethylene action, improves the life of fruits, cut flowers and potted plants. *Plant Bioregulators in Horticulture* 394: 337-346.

Serek, M., E. C. Sisler and M. S. Reid 1967. Effects of 1-MCP on the vase life and ethylene response of cut flowers. *Plant Growth Regulation* 16(1): 93-97.

Song, H., X. Chen, Z. Yan, M. Yu, R. Ma and G. Yu 2006. Process for synthesis of 1-sodium methyl cyclopropene and products thereof. China, Jiangsu Province Academy of Agriculture Sciences. *CN 1721420A.*

URQUHART, J. 2016.World's first commercial MOF keeps fruit fresh. *Chemistry World.*

Xin, H. 2012. Method of producing controlled release emulsion of 1-MCP for increasing produce yield *CN 102648709 A.*

Xin, H. 2012. Preparation method of 1-methylcyclopropene (MCP) slow-release emulsion for crop drought resistance and yield improvement China. *CN 102648709 A.*

Yoo, S. K. 2012. Method of preparing 1-methylcyclopropene and applying the same to plants. *US 8314051 B2.*

Yuan, R. and D. H. Carbaugh 2007. Effects of NAA, AVG, and 1-MCP on ethylene biosynthesis, preharvest fruit drop, fruit maturity, and quality of 'Golden Supreme'and 'Golden Delicious' apples. *Hort Science* 42(1): 101-105.

Zhang, Q., H. Jiang, H. Wang, P. Guo and X. Wei 2011. Preparation method and application of MCP and its derivatives inclusion. China, Huazhong Agricultural University. *CN102273504 A.*

Chapter 24

Re-engineering Post Harvest and Supply Chain Management of Apple in J&K: Challenges and Opportunities

M. A. Mir

Department of Food Technology,
Islamic University of Science and Technology, Awantipora-J&K
E-mail: hod.ft@islamicuniversity.edu.in

1. Introduction

Horticulture is important sector in the economy of J&K State. There are around seven lakh families comprising thirty three lakh people who are directly or indirectly associated with horticulture. In J&K horticulture is Rs 7,000 crore to 8,000 crore industry. The industry can reach to Rs 25,000 crore to 30,000 crore by implementing high density plantation in apples.

The state produces many kinds of temperate/subtropical fruits due to ecological niches in various agro-climatic zones. The total area under horticulture in J&K during the year 2015-2016 was 3, 37,677 hectares among which 2, 41,182 under fresh fruits and 9, 6,495 hectare under dry fruits. The total fruit production estimates were 24, 93,999 MT comprising of 22, 17,584MT fresh fruits and 2, 76,415 MT dry fruits. The area under apple cultivation was 16, 1, 773 hectare with production of 19, 66,417 MT. About 67 per cent of the area under fresh fruits is occupied by apples in J&K where 88.7 per cent of the total fresh fruit production comprises of apples (Anonymous, 2015).

India is world's fifth largest producer of apples. The state of J&K contributes (67.1 per cent) of apple produced in India followed by Himachal Pradesh (25.7 per cent), Uttarakhand, (5.7 per cent) and others (0.50 per cent). Due to WTO agreements, competition in apple trading has increased as markets are flooded with apples from china, USA, Australia, Russia, Chile, New Zealand, Italy, Iran and Afghanistan.

During April-September 2015 India's apple imports stood at USD 153.6 million. In January 2016 Govt. of India has relaxed apple import norms and has allowed in bound shipment of fruits through sea port and air ports in Kolkata, Chennai, Mumbai and Cochin.

Horticulture sector is gaining momentum due to various centrally sponsored focused programmes of Horticulture Technology Missions covering aspects of production, processing and post harvest management technology. The state of J&K has achieved Horticulture Excellence award during the year 2011 in adoption of Horticulture Technology Programmes successfully.

Keeping in view the contribution of apple for employment and revenue generation, appropriate pre-harvest and post harvest technologies are to be adopted to meet the challenges for sustainability of apple industry.

2. Post Harvest in Apple

(a) Post Harvest Physiology of Apple

Apple is a climacteric fruit and continues metabolic processes of respiration, transpiration, ripening and senescence changes even after harvest. These changes are governed by storage conditions *i.e.* temperature and relative humidity of the environment. Respiration is a temperature dependent catabolic process with a Q_{10} value of 2.4 in case of apples. The process uses stored food materials leading to their depletion and consequently loss of quality. Shelf life of apple is inversely proportional to temperature after harvest. Increase of 10°C temperature after harvest reduces shelf life by two fold. Apples if kept for three days in the orchard lower the storage life by one month.Apples lose moisture by transpiration process through lenticels. Rate of water loss depends on temperature and relative humidity of environment, water loss in apples cause shrinkage.

Fruit ripening is the transformation of physiologically mature fruit from an unfavourable state of firmness, colour, and taste/aroma to a more favourable state for consumption. Ethylene triggers the ripening change, it regulates the expression of several genes which modulate the activity of various enzymes involved in process of ripening. These enzymes are responsible for softening of apples (pectin esterase, polygalacturonase) and those acting on complex carbohydrates to make them simple.

The metabolic changes catalyzed by enzymes, which take place during ripening in apple fruit, are accelerated by temperature. Hence, higher the post harvest handling temperature, rapid is the process of ripening which leads to quicker senescence changes in apple quality manifested in the form of mealiness, loss of crispness, lack of juiciness, shrinkage and final tissue death by spoilage causing microflora. Red Delicious cultivar of apple, which forms major portion of apple produce, undergoes very fast ripening changes. With the result, the fruit loses its crispness and becomes mealy after short time of harvest compared to other cultivars, thus the market quality becomes inferior resulting in economic loss to growers.

(b) Post Harvest Management of Apple

The term post harvest management includes many management decisions and processes that are involved in harvesting, handling, washing, grading, packing, storage, transport, processing and marketing of fruit. Post harvest handling is concerned with maintaining, rather than improving quality of apples after harvest. The management tools used by the apple growers and traders, storage operators and packers are focused on the following aspects:

☆ Reducing metabolic rates of processes (respiration, transpiration, ripening and senescence) that result in undesirable changes in colour, composition, texture, flavour and nutritional status.

☆ Reducing water loss that can result in loss of marketable weight, shrinkage, softening and loss of crispness.

☆ Minimizing bruising and preventing the development of physiological and pathological disorders.

The main function of post harvest management system is to maintain quality of apples from harvest till it reaches the consumer.

(c) Traditional Practices of Apple Handling in J&K

In spite of rapid strides in fruit production in the state, there is inefficient post harvest management system of apples, seen in apple growing belts of the state. Commercial apple cultivars of late season are harvested from September onwards up to middle of November depending upon the altitude. Paddy harvest season also coincides with apple harvesting and growers have hectic schedule of activities to take care of paddy as well as apple harvest. Apple after harvest are kept in heaps in the orchard area and covered with tarpaulin to protect from bad weather. Subsequently, at leisure apple are manually graded, packed and loaded in trucks for export to distant markets located outside the state. Delhi is 800 Km away from Srinagar, and Mumbai is about 1600 Kms. Due to nonexistence of on-farm storage facility or refrigerated storage (cold store) in apple growing zones of the valley, the growers are forced to send their produce outside valley before December or before snow fall, for marketing or for temporary storage in cold stores located at

Delhi or in the adjoining areas of Punjab. It has been observed that there is a huge pressure on transport facilities for handling of the produce, with the result a delay of 30-35 days from harvest to dispatch of produce outside valley is inevitable. During this period, fruit remains under uncontrolled conditions of temperature and is subject to quality deterioration at a rapid rate. The farmer receives non-remunerative returns owing to the poor quality of produce. Every grower tries to send his produce outside valley as a result disequilibrium between demand and supply reduces price of apples (Figure 24.1).

Figure 24.1: Traditional Practice of Heaping.

(d) Post Harvest Losses in Apple

Loss means any change in availability, edibility, wholesomeness or quality of fruit that prevents it being consumed by people. There are pre-harvest losses and post harvest losses in apples.

(i) Pre-harvest Losses

Pre-harvest losses are on account of

- ☆ Hail injury
- ☆ Frost injury
- ☆ Untimely rains/draught
- ☆ Floods Branch/tree bruise
- ☆ Packing/harvesting bruise
- ☆ Very high temperature

☆ Fungal infestation (Scab, soothy blotch, powdery mildew, blight *etc.*)

☆ Insect infestation (Sanjose scale, red mite, wooly aphis)

(ii) Post Harvest Losses

Factors responsible for post harvest losses are:

☆ Lack of knowledge of apple maturity standards. Not harvesting apples at correct stage.

☆ Improper picking practices

☆ Bruising during picking, washing, sorting, grading, packing *etc.*

☆ Rough handling of produce dropping from height while emptying the pick baskets in the sorting area.

☆ Carelessness in transportation.

☆ Improper container for field use

☆ Improper packing and poor choice of packing material. In CFB boxes use of Paddy straw

☆ Long lapse between harvesting and storage of fruit

☆ Lack of on farm precooling and storage facilities due to small holding size of apple orchards.

(iii) Apple Bruising

Bruising is main quality defect of apples. It may occur during harvesting, grading, packaging, and transportation. There are three types of bruising encountered during handling of apple *viz.* impact bruising, compression bruising, and vibration bruising. Bruising breaks the cellular structure beneath the skin, exposing cells to intercellular air and to the oxidation processes. The injured areas turn brown in colour. It has been estimated that 40 per cent of bruising occurs in field, 40 per cent during grading and 20 per cent during transport.

☆ Bruising has undesirable effect on fruit appearance as it increases fruit shrinkage and may lead to rots by molds.

☆ Proper method should be adopted to hand pick apples from spur and putting them in picking baskets.

SKUAST-K Division of Post harvest Technology under All India Coordinated Research Project (AICRP) sanctioned by CIPHET (Ludhiana) on post harvest loss assessment of apples have concluded that post harvest loss in apple ranges from 20.07 to 21.46 per cent respectively enumerated by observation and enquiry method in the field studies. The losses are reflected in Table 24.1.

However, in view of the various loses in apple post harvest management practices should be improved as indicated in Table 24.2.

Table 24.1: Per cent Losses in Apples at Various Stages

Sl.No.	Stage	Per cent Loss (per cent)	
		By Enquiry Method	By Observation Method
1.	**Farmers level :**		
	Harvesting	6.65	4.70
	Collection	1.08	
	Sorting	1.75	4.70
	Packaging	0.48	2.15
	Transportation	0.57	2.39
	Storage	1.83	
2.	Wholesaler Level	0.60	1.37
3.	Retailer Level	8.50	4.76
	Total	21.46	**20.07**

Table 24.2: Major Causes of Quality Loss in Apple

Major Cause	Method to Reduce Loss
Bruising	1. Gentle harvesting and handling
	2. Protective packaging *i.e.* Net packing of individual apples
Rotting by fungi	1. Good sanitation
	2. Cool storage
	3. Use of fungicides
	4. Maintain intact skin
Senescence	1. Cool/CA storage
	2. Prompt marketing
	3. Processing in stable form
Wilting	1. Maintain high humidity in surrounding environment

(e) Post Harvest Activities in Apple

(i) Harvesting of Fruits

The generally accepted commercial practice is to hand pick the apples. It is important to know the appropriate harvest dates for apple varieties. Apple picked too early are susceptible to shrivel, scald, and bitter pit. They also may not ripe properly after harvest. Apples picked too late may be in the respiratory rise, which will decrease their shelf life and lead to disorders such as flesh browning and break down. The state grows both early, mid and late season commercial apple cultivars. Among late season cultivars, Red Delicious, Golden Delicious, Royal Delicious, American Apirogue, *Ambri* and *Maharaji* are commercially grown in traditional orchards (Table 24.3).

Table 24.3: Harvesting Season of Apples

Season	Varieties	Harvesting Time
Early	Benoni	2nd week of July- 4th week of July
	Irish Peach	
Mid-season	Cox's orange pippin	August –September
	Florina	
	Gala Must	
	Razakwar	
	Rome Beauty	
	Red Beauty	
	Red Chief	
	Royal Delicious	
	Red Delicious	
	Golden Delicious	
	Red Gold	
Late	Rich-a-red	November
	Yellow Newton	
	American Apirouge	
	Chamure	
	Baldwin	
	Ambri	
	White dotted Red	

(ii) Tests for Determining Maturity

Maturing indices are important for deciding when a given commodity should be harvested to provide some marketing flexibility and to ensure attainment of acceptable-edible quality to the consumer. The following guides are used for determining maturity of apples:

1. Ease of separation from the tree
2. Color of the fruit
3. Seed color
4. Sugar/soluble solids
5. Starch-iodine test
6. Acid levels
7. Flesh firmness

(iii) Determination of Fruit Firmness

(a) *Principle of the Test*

As the fruit ripens, the cementing material between the cells (pectin) is hydrolyzed by pectolytic enzymes and becomes soluble, due to which fruit softens. It is a common experience that a mature fruit is felt harder on pressing by fingers in the hand, as it ripens it becomes soft. In this test, hardness of the fruit is measured by an instrument called penetrometer or pressure tester (Figure 24.2).

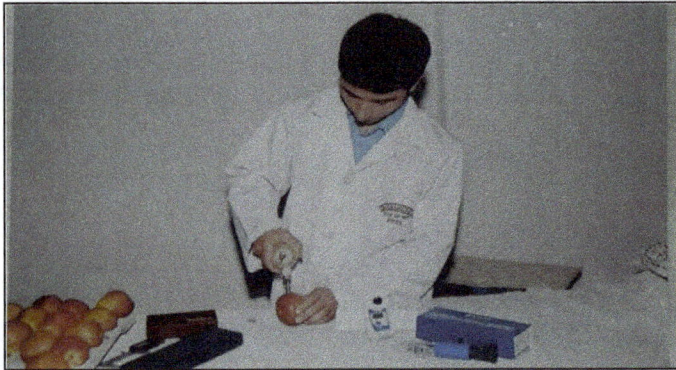

Fig 24.2. Determination of Fruit Firmness using Penetrometer.

(b) *Pressure Tester*

The Effegi model 327 pressure tester is provided with 2 types of plungers. Plunger with diameter of 5/16 inch is used for pear fruits while plunger with a diameter of 7/16 inch is used for apple pressure testing. It indicates fruit pressure in Ib./sq inch or in Kg./sq inch. The plunger is pressed against the flesh and when this gives way a pressure reading is given by the scale.

(c) *Test Procedure*

1. The apple is placed against a hard surface for testing and should never be hand held.

2. In use, thin circular patches of skin about the size of 25-paisa or 1.5 mm diameter are removed from the cheeks of fruit both the blushed and green sides. A potato peeler is often used since the depth of cut can influence the instrument. The cuts are made half way between the stem and calyx end of the apple.

3. The plunger is placed against the pared flesh and the handle pushed until the plunger enters the flesh upto the depth ring approximately 0.6 centimeters (1/4 inch) from tip in 2 seconds.

4. Fruit pressure is read directly from the scale on the tester.

(d) Precautions

1. Calibrate the instrument before test.
2. Test the fruit of the same size. Longer apples are softer then smaller ones, which give accurate comparisons.
3. Use proper plunger (7/16" for apples and 5/16" for pears)
4. Insert the plunger only as far as the inscribed line on the tip. Going beyond the line will give a higher reading.
5. Shaded fruits taste softer then exposed to light.
6. Cold fruits taste harder than warm samples.
7. Fruits under dry land conditions taste harder than that raised under irrigated conditions.
8. Avoid blushed side, which usually reads too hard.
9. Do not taste russetted and culled apples/pears as former usually read hard.
10. Test at least 12 fruits taken from several parts of the block for each variety.
11. Single fruit samples are not representative of the entire orchard.
12. Testing should be carried out in the orchard at the same time each day in morning hours.
13. Fruits picked or stored for several hours are not suitable for testing.
14. The most important aspect of testing is the speed of applying force to the instrument. Higher readings will result from applying pressure too fast and it is recommended that speed be regulated by counting "one thousand one and one thousand two" as the plunger is inserted in the fruit.

(iv) Starch-Iodine Test

Starch – iodine test is used as a field evaluation aid for apple maturity determination.

(a) Principle

The principle of the procedure is that iodine will give a blue black stain to the starch, which has been deposited during the vegetative growing period in the fruit. During the maturation process this starch is converted to sugar. This conversion occurs in the core area first and continues outward into the flesh. The pattern of starch disappearance is specific for each variety. For example in Red Delicious starch appears in a fairly even expanding ring, while Golden Delicious shows an uneven pattern.

(b) Preparation of Iodine Solution

An iodine solution can be prepared by mixing 15 g potassium iodine in 150 ml distilled water and dissolving 6 g iodine crystals in it. After all iodine has dissolved (overnight) the final solution is made to 1000 ml with distilled water.

(c) Test Procedure

1. Cut the apple at right angle to the core approximately half way from the stem to the calyx end.
2. Pour the iodine solution in a petri dish or a saucer.
3. Dip the cut surface of the apple with pedicle attached to it in the iodine solution for a minute.
4. Drain the excess solution and let the surface dry for a few minutes and blue-black pattern will become apparent.

The starch-to-sugar conversion occurs in a uniform manner during the maturation period. The rating system used in the Apple Maturity Programme is on scale of 1 to 6 as follows (Figure 24.3)

1. Full starch (complete blue-black starch)
2. Core area clear of starch
3. Clear through the area including vascular bundles
4. Half flesh (cortex) clear
5. Starch just under skin
6. Free of starch

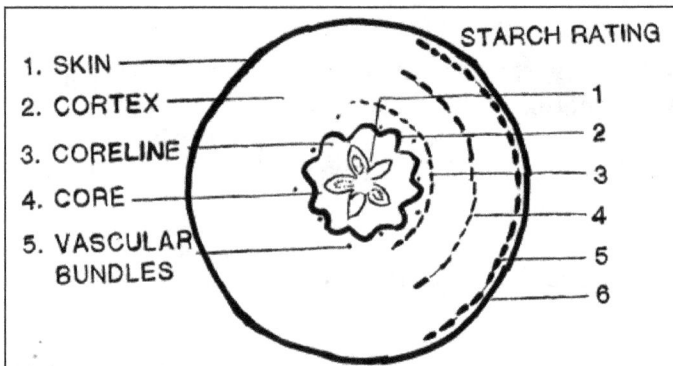

Figure 24.3: Cross Section of Apple.

Ideal starch index rating for harvesting Red Delicious cultivar of apples lies between 2.0 to 2.5 never greater than 3.0

(v) Sugar/Soluble Solids

Soluble solids are determined by refractometer using juice from apple. In apples, the level of sugar in expressed juice can give a useful assessment of maturity. Normally the fruit is squeezed and the sugar level is measured with a refractometer. This is very simple and quick and gives a direct reading in degrees brix, which is more correctly a measure of soluble solids. Since most soluble solids in apple are sugars, thus test gives a good indication of maturity status of apples.Refractometer is in the range of (0-32). Brix is used to measure total soluble solids.

(vi) Precooling of Apple

Precooling is first step in good temperature management. Delays at high temperatures between harvest and start of precooling are certain to increase deterioration. A delay of 3 days in the orchard or in a warm packing shed may shorten storage life as much as 30 days even if they are then stored at (-1°C).

Apples should be cooled as quickly as possible after harvest. A delay of 1 day at 21°C after harvest takes 7 to 10 days of potential storage life at 0°C. There should be rapid heat transfer from apples to a cooling medium from 20 minutes a less to 24 hours.A desirable goal for precooling apples should be core temperature of fruit in the centre of stacks to drop to 0°C to 0.6°C in 2-3 days.

(a) Advantages of Pre-cooling

1. Retards ethylene production.
2. Inhibits water loss.
3. Restricts enzymatic and respiratory activity
4. Inhibits growth of decay producing microorganism.
5. Reduces refrigeration load by removing field heat of apples.

3. Sorting/Grading

The separation based on single parameter, it may be size, shape or colour is called sorting while separation based on multiple properties *i.e.* size, shape and colour blemishes is called grading. Table 24.4 represents the revised grade of Red Delicious variety of apple based on size, colour and surface blemishes.

(a) Principles of Sorting and Grading

1. Out grading may be done in field to remove small, large and blemished fruit.
2. Understanding market requirements.
3. Uniformity in color, size, weight and stage of maturity.
4. Skilled and experienced staff should be well aware of quality characteristics of different varieties.
5. Quality defects such as hail damage, bruise marks, sooty blotch, fly speck, russetting, scab, Sanjose scale, unhealed crakes, sun burn, shriveling and skin puncture reduce the apple appearance and storage life.
6. Proper care during grading should be exercised along with size and color criterion in apple.

Table 24.4: Revised Grade of Quality of Red Delicious Variety of Apple

Grade Designation	Definition of Quality Special Requirements					
	Defects					
	Skin Defects					
	Size Diameter per Fruit in mm	Colour Development per Fruit per cent of Surface Area (minimum)	Blemish Diameter per Fruit in mm (maximum)	Russetting per Fruit per cent of Surface Area (maximum)	Other Skin Defects per Fruit Aggregate Diameter in mm (maximum)	Flesh Defects not to Exceed per cent of total Fruits per Package
1	2	3	4	5	6	7
Super large*	81 and above	65.0	0.0	Permitted in stem cavity	0.0	0.0
Super medium	71-80 mm	65.0	0.0	-do-	0.0	0.0
Super small	55-70 mm	65.0	0.0	-do-	0.0	0.0
Special large**	81 mm and above	45.0	5.0	Maximum surface area permitted 10 per cent	5 mm	Slight incidence tolerated upto 2 per cent
Special medium	71-80 mm	45.0	5.0		5 mm	-do-
Special small	55-70 mm	45.0	5.0		10 mm	-do-
Fancy large	81 mm and above	30.0	10.0		10 mm	Fruits with externally visible patches tolerated upto 5 per cent
Fancy medium	71-80 mm	30.0	10.0		10 mm	
Fancy small	55-70 mm	30.0	10.0		10 mm	
Processing grade	55 mm and above	Less than 30 per cent or greater than 90 per cent	–	–	–	–

(i) General Requirements (Common for all grades)

☆ The apple shall be fruit obtained from the plant botanically known as *Malus domestica* of family Rosacease

☆ Be carefully handpicked, sound, clean, firm, intact, fairly well formed and developed.

☆ Be reasonably uniform in shape, size, colour and have taste, texture and flavor characteristics of the variety, slight abnormal shape not effecting appearance of the fruit shall be allowed.

☆ Have reached the stages of development which will ensure the proper completion of repining process be free from skin punctures, shriveling bacterial and fungal disease, sanjose scale. Off flavor and or any damage that appreciably effect the appearance and quality of fruit except to the extent specified under the special requirements.

(b) Advantages of Sorting and Grading

1. Fruits properly graded according to well prescribed standards will help the growers to get premium price in competitive market.

2. The growers are assured of stable market for their produce in time when product is in over supply.

3. Buyers get maximum joy and satisfaction in purchasing the graded fruits of uniform quality.

4. Growers can save cost of packaging, transportation/storage charges in not sending low-grade fruits outside valley.

5. Fruits properly graded will fit well in box/tray pack and shall not suffer transit damage.

6. Grading builds reputation of growers and confidence for repeat transaction by buyers.

7. Nowadays in pack house mechanical grading is done where apples are graded on size/weight basis and laser colour sensor are used for color sorting.

(c) Packaging

Traditionally apples are packed in wooden shocks cushioned with paddy straw and newspaper as liners which are being replaced by corrugated fibre boxes. By various intervention approaches more than 60 per cent of apples are now a days packed in CFB (Figure 24.4). The advantages of using CFB for apple packing are:

1. The boxes are light in weight

2. They cause less bruising damage to apples.

3. They are attractive, easy to handle and print.

4. They reduce freight cost.

(i)   (ii)

Figure 24.4: Packaging of Apple (i) Corrugated Fiber Board; (ii) Net Packing.

5. They are made from cheaper wood and cellulose waste and thus are recyclable.

6. They are free from fungal infections

7. They can be easily palletized.

8. They bring good returns to growers.

4. Shelf Life Extension

The following studies on various post harvest treatments on apple have been conducted at SKUAST-K Division of post harvest technology

(a) Use of Wax Coating Materials

Post harvest application of wax coating materials Stay-fresh, Sta-fresh-960 (1:1) and virosil-agro (2.5 per cent) for 10 min proved to be effective in reducing starch hydrolysis, physiological weight loss and texture breakdown in Red Delicious apples. The treated fruits remained in fair to good quality up to 65 days compared to 45 days in control lot. The post harvest treatments of wax coatings helped to maintain the quality attributes over a span of +20 days under ambient conditions (3.2-18.8°C and RH 53.5-80 per cent) without deleterious effect on sensory quality (Mir *et al.,* 2004). Standardization of lac based wax coating materials revealed that $SHOO_3$ formulation when applied at 10 per cent was superior in retaining all quality attributes *viz.* fruit firmness (Ibs/sq. inch), juice yield (per cent), TSS, total titrable acidity (per cent) of apples under both the conditions of ambient and refrigerated storage. Shelf life of Red Delicious fruits treated with $SHOO_3$ at 10 per cent formulation was increased by 11 per cent and 9 per cent under refrigerated (180) days and ambient (90) days conditions of storage, respectively. Whereas, shelf life of Ambri fruits treated with same wax formulation increased by 10 per cent and 6.8 per cent under refrigerated (180) days and ambient (90) days conditions of storage, respectively.

(b) Post Harvest Calcium Chloride Treatment

The calcium content of apples influences fruit firmness and storage life. Studies conducted have revealed that post harvest calcium treatment (4 per cent calcium chloride dip for 5 mins.) helped to retain fruits firmness of Red Delicious apples 14lb/inch during ambient storage of 60 days and the firmness values of 11.8 lb recorded after 110 days of refrigerate storage (1°C). The calcium chloride treated fruits had lesser physiological loss in weight during storage.

(c) Irradiation Technology for Shelf Life Extension of Apple

Four commercial varieties of apple *viz.,* Red Delicious, Golden Delicious, Royal Delicious and *Ambri* harvested at proper maturity stage were subjected to Gamma irradiation doses in the range of 0.1 to 0.5 KGy (Hussian *et al.,* 2008).The highlights of the study were as under:

1. The fruit response to irradiation was cultivar dependent.
2. Gamma irradiation doses of 0.2, 0.3, 0.4, and 0.5 KGy significantly helped in overall quality retention of Ambri, Golden Delicious, Royal Delicious and Red Delicious apples under both ambient and refrigerated storage conditions.
3. Based on the weight loss during storage, the above doses can be applied to extend the shelf life of the tested apple varieties by about a period of (30) days under ambient storage conditions and a (90) days extension in shelf life is achieved when the above irradiation doses are applied in combination with refrigerated storage.
4. The post harvest wax coating dip treatment of apples (lac based SH-002 10 per cent for 5 minutes) in combination with Gamma irradiation doses of 0.4 KGy or 0.3 KGy applied to *Ambri,* Golden Delicious, Royal Delicious and Red Delicious extended the shelf life of apples upto (60) days under ambient conditions without affecting the other quality parameters.
5. The irradiation treatments significantly reduced the yeast and mould count of the apple varieties from hygienic point of view. The minimum yeast count of 2×10^3 CFU/g of sample was recorded in case of apple varieties treated with 0.5 KGy, whereas maximum yeast and mould count of $23 \times 10^3, 25 \times 10^3, 20 \times 10^3$ and 18×10^3 CFU/g sample was recorded in case of control samples of Ambri, Golden Delicious, Royal, Delicious and Red Delicious apples after (90) days of storage under ambient conditions, respectively.

(d) Use of 1-Methyl Cyclopropene(1-MCP) Application in Apple

Application of 1-MCP as Smart Fresh Technology is the cheapest means for maintaining the quality of apples. 1- MCP is an innovative product that blocks the action of ethylene in plants and harvest fruits. Its mode of action is *via* a preferential attachment to the ethylene receptor, thereby blocking the effects of ethylene from both internal and external sources. Treatment of fruits with 1-MCP is a simple technique that does not require sophisticated instruments and can be adopted by growers in remote areas without any difficulty. The work was initiated with overall objective of enhancing quality of fresh Red delicious apples to the consumers by prevention of mealiness or textural breakdown under APEDA sponsored project.

Physiologically mature Red Delicious apples harvested after 158 days from full bloom were treated with 1-MCP SmartFresh™(1ppm) for 24h under small scale airtight plastic tent at 14±2°C. The treated fruits along with untreated fruits were subsequently stored under two storage conditions *i.e.* non-refrigerated (12±2°C) and refrigerated (0-2°C) conditions for monitoring quality changes. The studies indicated (Tables 24.5–24.7) that quality attributes like fruit firmness, juiciness and TSS and acidity were retained in treated samples up to 110 days while as to untreated apples retained the same quality parameters up to 30 days, thus the shelf life of treated apples was increased by 80 days under non-refrigerated conditions (12±2°C). Similarly, under refrigerated conditions 1-MCP treated fruits retained maximum quality attributes up to 270 days, while as untreated fruits retained quality attributes up to 180 days thus storage life was increased by 90 days. Post cold storage deterioration in 1-MCP treated apples was very slow as compared to untreated fruits. Maximum quality parameters in 1-MCP treated apples were retained up to 14 days of ambient storage (22±2°C) after refrigerated storage of 270 days, while untreated fruits retained quality parameters up to five days (Mir *et al.*, 2013).

Table 24.5: Effect of 1-MCP on Quality of Red Delicious Apple Stored under Non-refrigerated Conditions of Storage

Treatment	Fruit Firmness (lbs.)				
	Days after Storage				
	0	35	80	110	Mean
1-MCP	16.85	16.37	14.88	14.26	15.59
Control	16.85	10.60	9.27	8.81	11.38
LSD at 5 per cent	NS	0.393	0.382	0.403	
S.E.	0.366	0.386	0.366	0.407	
Treatment	Juice Yield (per cent)				
	Days after Storage				
	0	35	80	110	Mean
1-MCP	69.0	62.00	61.00	56.00	62.00
Control	69.0	53.00	43.00	41.00	51.50
LSD at 5 per cent	NS	0.360	0.510	0.510	
S.E.	0.65	0.325	0.65	0.651	

Treatment	Total Soluble Solid (°Brix)				
	Days after Storage				
	0	35	80	110	Mean
1-MCP	10.40	11.30	11.70	11.20	11.15
Control	10.40	12.10	11.00	9.80	10.82
LSD at 5 per cent	NS	0.393	0.403	0.403	
S.E.	0.386	0.386	0.407	0.407	
Treatment	Titrable Acidity (per cent as malic acid)				
	Days after Storage				
	0	35	80	110	Mean
1-MCP	0.217	0.198	0.167	0.147	0.18
Control	0.217	0.178	0.138	0.109	0.16
LSD at 5 per cent	NS	NS	NS	NS	
S.E.	0.400	0.400	0.400	0.400	
Treatment	Starch Rating (1-6 point scale)				
	Days after Storage				
	0	35	80	110	Mean
1-MCP	2.77	5.20	6.00	6.00	5.24
Control	2.77	6.00	6.00	6.00	5.47
LSD at 5 per cent	NS	0.400	NS	NS	0.399
S.E.	0.402	0.402	0.401	0.401	0.399
Treatment	Physiological Weight Loss (per cent)				
	Days after Storage				
	0	35	80	110	Mean
1-MCP	0.69	1.08	1.61	1.72	1.85
Control	0.69	0.95	1.44	1.62	3.51

Treatment	CPLW (per cent)	Spoilage (per cent)
	Up to 110 days	Up to 110 days
1-MCP	6.59	0.0
Control	7.99	20.0
LSD at 5 per cent	NS	0.406
S.E.	0.388	

NS: Not Significant.

(i) Effect of 1-MCP on CA Stored Apple

Red Delicious apples treated with 1-MCP (1PPM) and stored under CA condition, along with control lot for period of (225) days. Data presented in Table 24.8 reveals that 1-MCP treatment had significant effect on retention of all quality parameters during (225) days of C.A storage. 1-MCP treated fruits stored under 26±1°C RH 76 per cent during post C.A storage maintained crispness up to 12 days at ambient storage, whereas, untreated fruits were mealy after 4 days of CA

Table 24.6: Effect of 1-MCP on Quality of Red Delicious Apples Stored under Refrigerated Conditions of Storage

Treatment	Fruit Firmness (lbs.)				
	Days after Storage				
	0	90	180	270	Mean
1-MCP	16.85	14.95	14.75	14.51	15.26
Control	16.85	13.68	12.45	9.70	13.70
LSD at 5 per cent	NS	0.39	0.41	0.44	0.41
S.E.	0.366	0.38			
Treatment	Juice Yield (per cent)				
	Days after Storage				
	0	90	180	270	Mean
1-MCP	69.0	58.00	56.00	54.00	59.25
Control	69.0	54.00	48.00	42.00	53.25
LSD at 5 per cent	NS	0.38	0.42	0.45	0.39
Treatment	Total Soluble Solid ($^\circ$Brix)				
	Days after Storage				
	0	90	180	270	Mean
1-MCP	10.40	10.80	11.20	11.80	11.05
Control	10.40	13.60	12.40	11.40	11.95
LSD at 5 per cent	NS	NS	NS	NS	NS
Treatmet	Titrable Acidity (per cent as malic acid)				
	Days after Storage				
	0	90	180	270	Mean
1-MCP	0.217	0.167	0.161	0.154	0.174
Control	0.217	0.117	0.112	0.102	0.137
LSD at 5 per cent	NS	0.34	0.34	0.34	0.33
Treatment	Starch Rating (1-6 point scale)				
	Days after Storage				
	0	90	180	270	Mean
1-MCP	2.77	6.0	6.0	6.0	5.19
Control	2.77	6.0	6.0	6.0	5.19
LSD at 5 per cent	NS	NS	NS	NS	NS
Treatment	Physiological Weight Loss (per cent)				
	Days after Storage				
	50	100	200	270	
1-MCP	0.36	0.44	0.94	1.27	
Control	0.47	0.58	0.98	1.29	

Treatment	CPLW (per cent)	Spoilage (per cent)
	Up to 110 days	Up to 110 days
1-MCP	4.81	4.0
Control	5.19	14.0

NS: Not Significant.

Table 24.7: Effect of 1-MCP on Quality of Red Delicious Apples during Post Cold Storage (22-25°C)

Treatment	Fruit Firmness (lbs/sq.inch)						
	Days after Storage						
	90+7	90+14	180+7	180+14	270+7	270+14	Mean
1-MCP	14.75	13.93	14.09	13.25	14.08	12.92	13.83
Control	12.50	11.31	11.90	Mealy	Mealy	Mealy	10.45
LSD at 5 per cent	0.39	0.39	0.40	0.43	0.45	0.42	0.43

Treatment	Juice Yield (per cent)						
	Days after Storage						
	90+7	90+14	180+7	180+14	270+7	270+14	Mean
1-MCP	54.00	54.00	54.00	52.00	53.00	52.00	53.16
Control	50.00	46.00	48.00	Mealy	Mealy	Mealy	

Treatment	Total Soluble Solids (°Brix)						
	Days after Storage						
	90+7	90+7	90+7	90+7	90+7	90+7	Mean
1-MCP	11.80	11.80	10.80	11.00	10.60	10.80	11.13
Control	11.60	12.80	11.60	11.20	11.00	11.40	11.60

Treatment	Titrable Acidity (per cent as malic acid)						
	Days after Storage						
	90+7	90+7	90+7	90+7	90+7	90+7	Mean
1-MCP	0.156	0.147	0.152	0.144	0.150	0.134	0.147
Control	0.154	0.137	0.136	0.124	0.112	0.108	0.13

NS: Not Significant.

storage. The treated fruits tried second higher sensory scored after 12 days of storage (Mir, 2009).

(e) Controlled Atmosphere Storage of Apple

Apple is the predominant horticultural commodity stored under CA conditions. The objective of CA storage is to lower oxygen and increase carbon dioxide concentrations in storage atmosphere to levels that will maintain fruit quality by decreasing respiratory metabolism and reducing ethylene production and action, but not to the levels that induce fermentation or other damaging events. Post harvest Technology (PHT) Division of SKUAST-K has for the first time in the

Table 24.8: Effect of 1-MCP on Quality of Red Delicious Apples Stored under CA Conditions

Sl.No.	Quality Parameter	Days after Storage	Control	1-MCP Treated
1	Fruit Firmness	0	12.92	12.92
	(lbs./sq. inch)	225	8.90	12.20
		225^{+4*}	7.60	9.80
		225^{+8*}	7.50	8.50
		225^{+12*}	6.00	7.80
2	TSS (per cent)	0	12.8	12.8
		225	14.4	17.0
		225^{+4*}	13.8	16.7
		225^{+8*}	13.5	16.0
		225^{+12*}	12.7	15.5
3	Acidity (per cent)	0	0.27	0.27
		225	0.18	0.24
		225^{+4*}	0.15	0.21
		225^{+8*}	0.13	0.18
		225^{+12*}	0.11	0.16
4	Juice	0	57.0	57.0
	Yield (per cent)	225	50.0	55.0
		225^{+4*}	45.0	51.0
		225^{+8*}	mealy	49.0
		225^{+12*}	mealy	47.0
5	Rot (per cent)	225^{+12*}	40	25.0
6	CPLW (per cent)	225^{+12*}	7.60	6.25

* After CA Storage.

country developed know-how for controlled atmosphere storage of apples. The experimental CA storage facility has been established under World Bank aided project at SKUAST-K Shalimar, which was operational since September 1985 (Anonymous, 1988).

(i) CA Storage of Apple var. Red Delicious

Red Delicious, the commercial of apples is a poor keeper and undergoes rapid softening under ambient storage conditions. Storage studies conducted during two years period (1985-87)indicated that this variety can successfully be stored upto 231 days (about 8 months) under CA conditions at (1±0.5°C) temperature with gaseous regime of CO_2(1 per cent) and O_2 (5 per cent). The fruits maintained orchard freshness and superior quality attributes over the prolonged period. The fruits stored under ambient conditions lost fruit firmness within 60 days of storage.

Table 24.9: Qualitative Analysis of Red Delicious Apple after 231 Days

Parameters	At Harvest	C.A	R.A	Ambient	CD at 5 per cent
Pressure (lbs)	17.13	12.37	9.4	5.11	0.322
T.S.S (. per cent)	14.15	16.62	14.89	16.46	NS
Acidity (per cent)	0.240	0.211	0.134	0.168	0.002
Vitamin C (mg/100g)FW	6.00	6.004	3.912	2.08	0.1240
Organoleptic evaluation (1 to 9 rating)		8.16	6.23	2.70	1.99
Remarks			**Developed off flavour in March**	**Developed Mealiness in November**	

(ii) C.A Storage of Apples (Varietal mix)

During study period of 1987-89 CA storage evaluation studies were extended to Ambri Cv. Stored alone or in combination with Red Delicious, Maharaji and American Apirouge CVs at 1.0±0.5°C with oxygen content (3-5 per cent)and CO_2 (1 per cent) in the storage chambers. The studies indicated that Red Delicious can successfully stored in combination with Ambri or American Apirouge cultivars at (1.0±0.50°C) at this temperature Maharaji variety developed low temperature injury which was mitigated when temperature or Maharaji cultivar was raised to (1.5±0.5°C).

Table 24.10: Quality Parameters of some Commercial Apple Cultivars after 238 Days of Storage in CA and Ref. Conditions (1988-89)

Test Variety	Initial Quality Values			CA Storage			RA storage		
	Fruit Firm-ness (lbs)	TSS (per cent)	Acidity (per cent)	Fruit Firm-ness (lbs)	TSS (per cent)	Acidity (per cent)	Fruit Firm-ness (lbs)	TSS (per cent)	Acidity (per cent)
Red Delicious	14.86	12.52	0.161	13.99	1.08	0.084	12.96	11.30	0.083
American Apirouge	19.33	12.30	0.301	16.03	11.3	0.088	14.49	10.92	0.071
Maharaji	21.00	13.60	0.790	13.25	906	0.247	12.46	8.66	0.237
Ambri	15.33	12.36	0.140	15.00	14.0	0.129	13.16	12.70	0.118

CA storage temperature (1.0±0.5°C), CO_2 (1 per cent) and O_2 (3-5 per cent).

The comparative evaluation of various storage systems for prolongation of storage life of commercial apples cultivars have explored the possibility of regulation market supplies, wherein distress sales in the event of gluts can be reduced, consumers can be assured of good quality fruit over a longer span of time, and farmers can expect to receive better returns for their produce. Commercialization of CA Technology on co-operative basis can uplift the rural economy of state.

5. Supply Chain Management

J&K Govt. has established separate Department of Horticulture Planning and Marketing for development of efficient marketing system, facilitating reasonable price to producer and fair price to consumers. The department has three terminal markets *viz.* Sopore Fruit Market, Parimpora Fruit Market at Srinagar and *Narwal Fruit Market* at Jammu. The department has established 17 satellite markets located in each district. The department provides market intelligence through print and electronic media.

(a) Cold Chain

The fruit quality after harvest is subjected to deterioration in quality depending on time and the environment conditions,poor handling practices by any one in custody of goods can have a damaging effect on fruit quality value. Cold chain refers to the procurement, warehousing, transportation and retailing of food products under controlled temperatures. Cold chain is a supply chain based business model, relying on temperature controlled logistics.The cold supply chain assures the physical and quality parameters of perishable goods, in course of their passage from producer to consumer. Figure 24.5 depicts cold chain vision for quality apple marketing.

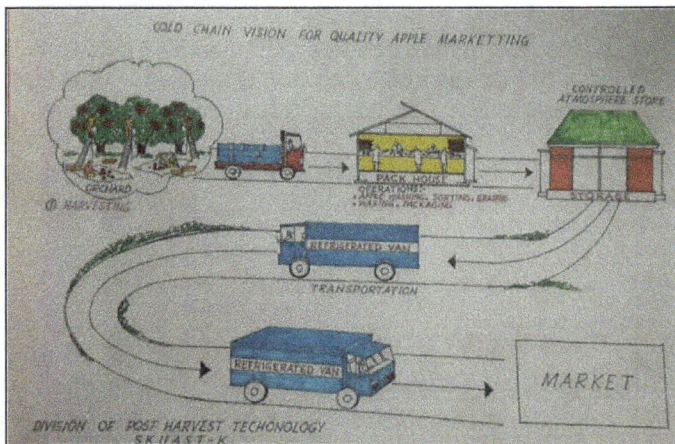

Figure 24.5: Cold Chain Vision for Quality Apple Marketing.

To benefit from cold chain, the supply chain operator needs to prepare the harvested produce for onward travel from farm gate to market. Integrated pack house is the initiator of cold chain. It is the nervecentre where multiple supply lines are decided, prepared, for and triggered.The function of pack house involves washing, sorting, grading, retail packaging and labeling, pre cooling, climate controlled storage and transport. These procedures make it possible for fresh produce to be supplied more efficiently.

(b) Integrated Cold Chain for Apple in J&K

The success of storage of apples under experimental CA conditions at SKUAST –K (J&K), has proved beneficial in motivating the policy planners and entrepreneurs to set up CA storage units in Kashmir on commercial scale. Liberal funding has been granted by Govt. of India through APEDA, NHB and MOFPI as a result a chain of integrated CA storage units have been established or under construction at Industrial Estate Rangreth and Industrial Growth Center Lassipora(Pulwama) J&K State (Table 24.11).

Table 24.11: Controlled Atmosphere Facility in J&K

Sl.No.	C.A Facility	Location	Capacity
1.	M/S FIL Industries Ltd (Kohinoor)	Industrial Estates Rangreth	5000 MT
2.	M/S Fruit Master Agro Fresh	Food Park Lassipora	5000 MT
3.	M/S Harshena Naturals	Food Park Lassipora	5000 MT
4.	M/S Shaheen Agro	Food Park Lassipora	8000MT
5.	M/S Kashmir Premier Apples	Food Park Lassipora	5000 MT
6.	M/S Valley Fresh	Food Park Lassipora	5000MT
7.	M/S Queewah Group	Food Park Lassipora	5000 MT
8.	M/S Golden Apple	Food Park Lassipora	3000MT
9.	M/S Chowdhary CA Store	Industrial Estate Zainakote	5000 MT
10	M/S H.N Agri Serve Pvt.Ltd	Food Park Lassipora	5000MT
11	Shopian Orchard	Food Park Lassipora	3000MT
12	Nedous	Food Park Lassipora	8000MT
13	Say Infra and Storage Pvt Ltd	Food Park Lassipora	3000MT
14	Peaks Agro warehousing Pvt.Ltd	Food Park Lassipora	5000MT
15	Alamdar Cool Store	Food Park Lassipora	5000MT
16	Mir Agro Industries	Food Park Lassipora	5000MT
17	Kashmir Fruit Preserver	Food Park Lassipora	5000MT
18	Green Tree	Food Park Lassipora	5000MT

In these integrated units Hi-Tech systems for washing, sorting, grading, packing, CA storage and refrigerated transportation of apples are ensured.

(i) Process Technology

The process involves collection of the harvested apples in plastic crates from orchards and their transportation to pack house. In the pack house apples are washed, sorted, graded, waxed, dried packed and stored in CA chambers under controlled oxygen/carbon dioxide and low temperature conditions.

(ii) Machinery Required

The washing, sorting, grading and waxing line (Figure 24.6) consists of dumping tank, sorting table, cull belt reject conveyor, washer, sponger, singulator grader, wax application machine and drier.

Figure 24.6: Internal View of Pack House for washing, Grading, Waxing and Packing of Fruit and Vegetables (*Division of Post harvest Technology, SKUAST-K Shalimar*).

The functional details of various units are as under:

(i) Dumping tank: the fruits after reception form field are put in to the dumping tank fitted with water pump for circulation of water inside the tank. The apples float on water and are elevated to inspection unit.

(ii) Sorting table: it consists of PVC rollers which are of food graded material and the contact parts made of stainless steel and structure of mild steel. Rollers move on their own axis and at the same time shall move forward.

(iii) Cull belt/Reject conveyor: A conveyor belt is placed on the side of shorting table so that the worker can sort out the rotten or damaged fruit from the sorting table and place on this conveyor. The belt of this conveyor is in the opposite direction of the sorting table so that the rotten fruits placed over the cull belt conveyor be carried backward and collected in plastic crates at the end. The belt is of food grade material.

(iv) Washer: The washer is fitted with number of Nylon bristle rollers rotating on their own axis and the fruits are fed on to the bristle rollers from the sorting table. Sprinkler arrangement is provided on the top of moving fruits. Water is sprayed on the fruit while they move forward slowly. The bristle roller gently rub the surface of fruits and clean it properly. The dirty water is accumulated on the water tank fitted below the machine and discharged out continuously. The washing process is automatic and the machine is connected with a constant water supply point.

(v) **Sponger:** The fruits after washing in the washer are fed on to the rollers made of high sponge material. These rotating sponge rollers absorb the water from the surface of fruits. Squeezing rollers are provided in between the sponge rollers to squeeze water from them.

(vi) **Singulator:** It is equipped with PVC coated food grade material rubberized belt. From the inclined roller conveyor, fruits are directed to the horizontal belt. Conveyor for proper singulation of the fruits before feeding to the grader.

(vii) **Grader:** Grader is fitted to grade the apples on the basis of size. Rollers and other contact parts of the grade are made of food grade material.

(viii) **Wax application machine:** This machine fulfills the purpose of waxing the fruits. Hair rollers are provided with this machine which polishes the fruits after the application of liquid edible wax. The wax storage tank made of acrylic sheets is also provided. All the contact parts are to be fabricated in stainless sheet. The wax in the form of liquid is to be sprayed by wax spraying gun from the top on the moving and tumbling fruits on specially designed rollers with the help of jets and suitably designed pump with adjustment to control the flow of wax. Bristle rollers are provided for the uniform application of wax on the fruits.

(ix) **Dryer/drying unit:** It is provided for drying of waxed fruit. The dryer comprises of PVC rollers on which the fruits are conveyed after being waxed. It is fitted with electric heaters and infra lamps. Drying of fruits occur in the dryer by hot air at controlled temperature of 30-50°C. It is provided with air curtains, hot air circulation fan, electrical heaters, IR lamp, temperature controller, thermostat *etc.* the contact parts are made of stainless steel.

(c) Socio-economic Benefits of Cold Chain

The introduction of cold chain by entrepreneurs has revolutionized the apple trading concept among Kashmir fruit growers. The following mentioned benefits have been observed by adoption of cold chain technology in J&K state.

☆ Apple after picking are washed, sorted, graded, waxed and stored under CA conditions up to 6 (Six) months or more till market prices are favorable.

☆ There is no mad rush for sending apples outside state during peak seasons, which has helped to control the price line in favour of producers.

☆ The quality of apples stored under CA conditions is superb with respect to texture, Juiciness/sweetness attributes.

☆ Apples are packed in attractive packs after machine grading which helps in better sale proceeds.

☆ The member growers can avail credit facilities from CA management and production related inputs and advice.

(d) Protocols for Scientific Post Harvest Management of Apples for Marketing

1. Harvest apples at proper stage of maturity.

2. Test quality of your apples by use of starch-iodine test, fruit pressure tester and refractometer for making harvest maturity decisions.

3. Train your harvest crew properly before start of harvesting. Adopt correct technique of fruit picking.

4. Gently pluck the fruit from the tree, avoid damage to fruit and future buds.

5. Use plastic cushioned pick baskets for apple picking.

6. Do not overload the pick baskets.

7. Make arrangements of shading the fruit collection spots before harvest in the field.

8. Collect the harvest produce under shade in plastic crates.

9. Allow pre-cooling of fruit and avoid direct exposure of apples to the sun.

10. Do not make big heaps and cover apples with tarpaulin or plastic sheets in the field to avoid generation of heat of respiration.

11. Discard fruits with disease spots, puncture marks and insect attack.

12. Arrange packaging material well in advance of harvest season.

13. Grade your apples properly to obtain premium price in the market.

14. Avoid topping practices. Do not mix grades.

15. Switch on to newer attractive eco-friendly packaging material *i.e.* CFB boxes and molded trays.

16. Avoid use of paddy straw in CFB boxes. Use trays.

17. Try packing in small gift packs which fetch better price in the local market.

18. Take care while fixing nails on wooden boxes.

19. Do not overload boxes in the trucks.

20. Dispatch fruit to different markets as per demand of market to avoid glut and low prices in fruit *mandies*.

21. Try your best to minimize the time lapse between harvest and cold storage of fruit.

22. Avoid bruising of tender fruit during harvesting, grading, packing, storage and transportation of fruit.

23. Store your apples in the farm storage structure till favorable market information is received.

24. During home storage make inspection of the fruits regularly and discard rotten and diseased fruits.

25. You can store apples for more than six months in controlled atmosphere stores located at Industrial Estate Rangreth, or at Industrial Estate

Lassipora (Pulwama). Make contacts in advance with the firms for booking the storage space.

26. Avail the market information from Radio and TV provided regularly by Directorate of Horticulture Planning and Marketing J&K Govt. regarding dispatches, arrivals and prevailing rates of apple varieties in different fruit markets of India.

27. Collect unmarketable low grade fruits in plastic crates and deliver it nearby to community processing centers for making jams/juice *etc.*

28. Prepare jam and juice from unmarketable fruits in your home for your family to avoid wastage.

29. Try to brand your produce for better marketing.

30. Growers can establish "Fruit Pack Houses" for scientific washing, waxing, grading, packing and cold storage/CA storage of apples/other fruits on cooperative basis.

6. Challenges and Opportunities

(a) Low Productivity and Poor Quality

The apple productivity in J&K is about 10-13 tons/hectare, which is much lower than the yield of advanced countries like Belgium (46.22 tons/ha), Denmark (41.87 tons/ha), Netherlands (40.40 tons/ha). The low productivity has been attributed to various factors *viz.* monoculture of few varieties, climatic constraints, poor quality planting material, lack of pollinizers, poor canopy management, low planting density, faulty plant nutrition, poor soil and water management, incidence of pests/diseases (Banday, *et al.,* 2010).For improving productivity and quality of apples Govt. has launched High Density Apple orchard scheme in J&K in Feb 2017 and declared the 2017 year as "Apple Year". Govt. will provide 50 per cent subsidy for high density apple plantation to the orchardists.

(b) Transportation

J&K is hilly state and the orchard sites are located mostly on uneven topography. The orchards are not linked by metalized lanes/roads. After harvesting from the orchard, fruit is to be shifted to collection centers and lot of mechanical bruising occurs to fruit.

1. There is no State transport policy for fruit shipment in the State.During peak seasons growers face lot of problems, due to shortage of sufficient vehicles. As a result extra freight-charges are claimed by transporters.

2. There is overloading of trucks with fruit boxes, and is absence of any traffic checks, the quality damage is encountered in fruit due to traffic jams for 4-5 days or more.

3. The Jammu-Srinagar National Highway (NH-1A) is not operational throughout year due to landslides as result trucks loaded with fruit remain

stranded days to gather. This type of problem is quite experienced during winter months, due to avalanches/snowfall in the region. As a result of which perishable fruits worth huge amount are lost each year.

(c) Utilization of Low Grade Apple

About 30 per cent of the apples produced in the state are of low grade which can be processed for value addition, rather than for direct sale. This requires processing capacity of 3 lakh tones per annum. The state has processing capacity of only 70,000 tons which is not adequate.

(d) Employment Opportunities in Post Harvest Management

Apple production can prove a sustainable enterprise, provided production technology and post harvest technology/management are given equal priorities in policy planning decisions. Ample self-employment opportunities in the field of post harvest management of apple do exist, some of the areas are as under.

(i) Establishment of Integrated Pack House Facilities

Educated unemployed youth can avail the central sponsored incentive schemes for establishing Cold-Chain units. There is enough demand of storage space for of apples in the state. A large number of skilled manpower can get Jobs in these units as supervisors, grading, packing or storage operators and supply chain managers.

(ii) Establishment of CFB Making Units

Apples are to be packed in CFB boxes to compete in the national/international market. These CFB making units are sanctioned by SIDCO (State Industrial Dev. Corporation) J&K to the entrepreneurs in Food Parks, Lassipora, and Khanmoh. Apple tray making units can also be established.Khadi and Village Industries has announced 40 per cent assistance for setting up of CFB making units in the state in Feb 2017.

(iii) Establishment of Apple Processing Units

Apples unfit for export/low grade apples are procured by Govt., under MIS scheme and supplied to apple processing units to be used for making apple juice concentrate, or other value added products like, juice, jam, sauce, chutney, fruit toffees/fruit bars. Trained youths can find employment opportunities in such units.

(e) Export Potential

J&K has been designated as Agri. Export Zone for apple and walnut. There are enough opportunities to brand our apple varieties for export to Bangladesh, Sri Lanka, UAE, Bahrain, Singapore, Nepal, *etc.* Efforts are needed to meet the market demands of these countries by market studies and then streamlining the apple trade by proper decision making process by apple trade organizations.

(f) Financial Support for Post Harvest Management Activities

(i) Under Horticulture Mini Mission–III

The areas covered in post harvest management are establishment of marketing infrastructure for horticultural produce in Govt./Private/Co-operative sector. Financial support under Horticulture Technology Mini Mission III is provided for establishing following activities on 50 per cent subsidy basis:

1. On form collection and sorting unit (pack house)
2. Precooling unit
3. Cold storage unit
4. Integrated C.A chamber with facilities for precooling, cleaning, sorting and grading.
5. C.A storage units
6. Reefer vans/containers
7. Primary Mobile Minimal Processing units
8. Evaporative/low energy cool unit
9. Terminal markets
10. Whole sale markets
11. Rural markets/Apni *mandi*
12. Retail markets/out lets
13. Platform with cool chamber
14. Functional Infrastructure
15. For collecting, grading
16. Quality analysis
17. Market extension
18. Quality awareness
19. Market led extension activities

(ii) Government, Support to Cold Chain Development

Government of India, Ministry of Agriculture has established National Center for Cold Chain Development in the country under mission for integrated development of Horticulture. A financial assistance of (35-50 per cent) for cold-chain projects during 2016-2017 has been announced. The scheme is helping entrepreneurs in establishing modern pack houses with pre coolers, cold rooms, reefer vehicles, reefer containers and ripening units.

(iii) Market Intervention Scheme

For quality assurance in apple marketing scientific grading is must. It has been observed that some growers resort to topping practices, where in they pack good quality fruit in top layers, and subsequent lower layers are packed with mixed (inferior) quality fruit as a result, whole pack loses its market value.J&K Government has promulgated Market Intervention Scheme (MIS) for apples where in (C) grade apples are procured from growers for supplying to apple processing units. During year 2015, a target of procuring 9000 MT of apples under MIS was proposed by J&K Government for processing industry to make apple juice concentrate and other products.

(iv) J&K Bank Support to Apple Industry

For promotion of production and post harvest infrastructure development in the State, J&K Bank has launched Apple Advance Scheme and Apple Insurance Scheme for apple growers.

(g) Interventions Needed

(i) Road Connectivity

Most of the orchards sites are not connected by metalized roads and the harvested fruits at time have to be carried on ponies to road side. Due to bad roads, bruising of fruits takes place. There is need for connecting orchards with better roads and four lanning of National Highway.

(ii) Fruit Grower's Awareness Programmes

Post harvest management concept is not properly understood by the growers. Efforts are to be made to educate fruit/vegetable growers about scientific handling, grading, packaging and storage techniques, so that losses are reduced right from farm level. It has been observed that due to mere negligence, huge losses are occurring which are preventable. Knowledge of post harvest management practices can be communicated to fruit and vegetable growers by organizing seminars, workshops, field visits and other special training programs by university or development departments.

(iii) Entrepreneurship Development for Educated Unemployed Youth

Special training programs needs to be organized for 3-6 months for educated unemployed youths in horticultural entrepreneurship to start CA storage and agri-processing activities. Training modules comprising of theoretical as well as practical lessons needed to be developed for product specific activities. These skilled trainees can start production activities in food parks after proper DPR formalities and can get financial assistance from Ministry of Food Processing Industry (MoFPI),National Horticulture Board (NHB), Agricultural and Processed Food Products Export Development Authority (APEDA) and State Industries Department.

(iv) Transport Policy

J&K being horticulture state doesn't have perfect and sound transport policy for fruit shipment. At peak harvest season, adequate number of trucks are not available. Trucks are not refrigerated or specifically designed, due to which fruit quality is deteriorated during transportation. Efforts at government level are to be made for making adequate transport arrangements at reasonable cost during fruit harvesting time. Freight charges are to be rationally announced well in time.

(v) Marketing management

The state has separate Department of Horticulture Planning and Marketing which has established fruit *mandies* in each district were growers and buyers are meeting for trade negotiations. Adequate arrangement for storage, housing and internet facilities are to be ensured at *mandi* sites. The growers must get timely information on internet regarding fruit dispatches in various markets of India and price structure so that transparency in trade is ensured. The modernization of fruit *mandies* and the existing market yards should be upgraded with improved facilities.

1. Improved auction platforms.
2. Transit storage/CA/RA facilities.
3. Price/information displaying mechanism
4. Waste disposal.

(vi) Role of NGO's and Women's Co-operatives

Women's Co-operatives and NGO can contribute in setting up processing units at community level to meet the local demand of processed fruit and vegetable products.

(vii) Creation of Processing Facilities

Processing facilities for fruits and vegetables are in-adequate in the state. The state produces approximately 4 lakh MT of apples (low grade) *i.e.* 30 per cent of the total production. However, the existing capacity is approximately 30,000 MT out of which 10,000 MT is owned by JKHPMC.

(viii) Modernization of Existing Processing Units

The existing processing units run by local entrepreneurs need to be modernized.

(ix) Relaxation of Toll Tax on Cardboard Boxes and VAT on Molded Trays

Corrugated Fibre Boards (CFB) are imported from Punjab, Delhi and Gujarat and toll tax on import of CFB has been enhanced from Rs 80 to 175/Quintal w.e.f. Feb 2017, which needs to be relaxed. Similarly, toll tax on import of raw material for manufacture of CFB by local unit holder makes local made CFB costlier. Imported molded trays used for apple packaging are charged @ 14.5 per cent VAT which is making the packaging in CFB very expensive.

(x) Minimum Selling Price

In order to safe guard interests of growers minimum selling price of apple varieties should be worked out by competent authorities.

(xi) Popularization of 1-MCP Technology in the Field

Laboratory scale trails have indicated good response of 1-MCP for enhancement of shelf life in apples, the same needs to be popularized for commercial application under proper supervision and guidance.

(xii) Buyers-Sellers Meet/Apple FEST

To promote the horticulture sector in the state APPLE FEST should be regularly organized by confederation of Indian Industry to host the exhibition on pre and post harvest technology.

References

Anonymous 1988.*Controlled atmosphere storage of apples in Kashmir* (A preliminary report).Sher-e-Kashmir University of Agricultural Sciences and Technology, Divison of Post harvest Technology, J&K.

Anonymous 2015. *Area and production statement of fruits in J&K state*. Department of Horticulture, J&K Govt.

Banday, F. A., Sharma, M.K and Ahsan Hafiza 2010. Apple production in Kashmir: Challenges and Strategies. *SKUAST Journal of Research* 12:169-181.

Hussian, P. R., Dar, M. A., Meena, R. S., Mir, M. A., Shafi, F., Wani, A. M 2008. Changes in quality of apple (*Malus domestica*) cultivars due to γ-irradiation and storage conditions. *Journal of Food Science and Technology* 215(1): 44-49.

Mir, M. A., Rasool, A., Beigh, G. M., Rather, A.H. Shafi, F and Hussian, R 2004. Effect of post harvest application of wax coating materials and calcium chloride on fruit qualitychanges in "Red Delicious" apples. *Applied Biological Research* 6:1-6.

Mir, M.A 2009. *44th Research Council Meeting Rabi-2009-10*. Directorate of Research, Sher-e-Kashmir University of Agricultural Sciences and Technology, Shalimar (J&K) Nov -11-12th

Mir, M. A., Mushtaq, A. B and Fouzia, S 2013. Effect of 1-MCP on quality attributes of Red Delicious apples under non-refrigerated and refrigerated conditions of storage.*Research on Crops* 14(2): 522-529.

Chapter 25

Apple Processing: Global Scenario and an Indian Overview

F. A. Masoodi, Furah Naqash and Sajad A. Rather

Department of Food Sciences and Technology,
University of Kashmir, Srinagar
E-mail: masoodi_fa@yahoo.co.in

1. Introduction

Apple has been grown by mankind since the dawn of history. This is mentioned in early legends, poems, and religious books. The "fruit" that the Bible says Adam and Eve ate in the Garden of Eden is believed by many to have been an apple. The ancient Greeks had a legend that a golden apple caused quarreling among the gods and brought about the destruction of Troy. The Greek writer Theophrastus mentions several cultivars grown in Greece in the fourth century B.C. Apple trees were grown and prized for their fruit by the people of ancient Rome. The apple species *Malus pumila*, from which the modern apple developed, had its origin in south-western Asia in the area from the Caspian to the Black seas. The Stone Age lake dwellers of central Europe used apples extensively. Remains of apples were found in their habitations show that they stored them fresh and also preserved by cutting and drying in the sun (Root and Barrett, 2005). The apple was brought to America by early colonists from Europe. Apple is a non- climacteric fruit, is commercially grown in the temperate regions of the world in the latitudes between 30°and 60°north and south. It adapts well in climates where the average winter temperature is near freezing for at least 2 months. The flowering starts in the spring when trees bear white blossoms that look like tiny roses, which grow into ripened fruit in about 120-170 days.

Annual apple production at the global level is about 76.4 million metric tons for the year 2015-2016, with the production market dominated by China at 43 million metric tons followed by European Union and U.S. at 12.2 and 4.5 million metric tons respectively. Of the total produce, major proportion is for fresh consumption (85.07 per cent), whereas 14.39 per cent of apple produced is utilized for processing. In the processing market, China and European Union (EU) lead closely at 4.0 and 3.8 million metric tons respectively, which nears to about 36 per cent and 34.95 per cent of the total processed lot. U.S. is also a major player in the processing, however, at a relatively lower percentage than China and EU at 12.63 per cent (Foreign Agricultural Service, USDA, 2106). Many underlying factors lead to the global emergence of the leading and lagging players in the market of production and processing. For the marketing year 2015-2016, about 11 million MT were processed into different products. Table 25.1 provides the country wise listing of the apple process.

Table 25.1: Major Countries Involved in Apple Processing (2015-2016)

Country	Processed Apple (1000 MT)
China	4,000
European Union	3,852
United States	1,392
Chile	320
Russia	335
Argentina	293
South Africa	200
New Zealand	141
Japan	135
Canada	135
Turkey	100
Brazil	20

Source: USDA, Foreign Agricultural Service analysis of customs data from Production, Supply and Distribution online, Inc.

Apples are processed into a variety of products, although apple juice is the major processed product. Apples are also processed into canned, frozen, and dehydrated apple slices and dices, and several styles of apple sauce. Juice is processed from apples that are unsuitable for use in peeling operations. Fresh apple juice is the product of sound, ripe fruit pressed and packaged without the addition of any preservatives. Apple juice character is linked to the cultivar and maturity stage of the fruit used to make the product. Regardless of the cultivar, only sound, ripe fruit showing no signs of decay should be used. Apples picked from the ground are to be avoided for processing into juice due to the musty or earthy flavor they pick up. Immature apples lack flavor and over-mature ones are difficult to press, clarify, and filter.

The cultivar used in processing is dictated to some degree by the quality of the product to be produced. Many of the apples that have some imperfections, such as skin blemishes or off shapes rendering them undesirable for the fresh market, are utilized by processors. These are perfectly good quality apples and are in high demand. In the United States, an average of about 20 per cent of the Delicious and other fresh market apples are processed. Varieties such as Golden Delicious, Rome Beauty, Granny Smith, McIntosh, and others may have more than 20 per cent of the volume diverted to processing. Delicious apples that are firm, sweet, and juicy yield a good volume of high quality juice. Although sauce can be produced using Delicious apples the product would not be of good quality, particularly in relation to texture and color. The apple sauce yield is less with the Delicious apple due to the thicker skin that results in greater loss during peeling. Golden Delicious on the other hand not only makes a good quality juice but also produces a high quality sauce and sliced processed product. Cultivars utilized in processed products are determined by availability of the raw product, quality of the product produced, and market demand from the region grown (Root and Barrett, 2005).

Apple juice is the second most consumed fruit juice in the world, with total export value of single-strength and concentrated juice of US$3936 million in 2009 (FAO, 2010). In 2011, 394,400 MT of apples were processed into juice (AMS, 2011). In 2010, Americans consumed 8 L/yr. of apple juice, accounting for 31per cent of the total juice consumption (Corcuera *et al.,* 2014). Annual global apple juice production is around 1.1 m tonnes per year. The main producers are China (with 57per cent of global production) followed by Poland (6 per cent), the US (5 per cent) and Italy (3 per cent). China dominates the export market with more than 80 per cent of its apple juice concentrate production going to the export market. In the EU, the main exporter is Poland.

China is the world's top apple juice producer reaching 1 million tons in 2007. Strong global demand and a shortage of juicing apples in China have forced Chinese producers to tap into their higher priced fresh market. Crushing companies are paying double prices to purchase apples. In addition to China, Argentina, the EU-27, Chile, and Brazil also supply juice to the world market. Argentina typically exports nearly all of its production, most of which goes to the United States. Chile's production also reflects foreign demand but like many producing countries, processing-apple availability is tight. In the United States, a small portion of apples are grown just for juicing. Most juice apples are culled fruit from fresh packing lines. During 2006/07, about 64 per cent of all apples went into the fresh market, the rest, or about 36 per cent were processed. Of the processed apples, 45 per cent went to the juice and cider market or about 16 per cent of total apple production. About 3.4 billion pounds of apples were processed during 2006/07 for products including canned, juice, frozen, dried, and fresh slices (FAS, USDA 2008). Table 25.2 represents the percentage of apple utilized as fresh and processed products.

Table 25.2: U.S. Apple Utilization as Percentage of Total Yearly Production

Product	2004	2008
Fresh	63.5	64.5
Processed	36.5	35.5
Juice and Cider	18.0	15.7
Canned (sauces)	12	12.2
Frozen	2.5	2.1
Dried	1.9	2.1
Fresh Slices	0.4	1.3
Other	0.7	1.3

Source: Data obtained from USDA, (National Agricultural Statistics Service) NASS, Noncitrus Fruits and Nuts Summary: 2007 and 2009.

Factors like improvement in production practices, government policies, processing and marketing infrastructure, and share at the global level could positively as well as negatively influence in shaping the global position.

2. China: The Global Leader

China has emerged as a leading exporter of processed food products, the underlying reason of which has been capital investment in processing facilities. The export of Apple Juice Concentrate (AJC) for instance, is exemplary of the nexus between investment and labor-intensive agri-exports. Expansion of apple orchard area no doubt crowned China to be the world's leading producer but decreased the prices as well. However, combined government, private and foreign investment helped to construct a network of apple juice processing plants that made apple juice concentrate into an export oriented commodity. Further, expansion of the AJC industry was brought about by penetrating into economically weaker north-western region in search of additional apples by guiding investment and recruiting apple growers. China gives tough competition to other countries in terms of its raw material supply, and could hold monopoly in the AJC market.

China first emerged as a significant exporter of apple juice concentrate in the late 1990s when it accounted for about 20 per cent of world trade in AJC. Major exporters at that time included Poland, Germany, and other European Union countries, and Southern Hemisphere countries that supplied products during the Northern Hemisphere's off season. At present however, China dominates the global export market also. U.S. apple growers view China's AJC exports as a threat. U.S. apple industry leaders cite the rapidly rising volume and low cost of Chinese apple juice concentrate as the main force putting downward pressure on prices of juice and apples.

(a) Factors Advantageous to China

(i) Raw Material

Abundant, inexpensive fresh apples are the key to China's competitive advantage in producing apple juice. China began boosting apple production in the early 1990s as farmers sought to grow commodities that would bring higher returns than traditional field crops. In the early 1990s, China and the United States produced a similar volume of apples, each accounting for nearly a fifth of world apple production. After two decades of growth, China produced about seven times as many apples as the United States in 2009/10 and now accounts for roughly half of world apple production Also during the 1990s, Chinese consumers diversified their diets to include more fruit, but not fast enough to absorb a fourfold increase in domestic apple production. Supply outpaced demand and pushed down Chinese apple prices.

In 1991, the average farm price of apples in China was the equivalent of about $.25 per kg, roughly one-half of the U.S. price. The price in China reached a low point of $.10 per kg in 2000, about one-fourth of the U.S. price that year. Chinese apple prices began rising after 2003, coinciding with the juice export boom. The average reached $.30 per kg in 2008- three times the price in 2000 but still about one-half the average U.S. price at the time (Figure 25.1). Depressed apple prices in the 1990s encouraged agricultural officials in China to look to juice exports as an outlet for the apple supply. Low apple prices also generated interest among juice processors; according to Chinese industry sources, each kilogram of juice concentrate requires about 7 kg of apples as raw material and apples constitute 60-70 per cent of AJC production costs.

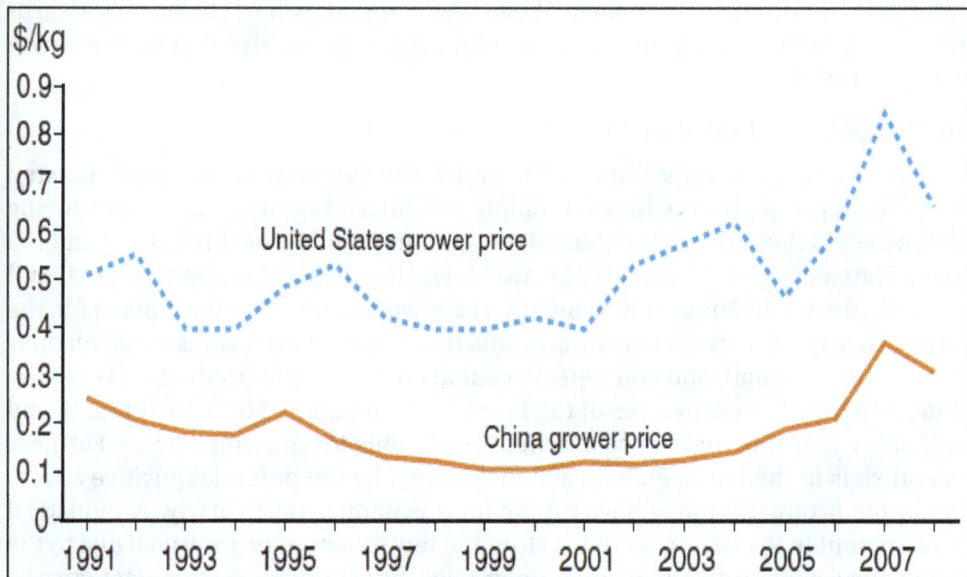

Figure 25.1: Decreasing Apple Prices in China (*Source*: USDA, ERS).

In China, few apples are grown specifically for juice processing. The AJC industry offers an outlet for the abundant supply of apples that are too small, misshapen, or off-color to sell on the fresh market and would otherwise be discarded or used as animal feed. Typically, 20-30 per cent of an apple crop is below fresh-market standards. USDA estimates indicate that processing accounted for only 5 percent of China's apple use in the 1990s, but the share rose to 15-25 per cent afterwards. AJC exports are now consuming a substantial share of China's apple crop.

(ii) Cooperative Intervention

The transformation of apples into juice cannot take place without investments in processing plants and equipment. Thus, capital investment in processing capacity has been instrumental to the boom in China's apple juice exports. Apple growing is a labor-intensive process conducted by an estimated 10 million small farmers who have little education and minimal financial resources. Since the 1990s, the industry has expanded rapidly through a diverse mix of government, private, and foreign investment, loans from Chinese banks, and infusions of capital from stock market listings in China, Hong Kong, and overseas. Joint ventures of foreign and government investments laid the basis for the establishment of the apple juice industry, and ensured to keep the industry growing and output oriented.

According to industry reports, China's apple juice concentrate production capacity rose from about 3,000 MT per year in the 1980s to 20,000 MT in 1995. As new companies entered and expanded, capacity shot up to 370,000 MT in 2001, and it expanded 44 per cent in 2004/05 alone. Recent industry reports indicate that capacity now far exceeds export demand. With heated competition and maturation of the domestic market, some AJC companies have sought to forward-integrate into beverage products. Consumer beverage companies have begun backward-integrating by forming joint ventures with processors or developing their own supply networks.

(iii) Geographical upliftment

Geography is an important factor in the development of the juice industry. Apple juice production is located mainly in China's two main apple-producing regions: the Bohai Gulf region (Shandong, Liaoning, and Hebei Provinces) and the Loess Plateau region (Shaanxi, northwestern Henan, southwestern Shanxi, and parts of Gansu and Ningxia Provinces). The government's strategic plans for the apple industry call for concentrating production in these two regions based on their temperature, rainfall, and soil type. Recent growth in apple production is mostly in the northwestern provinces of the Loess Plateau region. Studies of climate and soils show that this region has excellent conditions for growing apples. Farmers and officials in the Loess Plateau are encouraged by the potential positive effects that apple production may have on the local economy, particularly as industrial development in the region has been slow and few other crops grow well due to the region's arid climate and hilly topography. Analysis of Chinese customs statistics by region shows that most of China's apple juice exports come from a belt of countries in the Loess Plateau and eastern Shandong and Liaoning Provinces, which reflects the geographic concentration of processing investment.

In 2000, the Chinese government formally launched a "Develop the West" (*Xibu Da Kaifa*) program to steer investment to the less-developed western provinces, including the Loess Plateau (Ma and Summers). The program includes direct government investment, financial transfers to western provincial governments, and such measures as subsidized loans, tax breaks, and access to infrastructure projects to "guide" private investment to western provinces. The strategy emphasizes developing industries based on local resources ("industries with special characteristics"). The Ministry of Agriculture issued a set of strategic plans ("advantaged regional layout plans") that identified apples as one of China's most internationally competitive crops and called for concentrating production in the most efficient production regions, including the north-western provinces. In 2006, the Ministry of Commerce formulated a plan for promoting agricultural exports that encouraged companies to develop "export bases" in inland areas.

(iv) Government Role

Integrating juice companies with small farmers was adopted as part of the government's development strategy. Shaanxi's development plan was to "follow the large company development road" by cultivating "dragon head enterprises" that lead farmers into the market (China Ministry of Agriculture Development Plan Office, pp. 386-87). Dragon head enterprises (the term is sometimes translated as "leading" or "flagship" enterprises) are companies identified by central, provincial, or local governments as having potential to help farmers by providing them with a market for their products and disseminating technical information. In return, companies receive tax breaks, preferential bank loans, access to land, assistance raising capital, and government help meeting export standards. Dragon head enterprises have been identified as the key players in the fruit industry along with the support measures that include low-interest loans, subsidies for storage and cold-chain facilities, and export incentives such as value added tax rebates for exports.

(b) Bottlenecks

Weaknesses in the Chinese apple juice industry could be of potential benefit to the competitor countries globally. A weakness of Chinese juice in the international market is its low acidity, which reflects the predominance of sweet apple varieties like Fuji and Guoguang used as raw material. Sweet juice from China must be mixed with other juices and ingredients to make final products with high acidity demanded by consumer markets in North America and Europe. A number of industry representatives in China have voiced concerns over the lack of high-acid apples. In Shaanxi Province, high-acid varieties account for less than 10 per cent of apple area, and juice processors there procured less than half the volume of high-acid apples needed to fulfill their contracted juice sales in 2005. Lacking options, processors use sweet apple varieties intended for the fresh market that yield juice with low acidity; the resulting juice concentrate may sell at a 40 per cent discount or be rejected by customers. The lack of high acid apples has been identified as a bottleneck for the industry's growth. While the price for apples purchased by juice companies has been rising, apples sold for the fresh market still sell for higher prices. Consequently, the expansion of high-acid apple plantings has been slow because farmers prefer to grow sweet varieties that can be sold on the fresh market. Some companies and officials are promoting planting of orchards of high-acid apple. Varieties specifically for use in juice processing and the government's strategic plan for the apple industry encouraged production of these varieties (China Ministry of Agriculture). However, farmers are reluctant to plant apples primarily for juice processing. Juice processors use predominantly defective or fallen fruit intended for the fresh market. While pasteurization kills bacteria, Chinese industry experts warn that use of poor quality fruit can yield juice with a brown color or poor taste, or juice with high levels of patulin (a mycotoxin produced by molds) that is rejected by buyers.

While China's industry supplies a large share of the world's apple juice concentrate, the industry's reliance on price for its competitive position leaves it in a tenuous long-term standing. Chinese juice suppliers have not developed strong brands or consumer product lines, and the industry is predominantly a supplier of low-cost generic ingredients to multinational consumer product companies. This leaves the Chinese industry vulnerable to wide swings in demand resulting from the purchase decisions of price-sensitive downstream users. The sudden boom in demand for Chinese juice followed by a sharp reversal during 2007-08 illustrates the vulnerability of the industry to shifts in demand for its product. Juice from China competes with apple juice from other countries and other types of juice and ingredients, and it may be affected by swings in consumer preferences toward non-juice beverages (Gale *et al.,* 2010).

3. Status of Apple Processing in India

India ranks fifth in the global production of apple at 1.9 million metric tonnes, contributing to about 2.5per cent of the global share (FAS, USDA 2016). For the year 2013-2014, the production of apple recorded was 2585000 metric tonnes (MT), ranking sixth among the fruit production, and productivity of 21.8 MT/Ha ranking third. Apple occupies a meager share of 2.8per cent of the total fruit crops produced in India. Major apple producing states of India are Jammu and Kashmir, Himachal Pradesh, Uttarakhand, and Arunachal Pradesh. The apple production is divided into early, mid and late season varieties. The productivity of apple increased from 8.6 MT/Ha in 2001 to 10.2 MT/Ha in 2014. Jammu and Kashmir ranks first in the production of apple followed by Himachal Pradesh (Indian Horticulture Database, 2014).

In terms of processing apple, India has remained behind, as it does not find a listing in the players in this field. India produced 4500 tonnes per annum of AJC and this is equivalent to about 0.64 per cent of the total world production.

Table 25.3: Area, Production and Productivity of Apple in India

State	2001-02			2012-13		
	Area (ha)	Production (000 MT)	Productivity (MT/ha)	Area (ha)	Production (000 MT)	Productivity (MT/ha)
Jammu and Kashmir	157.28	1348.2	8.6	160.9	1647.7	10.2
Himachal Pradesh	106.23	412.4	3.9	107.7	738.7	6.9
Uttarakhand	33.76	123.2	3.7	30.0	77.5	2.6
Arunachal Pradesh	14.07	31.0	2.2	14.3	31.9	2.2

Source: Indian Horticulture Database, 2014.

In a report tilled, "Impact Evaluation Report of Horticulture Mission for North East and Himalayan States (HMNEH)" published by Agricultural Finance Corporation Ltd, it is mentioned that the horticulture processing sector is highly underdeveloped. In India, less than 2 per cent of horticulture produce is currently processed as compared to 65 to 80 per cent in the developed countries. There is a need to encourage setting up of horticulture based processing units at different levels with due regard to the product quality and safety measures. Similarly, value addition is only 7 per cent in India as against 88 per cent in countries like UK. China's export competitiveness arises from low costs and a growing processing industry (Murtaza, 2015).

4. Status of Apple Production and Processing in J&K

In India, every state has its own natural advantage. Still due to the non-alignment of strategic goals many states have not been able to drive the advantage of liberalization and globalization advantage in last twenty years. Apart from its scenic beauty the state has a strong advantage of natural endowment towards agro based industries due to its saffron, walnut, apple and tourism. Somehow, the state falls short of capitalizing the resource advantage due to lack of strategic alignment of asset of enterprises, and institutional framework. The strategic geographical location of Jammu and Kashmir is a blessing for the state's economy. Due to its geographical location, climate and soil type, the state is the producer of rich variety of fruits and vegetables. The state is the highest temperate fruit producing state of India. However, the geographical terrain and climate also create disadvantages for the state in terms of connectivity and other factors (Malik Zahoor, 2013). The total area and production under fresh fruit cultivation in Jammu and Kashmir for the year 2012-13 was 236,780 hectare and 1,524,593 metric ton respectively. Out of this the area and production of apple cultivation has been 157280 hectare and 1348149 metric ton respectively (Anonymous, 2013). Nearly about 70 per cent of the population in the state derives its livelihood directly or indirectly from the agriculture sector. There are about thirty lakh people in the state comprising of about six lakh families who are directly or indirectly associated with horticulture sector (J&K Economic Survey, 2012-13). The fruit industry is the second most important industry after tourism in Jammu and Kashmir (Buyer-seller meet and

PROCESS FLOW DIAGRAM of Apple Juice Concentrate

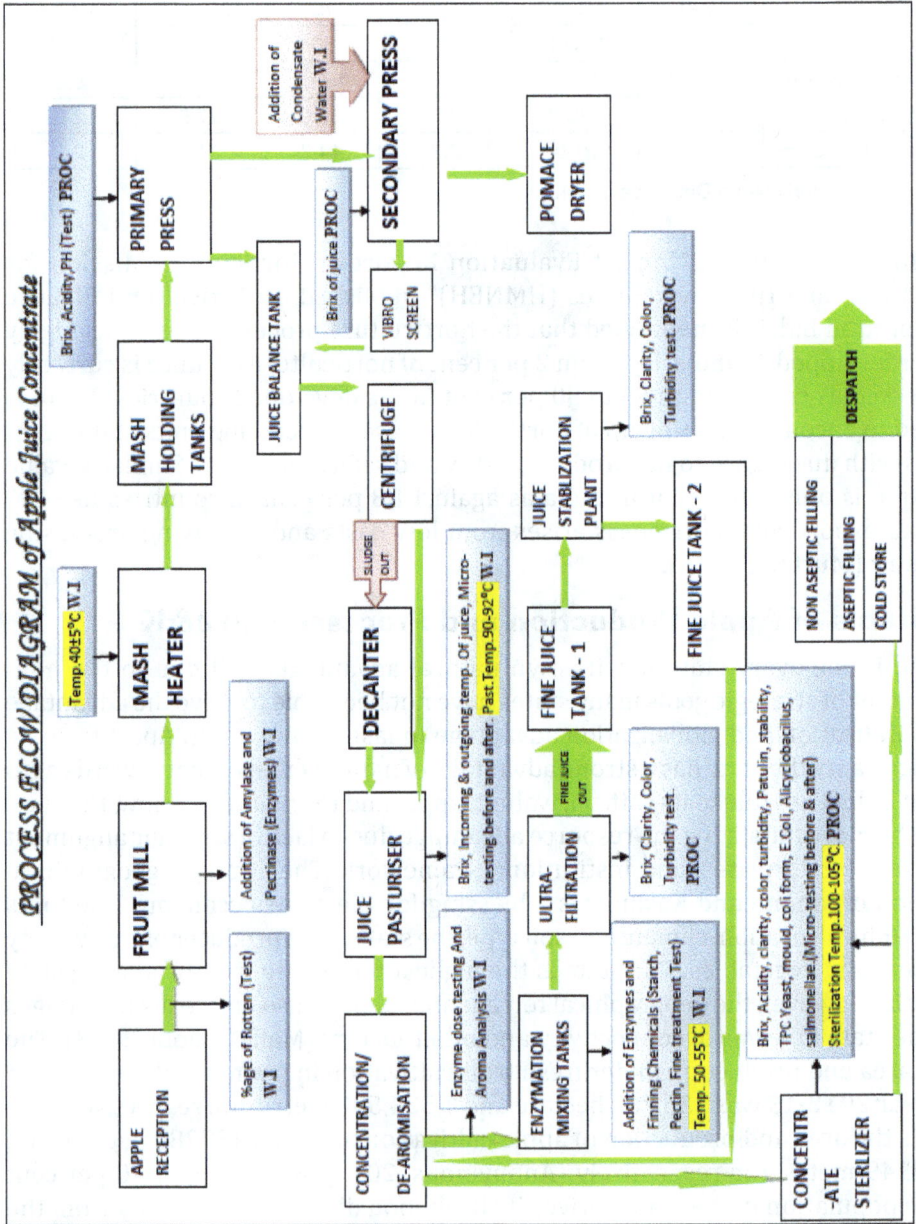

Figure 24.2: Process Flow Diagram for Production of Apple Juice Concentrate.

Conference, 2007). The annual exports of apples from India is about INR 400 million, out of which nearly fifty per cent of apples come from Jammu and Kashmir which also provides job opportunity to about 1.2 million people directly or indirectly (Malik Zahoor, 2013). However, the processing of apple in the state despite the above-mentioned facts is at odds. Most of the processing activities centre on the production of apple juice concentrate. Mostly cull grade apples are used for the production of concentrate (Figure 24.2). Some of the contributing factors leading to such situation are discussed in the following sections.

5. Factors Affecting Apple Processing in J&K

(a) Government Policies

Good governance demands consideration of all possible means that lead to the overall development and betterment along with that of the economic fabric of the setup it rules. The implementation of schemes intended for promoting market activities is not going the way it should. There is weaker implementation of schemes like Market Intervention Scheme (MIS) in the state. This has helped outside juice industries to grow at the cost of local industry as culled and C-grade fruit is marketed outside the state which results in lesser availability for local processors. Agents with unclear motives have been luring state growers to sell their culled and C-grade apple to them at higher prices. The production of apple juice in Kashmir depends on the availability of culled and C-grade apple. The industry being in infancy, unit holders do not find themselves capable enough to pay higher prices to the growers against the raw material. The fruit growers therefore, sell their fruits outside. Experts from the state Federation Chamber of Industry have even expressed openly the need for government banning the export of culled apples from the state. They hold the opinion that the government of J&K lacked the dedication to support the fruit processing industry as compared to Himachal government, where they adopt absolute 'grading management techniques' restricting C-grade apple for processing only. The production in valley (10 tons per acre) as compared to the apple producing countries (26 metric tons per acre) is very low. Till these targets are attained, the government needs to support the growers with reasonable price for culled apples. The performance of the market intervention scheme has been disheartening as it has not been able to meet the targets of procurement of low grade fruits from producers because of many reasons. Suspension of the scheme for some time and reimplementation in some specific districts only has further compounded the problem.

MIS implemented in the valley under Technological Mission Scheme in 2002, used to provide C-grade fruit to the processing units at subsidized rates. MIS ensured the grading of Kashmiri apple and made it competitive in the outside market. In Himachal Pradesh on the contrary, MIS scheme is operational for the past 30 years, and the production of the state is less than J&K. Considering the geographical disadvantages, this scheme and additional schemes, incentives and facilities are needed. The suspension of scheme has adversely affected the business, as the processors now need to buy the fruit directly from the farmers at much higher rates (Greater Kashmir 2007, 2008).

Taking an example of the JKHPMC, set up in 1978 to give boost to the horticulture sector, the corporation is making losses over the past two decades despite having huge asset base across the length and breadth of the State. The recommendations made by committees constituted by the government from time to time for revival of HPMC need to be relooked at (A report published in Greater Kashmir, 2016).

(b) Seasonal Limitation

Apple is available from August to November with peak production in the month of September. The harvesting season of apple extends up to the month of November. The seasonality of fruits is a major cause of concern for any fruit processing unit. Concerns center around the suspension of activities for the majority of the year, as the apple season spans a little breadth of the year. These concerns could be tackled if the manufacturing facility is diversified for production, as per the seasonal availability of the fruits in the state.

(c) Raw Material

Apple contributes 79 percent of the fruit production in J&K (Directorate of Horticulture, 2016) and the chief varieties being Delicious, American, *Ambri, Maharaji, Kesari,* and *Hazaratbali.* However *Ambri* or *Amri* is the most popular and has a large round red and while sweet fruit, ripening in October and keeping its condition for a long time. This variety attracts maximum consumer's attraction due to its sweetness and handsome appearance. Unlike *Amri, Mohi Amri* has acid and redness. Another species known as *Kuddu Sari* in longer is shape and possesses more juice rather than acid but has short life. Though the cultivation of apple in India is concentrated in Jammu and Kashmir, Himachal Pradesh and Uttarakhand, yet, Kashmir enjoys the distinction of being hub of apple industry of the country. This is obviously so because the State has not only superiority over Himachal and Uttarakhand in the field of production but also in marketing (Mir, 2014).

In a recent publication, Bhat and Choure (2014) reported significant increase in the production of apples in J&K during 2004-2012, and indicates the importance of apple as outstanding commodity for fruit industry. The estimated marketable surplus (2011-12) of grade A and B apple is approximately 12 lakh MT. Surplus grade C apple available for processing is 3.22 lakh MT, of which, about 30,000 MT of grade C and culled apples are being processed into apple concentrate by processing plants in Kashmir region. Small quantity of apple is also processed into other products, like, jam, jelly and apple preserve. According to the J&K economic survey (2012-13), prefalls and culled apples (about 30 per cent of A grade, 40 per cent of B grade and 30 per cent of C grade) create a substantial quantity of 50 thousand tonnes as raw material, which needs to be explored for the apple processing industry (Tali, 2014). This estimate is as per the current production level, which is significantly low as compared to global yield standards. The apple processing sector of J&K has potential to create market for its product at regional, national as well as for the international level. The Jammu and Kashmir state has huge raw material base for the fruit and vegetable processing industry particularly to the apple processing industry. With respect to the other apple producing states, J&K state has grown well in terms of the area harvested and apple production (Table 25.4).

Table 25.4: Year-wise Comparison of CAGR (per cent) of Production of J&K with Major Apple Producing States of India

State	1996-97 to 2001-02	2001-02 to 2006-07	2006-07 to 2011-12
Arunachal Pradesh	-10.97	2.88	25.50
Himachal Pradesh	-8.95	8.24	0.49
Jammu and Kashmir	2.20	6.09	7.75
Uttarakhand	-20.67	15.77	-0.10

Note: CAGR; compound annual growth rate (Source: Compiled from Indiastat.com).

However, the state has not been able to capitalize this resource base for the development of the apple processing industry. In addition, irrespective of the fact that the state has higher growth rate than other states, it is still not at the forefront of processing. To compare the global scene, the production needs to be increased far beyond that is at present, so that a huge raw material availability guarantees comparable processing.

(d) Industrial Infrastructure

The state of Jammu and Kashmir offers huge economic opportunity for the fruit processing and preserving industry, with reference to apple based products. The ample availability of raw material, labor force availability, per capita income offers an avenue for advantageous sustainability to the state of Jammu and Kashmir in terms of employment growth, strategic development of apple processing industry, social development *etc*. However, the state has not been able to take the leverage of these resources available for the strategic development of fruit and vegetable processing, particularly apple processing industry due to various reasons. Despite the advantage of natural endowment the state of Jammu and Kashmir (J&K) has comparatively sluggish participation in industrial segments related to food processing industry as compared to that of Himachal Pradesh. The data is negative with respect to the industrial development, and is far behind in comparison to HP. The invested capital, product value, inputs, income and profit from industrial units in J&K has decreased whereas, it has surged in HP (Annual survey of Industries).

J&K had only 8 units operational in fruit and vegetable processing and out of these only two units are involved in apple processing (Table 25.5). As of 2010 there were only six medium and small scale units for food processing in J&K. At present there are various units which are using apple as a base for processing such as FIL Industries Pvt. Ltd., Jammu Kashmir Horticulture Produce Marketing and Processing Corporation Ltd (JKHPMC), Snow Valley Industries, Wiss Cart Pvt. Ltd., Gulbadan Apple Juice Factory, *etc.* As compared to the other states of India, which have apple production, J&K has improved upon the area of harvest with a cumulative growth of 7.4 and 7.75 per cent of production. The food processing industry offers tremendous opportunity for commercial exploitation of horticulture of the State but commercial processing is around 1per cent only due to lack of post harvesting and processing facilities as well as unscientific packaging.

Table 25.5: Year-wise Production of Apple Processing Units of J&K (In metric tons)

	Years				
	2008-09	*2009-10*	*2010-11*	*2011-12*	*2012-13*
Public Sector Unit (JKHPMC)	170	170	90	46	280
Private Unit	2500	1700	1950	2100	2300
Total	2670	1870	2040	2146	2580

The required supply chain management is missing, "C" grade apples of pre falls and culled apples accounting for substantial quantum of around 3 lakh metric tons which needs to be explored as raw material for processing industry. This requires processing capacity of 3 lakh tones per annum. The current capacity is barely 65000 tons and even this capacity is under-utilized in view of the misguided propensity of the growers to sell fruit without grading, sorting, and value addition (Malik and Choure, 2014).

(e) Awareness and Willingness to Process

It is pertinent to note that Kashmiri horticulturists have preferred to remain the suppliers of primary products rather than diversify into value-added finished products. Despite the fact, that every processor would definitely have cull grade apple, in the years' produce, they have not shown specific interest in diverting it towards processing or participating themselves in processing. The underlying reasons could be that either they are unaware of the benefits and returns associated with processing, or they lack the proper exposure towards getting involved in processing industry. This is not to be taken in isolation as only grower centric reason; the entrepreneurial effort in terms of making apple into processing is also not prominent. Growers, for instance would prefer easy returns in directly selling the low grade apple than to take pains in setting up facilities for processing, even though the returns may be better in the long run. In manufacturing, they need to find a balance between overhead costs and retail price increase.

6. Problems Associated with Processing

(a) Lack of Technical Support

Apple a major raw material available for food processing industry in the state of Jammu and Kashmir, is available from August to November with peak production in the month of September. The seasonality of apple is a major cause of concern for any fruit processing unit. The apple processing sector of Jammu and Kashmir has potential to create market for its product at regional, national as well as for the international level. Despite having advantageous cues available, there are also some core technical development issues which hamper the growth and development of this sector in the state of Jammu and Kashmir. The post harvest management of crops is very important for ensuring better results to the farmers. Post harvest management activities may include plucking, sorting, grading (Figure 25.3), waxing, designing, packaging, storage, processing, *etc.* The quality deterioration of the produce is fast if post harvest management is not arranged scientifically. Transportation is most important factor in the marketing of apples, which have to be carried from producing areas to the consuming markets. The main problem in this regard is lack of vehicles, vehicles not available in time, villages not linked with metaled roads, high transportation charges, non-availability of labour and no road up to 1 km from orchard site. Packaging is usually poor but its cost is high and become unbearable for small producers. The main reason of mismatch in supply and demand of apples is the inadequate cold and conditional atmosphere storage in the valley. In order to ensure regular supply of raw material to the processing unit and consumer market, it is imperative to set up a cold storage and these storage facilities shall go along way for processors to extend the processing period from the present 3 month to 6 months. The inadequate or lack of post harvest operational systems are directly or indirectly associated with the marketing of Kashmiri apples.

Figure 25.3: Grading of Kashmiri Apple.

In addition the existing technology is surrounded by many technical issues, such as advancement issues, obsolete techniques, and old machineries. Due to these reasons it has become difficult for the fruit processing industries to use an appropriate technologies and techniques to reduce the post harvest losses, time in operational activities and produce sufficient quantity of various products.

(b) Lack of Appropriate Processing C

The estimated marketable surplus of grade A and B apple in Kashmir is approximately 12 lac MT. Surplus grade C apple available for processing is 3.22 lac MT, of which about 30,000MT of grade C and culled apples are being processed into apple concentrate by processing plants in Kashmir region. Small quantity of apple is also processed into other products like jam, jelly and apple preserve (Hanan, 2015). The apple cultivar used in processing will be dictated to some degree by the quality of the product to be produced. Processing quality can be affected by decay, damage, maturity, firmness, colour, soluble solids, acids, and other chemical compounds, such as tannins, contained in the fruit (Downing, 1989). For example sugar, obviously is responsible for the sweetness of apple juice while acid (measured as malic acid) gives it tartness. Tannins are considered to promote astringency which should not be confused with tartness. In the present scenario in Kashmir valley apples that have some imperfections, such as skin blemishes or off shapes rendering them undesirable for the fresh market, are utilized by processors for development of various products such as juice, concentrate, *etc.* Delicious apples that are firm, sweet, and juicy yield a good volume of high quality juice. Although sauce can be produced using Delicious apples, the product would not be of good quality, particularly in relation to texture and color. The apple sauce yield is less with the Delicious apple due to the thicker skin that results in greater loss during peeling. Golden Delicious on the other hand not only makes a good quality juice but produces a high quality sauce and sliced processed product. Golden Delicious and Delicious apples are found to be suitable for cider production while Ambri and Yellow Newton are not (Verma and Joshi, 2006). Cultivars utilized in processed products are determined by availability of the raw product, quality of the product produced, and market demand from the region grown. Apples may be grown specifically for processing. Apples for processing should be harvested at optimum maturity for good fresh market storage and handling. Only in a few instances are apples harvested with the processed product in mind.

(c) Processing and Nutritional Loss

Almost everyone has heard the expression "An apple a day keeps the doctor away," reflecting the notion that apples and apple products are nutritious. Apples and apple products are sources of potassium, phosphorus, calcium, vitamin A, and ascorbic acid. Fructose, sucrose, and glucose are the most abundant sugars. In recent years there has been growing interest in the presence of polyphenolic antioxidants in various fruit and vegetable crops. Apples are a rich source of these beneficial phytonutrients that epidemiological studies have found to be associated with protection against aging diseases and cancers. However, earlier studies (van der Sluis *et al.*, 2001; van der Sluis *et al.*, 2002) have highlighted the effects of apple cultivar, harvest year, storage conditions and apple-juice processing methods on the concentration of polyphenolics. It has been reported that juice produced from some apple varieties (Jonagold apples) by either pulping or straight pressing had a significantly lower level of both polyphenolics and antioxidant activity. Antioxidant activity was found to reduce by 10 to 13 per cent that of in fresh apples by the juice-making process. It was determined that most of the polyphenolic antioxidants were retained in the pomace or press cake and were not extracted into the juice (van der Sluis *et al.*, 2002). These results have ramifications for apple juice processors interested in producing juice with higher nutritional value. It may be of interest to either market cloudy apple juice as a superior product or at least to utilize the pomace as a source of polyphenolic antioxidants.

(d) Competition

The high price of imported raw materials, lack of mechanized infrastructure and inadequate availability of skilled manpower has resulted in the low quality and high price of local apple processed products. Moreover, market uncertainties, handling and storage problems do also prevail and limit the expansion of apple processing sub-sector in the valley. By these constraints the Kashmiri apple processors are not able to compete in the national market. Further quality has a strong impact on the supply chain, so it leads to efficiency and less rejection by the customer. In the state of Jammu and Kashmir, there is a lack of quality standards to meet international quality for export, poor hygienic and safety standards, high quality degradation *etc.* In addition the local enterprise is facing fierce competition from the imported products.

(e) Apple Processing Waste

Most of the production of apple is used for table purposes but a portion is being processed into various products like juice, concentrate, *etc.* A conventional process removes 75 per cent of fresh weight of apple as juice and 25 per cent is the pomace (Shah and Masoodi 1994; Kaushal *et al.*, 2002). In India, total production of apple pomace is about 1 million tons per annum and only approximately 10,000 tons of apple pomace is being utilized (Shalini and Gupta, 2010). The large scale processing plants are located in Jammu and Kashmir and Himachal Pradesh, which produce huge quantum of apple pomace and is not being utilized at present but is dumped in the fields creating pollution problems because of fermentation and high chemical oxygen demand (COD) of 250-300 g/kg. Apple pomace being biodegradable in nature with high bio-chemical oxygen demand (BOD), disposal of apple pomace into the environment causes pollution, necessitating the efforts to find out the appropriate solution to this problem. In large scale apple processing industries, the wastes can be categorized into 2 types. The first type is the fruit discarded into the sorting belt due to its partially bruised/spoiled nature and named as belt rejection. The second type is the apple pomace obtained after juice extraction. The belt rejection apples are also dumped along with apple pomace as waste. Safe disposal of processing waste is very important to prevent environmental pollution. Apple pomace contains large amounts of water and is in a wet and easily fermentable form, therefore, causes serious disposable problems (Shalini and Gupta, 2010). A substantial cost is involved for disposal of such wastes. Pomace can be treated as an excellent example of waste food resource (Shah and Masoodi 1994). Apple pomace is an interesting resource of healthy molecules, such as phenolic compounds, pectin and fibre and serves various applications such as pectin recovery, enzyme production, animal feed, organic acids production, ethanol production, aroma compounds, natural antioxidants, biopolymers and pigments (Rabetefika *et al.*, 2014). Numerous attempts have been made by earlier researchers to utilize this waste, such as a source of dietary fiber (Sudha *et al.*, 2007), human food (Masoodi and Chauhan 1988; Masoodi *et al.*, 2002; Verma *et al.*, 2010; Huda *et al.*, 2014; Rather *et al.*, 2015, 2016), polyphenols (Lu and Foo, 2000), animal feed (Joshi and Sandhu 1996; Joshi *et al.*, 2000) and biofuels (Sandhu and Joshi 1997) without any further fractionation and purification. However, only a fraction of apple pomace is used due to rapid spoilage of the wet pomace. Sustainable food production and value-addition of wastes is the most important issue in the agro and food processing industry. Hence it can be concluded that apple pomace processing waste utilization is the necessity and a challenge to the food industry and needs first and foremost consideration to make the units economically viable through the recovery of value added by-products.

7. Strategies and Future Prospects of Apple Processing Sector in Jammu and Kashmir

If appropriate steps are taken at various levels of production by all the stakeholders, small and large scale industries and government, there will definitely be a positive trend in production, processing and the corresponding revenue. Following areas need immediate attention of researchers and policy planners to improve the scenario of apple processing in the country.

☆ It is claimed that J&K produces the highest proportion of apple in India. However, the production is far from matching global productivity standards, implying the need for increasing both production and productivity, which will in turn generate surplus for processing as well. Efficient use of technology as implemented by the global players needs to be looked at and adopted. Research emphasis for improving processing of apple could center around growing high yielding cultivars and cultivars with the desired quality parameters for different processed products. This would also reduce the glut during peak production season as well as post harvest losses. The importance of varietal difference in the market place or the preference of cultivar for specific processing purpose should be demonstrated.

☆ Apple marketing being complex phenomena requires special treatment and utmost care at present in the Kashmir Valley. Marketing of locally processed apple products needs to be studied in comparison with non-local products. It is not yet identified as to how far have the locally processed products, for instance apple juice concentrate penetrated the domestic, national, and international market. Brand establishment of Kashmiri apple juice in the market could first be bought by introducing it through promotional activities.

☆ Entrepreneurial training programmes aimed at dealing the problems should be organized for small entrepreneurs so that they can be imparted with necessary skills and technical knowledge. Moreover, technology up gradation, financial management, material management, manufacturing techniques of products, brand promotion, advertising the product are some important areas of training needed by entrepreneurs in this field. Training should be imparted regarding environmental management, business opportunities and guidance, processing of various products, labor management and procurement of raw materials.

☆ Revenue-centric opportunities could be grabbed by establishing processing unit's at large scale at district level reducing overall wastage to its minimum. Semi processing units could be established by government agencies/entrepreneurs nearer to apple producing areas. Small scale food parks can be developed at various center points of districts areas facilitating packaging, semi processing, grading and value addition of apple. Such food parks could also become the centres for imparting know how of marketing and communicating modern technologies through exhibitions. This could bring in and encourage local involvement, provided such practices are undertaken on sustainable basis.

☆ Processing facilities matching the latest standards of operation need to be adopted. The aim has to be maximum, yet acceptable output judged by quality standards. Obsolete technology needs to be shunned off and technology upgradation is the need of the hour for the existing processing plants. Mechanized sorting, grading, handling systems, cold storages, processing equipment's with appropriate technical standards should be facilitated.

☆ The fruit processing industries should adopt all sorts of best management practices like SPC (Statistical Process Control), SQC (Statistical Quality Control), TQM (Total Quality Management), *etc.* to make the processes error free and fool proof, which will subsequently result in final products with zero defects.

☆ Setting up/upgradation of quality control/food testing Lab, R and D and promotional schemes should be launched by the Government for enhancing the product quality upto the export standard. Improving the raw material standards and strengthening food testing network;

strengthening institutional framework to develop manpower for improving R and D capabilities to address global needs and expectations. Continued monitoring of products has to be done to ensure their confirmation with quality and safety considerations. Along with this, improvement and reformation in the products as per consumer needs could also help in maintaining the acceptability.

☆ Utilization of apple processing waste could give birth to a separate industry in itself, with a span of wide extension. The large quantity of apple pomace produced during apple processing suggests that the preparation of single product would not be economically feasible and production of all possible products needs to be explored. The commercial utilization of pomace shall ultimately be determined by economics of products and the cost of waste disposal coupled with pressure from environment protection agencies in implementing the laws (Kaushal *et al.*, 2002).

☆ A strong marketing support needs to be provided to local apple processing industry so as to make it sustainable in national and international markets. Identify market out lets for the product by the state and initiatives like minimum support price for processed apple products are some suggestions worth consideration.

☆ Technocrats with specific skills for processing of foods need to be encouraged to take apple processing as source of their livelihood. Special loans, subsidies and tax rates be provided to such entrepreneurs as incentives.

References

Agricultural Marketing Service (AMS) 2011. *National Apple Processing Report.* USDA Agricultural Marketing Service, Fruit and Vegetable Programs.

Anon 2013-14. Govt. of Jammu and Kashmir RFD (*Results-Framework Document*) for Horticulture (2013-2014)

Bhat, T.A. and Choure, T. 2014. Status and strength of apple industry in Jammu and Kashmir. *International Journal of Research 1(4):277-283.*

Buyer Seller meet and Conference on Apple, Temperate Fruits and Nuts of Jammu and Kashmir. 2007. J&K Institute of Management, Public Administration, and Rural Development, Srinagar.

Corcuera, J.I.R., Schneider, R.M.G., Barringer, S.A. and Urbina, M.A.L. 2014. Processing of fruit and vegetable beverages. *In :Food Processing: Principles and Applications* (editors Clark, S., Jung, S. and Lamsal, B.). 2nd edition. John Wiley and Sons Ltd.

Downing, D. L. 1989. *Processed apple products.* Van Nostrand Reinhold, New York.

Economic Research Service, USDA and Economic Survey, 2012-13, http://www.jandkplanning.com/images/Economic_Survey/17-horticulture. pdf

Food and Agriculture Organization (FAO) (2010) FAOSTAT. http://faostat3.fao. org/home/index.html.

Foreign Agriculture Service, USDA, 2016

Gale, F., Huang, S., Gu, Y. 2010. Investment in processing industry turns Chinese apples into juice exports. *Economic Research Service*, USDA.

Greater Kashmir News Paper, 2007, 2008 and 2016, Srinagar J&K.

Hanan, E. 2015. Entrepreneurship perspective for trade and management of horticulture sector in Kashmir Himalayan valley. *Inter. J. Social Sci. Mana.* 2(3): 284-289.

Huda AB., Parveen S., Rather S.A., Akhter R., Hassan M.2014. Effect of incorporation of apple pomace on the physico-chemical, sensory and textural properties of mutton nuggets. *Int. J. Adv. Res.* 2(4) :974-983.

Indian Horticulture Database. (2014). National Horticulture Board, New Delhi

Joshi V., Gupta K., Devrajan A., Lal B., Arya S. (2000). Production and evaluation of fermented apple pomace in the feed of broilers.*J. Food Sci. Technol.* 37:609-612.

Joshi, V.K. and Sandhu D.K. 1996. Preparation and evaluation of an animal feed byproduct produced by solid-state fermentation of apple pomace. *Bioresour. Technol.* 56:251-255.

Kaushal, N.K., Joshi, V.K. and Sharma RC 2002. Effect of stage of apple pomace collection and the treatment on the physical-chemical and sensory qualities of pomace papad (fruit cloth). *J.Food Sci. Technol.*39:388-393.

Lu, Y. and Foo, L.Y. 2000. Antioxidant and radical scavenging activities of polyphenols from apple pomace. *Food Chem.*68:81-85.

Malik, Z.A. 2013 Assessment of apple production and marketing problems in Kashmir Valley. *Journal of Economic and Social Development* 9:52-156.

Malik, Z.A., and Choure, T. 2014. Horticulture growth trajectory evidences in Jammu and Kashmir (A lesson for apple industry in India). *Journal of Business Management and Social Sciences Research* 3:45-49.

Masoodi, F.A. and Chauhan, G.S. 1988. Use of apple pomace as a source of dietary fibre in wheat bread. *J. Food Process. Preserv.* 22:255-263.

Masoodi, FA., Sharma, B. and Chauhan G.S. 2002. Use of apple pomace as a source of dietary fiber in cakes.*Plant Foods Hum. Nutri.* 57:121-128.

Mir, S.M. 2014. Problems of apple industry in J&K with special reference to Sopore town. *IJMSS*2: 33-46

Murtaza, Z. 2015. Horticulture and its role in the economic development (An empiricalstudy of Kashmir valley). *IJMSS*3:240-249.

Production, Supply and Distribution online, USDA.

Rabetafika, H.N., Bchir, B., Blecker, C. and Richet A. 2014. Fractionation of apple by-products as source of new ingredients: Current situation and perspectives. *Trends Food Sci. Tech.* 40:99-114.

Rather, S.A., Akhte, R., Masoodi, F.A., Gani, A. and Wani, S.M.2015. Utilization of apple pomace powder as a fat replacer in goshtaba: a traditional meat product of Jammu and Kashmir, India. *J. Food Meas. Char.* 9:389-99.

Rather, S.A., Masoodi, F.A., Akhter, R., Gani, A. and Wani, S.M. 2016. Effect of apple pomace powder on the physico-chemical and sensory characteristics of low fat Rista, a traditional meat product of India. *AnimSci J.* http://dx.doi.org/10.1111/asj.12684.

Root, W.H. and Barrett 2005. *Apples and apple processing*. In Processing fruits CRC Press.

Sandhu D.K. and Joshi V.K. 1997. Solid state fermentation of apple pomace for concomitant production of ethanol and animal feed. *J. Sci. Ind. Res.* 56: 86-90.

Shah, G.H. and Masoodi, F.A. 1994. Studies on the utilization of wastes from apple processing plants. *Indian Food Packer* 48(5):47-52.

Shalini, R. and Gupta, D. K.2010. Utilization of pomace from apple processing industries: a review. *J. Food Sci. Technol.* 47(4): 365-371.

Sheikh, S.H. and Tripathi, A.K. 2013. Socio-economic conditions of apple growers of Kashmir Valley: A case study of district Anantnag. *International Journal of Educational Research and Technology* 4(1): 30-39

Sudha, M.L., Baskaran, V. and Leelavathi, K. 2007. Apple pomace as a source of dietary fiber and polyphenols and its effect on the rheological characteristics and cake making. *Food Chem.*104:686-692.

Tali, A.H. 2014. Dynamics of horticulture in Kashmir, India. *Int. J. Curr Sci.*11: 15-25.

Van der Sluis, A.A., Dekker, M., de Jager, A. andJongen, W.M.F. 2001. Activity and concentration of polyphenolic antioxidants in apple: Effect of cultivar, harvest year, and storage conditions. *J. Agric. Food Chem.* 49: 3606-3613.

Van der Sluis, AA., Dekker, M., Skrede, G., Jongen, W.M.F. 2002. Activity and concentration of polyphenolic antioxidants in apple juice: 1. effect of existing production methods. *J. Agric. Food Chem.* 50: 7211-7219.

Verma, A.K., Sharma, B.D. and Banerjee R. 2010. Effect of sodium chloride replacement and apple pulp inclusion on the physico-chemical, textural and sensory properties of low fat chicken nuggets. *LWT Food Sci. Technol.* 43:715-719.

Verma, L.R. and Joshi, V.K. 2006. *Post harvest Technology of Fruits and Vegetables.* Indus Publishing Company New Delhi. Volume 2.

Chapter 26

Economic Viability of Apple Processing

F. A. Shaheen, S. A. Wani, Farheen Naqash and Haris Manzoor

Network project on Market Intelligence (ICAR),
School of Agricultural Economics and Horti-Business Management,
Sher-e-Kashmir University of Agricultural Sciences and
Technology of Kashmir, Srinagar
E-mail: fashaheen@yahoo.com

1. Introduction

Apples are processed into a variety of products, but by far the largest volume of processed apple products is in the form of juice. Apple juice is processed from apples that are unsuitable for peeling, such as "eliminator" apples, smaller than 57mm diameter, too small to peel, *etc.* Apple juice can be produced and sold in several forms. Fresh apple juice or sweet cider is juice of ripe apples, bottled or packaged with no form of preservation. This form needs to be sold at the orchard or at outlets close by. Even under these conditions it is important to pasteurize the juice to eliminate *E. coli* or other dangerous organisms. Processing of apple is important as a considerable part of the produce will always remain of too low grade to sell for fresh consumption. Processed apples into whatever product can often be better stored and/or have less weight or volume which reduces transport costs (Chengappa, 2004; Deepa, 2008; Jairath, 1996).

There are a number of procedures employed in apple juice production, depending upon the end product desired (Rosa, 1998). Figure 26.1 is a generalized flow scheme for producing some of these products.The Kashmir province of Jammu and Kashmir state specialized in temperate horticulture due to favourable agro-climatic niches produces fruits such as apple, pear, apricot, peach, plum, cherry and grapes besides other nut crops. The total production of fresh fruits during the year 2013-14 was reported 18, 55,000 metric tonnes (MT), out of which apple contributed about 78.51 per cent. The apple crop dominates the horticultural industry and has an important role in economic scenario of the state by contributing more than 4500 crores to State Gross Domestic Product (SGDP). Involving around half a million households, apple plays a key role in the rural economy of the state. Nearly 30 per cent of total produce of apple crop goes waste due to pre-harvest drop, making total annual quantum of such fruit about 0.25 million MT (Shah, 1999). The apples which are wasted due to pre-harvest drop, under development of colour, inferior grade and other reasons are utilized for the purpose of processing. These apples cannot be marketed as they give negative returns to growers. Due to non-availability of adequate processing facilities in the state, such fruits do not find an appropriate out let in the market. Though there have been multi-dimensional efforts to increase the production of apple in the state but processing sector has not received proper attention. As a sequel to present study, processing an important link in the value chain was also evaluated (Sidhu, 2005; Srivastava, 1989; Tripathy, 2006).

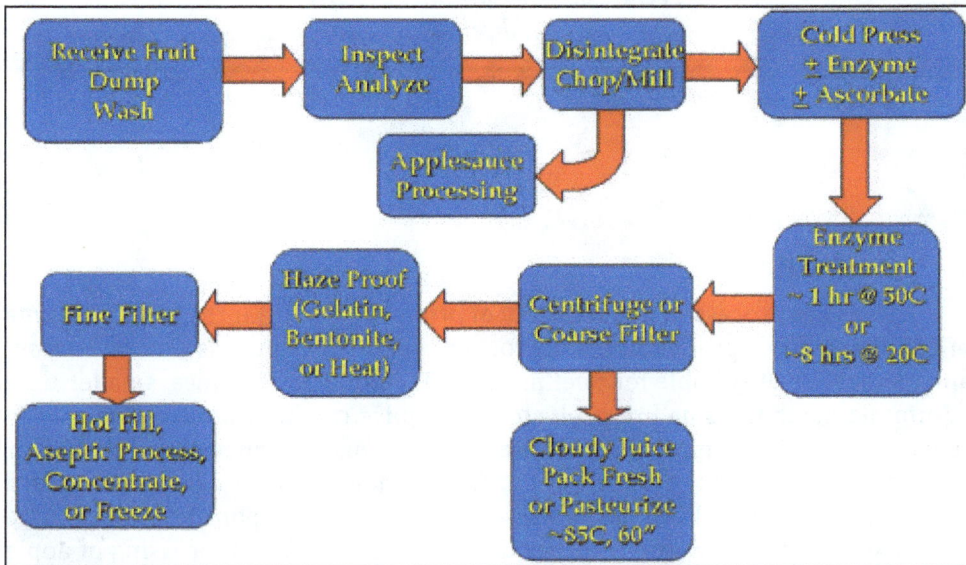

Figure 26.1: Generalized Flow Scheme for Processing.

As per the data with Ministry of Food Processing Industries (MoFPI), there are about 136 registered processing units in Jammu and Kashmir under Food

Processing Industry (FPI). The key sub-sectors in the state include grain-based processing industries (including *namkeen*, snacks *etc.*), fruit-based processing, floriculture related activities, spice processing, candy/pickles and other horticulture based industries(Shaheen and Gupta, 2004). APEDA has identified Kashmir as Agri- Export Zone for apple. Two major processing plants are presently operating in Kashmir with a total annual installed capacity of 70,000 MT to process raw apple culls. The processing plant, owned by Jammu and Kashmir Horticulture Processing and Marketing Corporation (JKHPMC) is located at the hub of apple producing area, *viz.* Sopore of Baramulla district. The plant with installed capacity of 10,000 M.T. was established by CADBURY, India Pvt. Ltd. in early eighties and was purchased by JKHPMC in nineties. The other processing plant with an annual installed capacity of 60,000 MT was established in the year 1999, by a private entrepreneur, *viz.* FIL Industries at Rangreth, Budgam.

The low grade apples *i.e.* Grade C are usually taken up for processing. Of the 18 processing plants set up only two are operational. The problems range from raw material scarcity, high cost of transport, law and order problems, competition from imports and unsuitable technology. The market intervention scheme run by government to procure low grade fruits from farms and supply the same at subsidised prices to processors was discontinued few years before leading to closure of some units. The current tendency of Grade C apples being mixed with better grades has to be discouraged by providing a market for these apples. Local processing plants for juice extraction (that avoid high transport cost of raw material) and centralised plants for making juice concentrates might provide a solution. But this needs to be backed up by a procurement chain that offers viable prices to farmers and ensures that landed cost of raw material is reasonable in the hands of processors. Some transport subsidy to bring the grade C apples to the plants might become necessary in the initial years when the investment cost overhang on new plants is still high. The potential for the establishment of 11 processing plants for a total capacity of 2.8 lakh tons in different districts (Table 26.1) (NABCONS, 2013).

Table 26.1: Potential of Apple Processing in Jammu and Kashmir

Sl.No.	District	Potential of Processing (MT)	Units Proposed
1.	Baramulla	183,000	40,000 MT- 4 units
2.	Kupwara	44,000	20,000 MT- 2 units
3.	Anantnag	17,500	10,000 MT- 1 unit
4.	Kulgam	20,000	20,000 MT- 1 unit
5.	Shopian	32,700	20,000 MT- 1 unit
			10,000 MT- 1 unit
6.	Pulwama	23,500	20,000 MT- 1 unit
	Total	320,700	40,000 MT- 4 units, 20,000 MT- 5 units and 10,000 MT- 2 units- Total 280,000 MT

Source: NABCONS, 2013.

For the present study mega apple juice plant, Budgam was purposively selected to evaluate the economics of apple juice processing in the value chain because from last two years JKHPMC plant is non-functional. The three years cross sectional data was collected from the processing plant, Budgam for the reference years 2011-12 to 2013-14, to avoid any abnormal production period (Naqash Farheen, 2015).

2. Processing Cost of Apple

The component-wise cost of apple processing (in lakhs) presented in Table 26.2 was studied under variable and fixed costs. More than 88 per cent of the total processing cost comprised of variable cost. The expenditures on raw material contributed about more than 84 per cent of the total variable cost followed by repairs and maintenance, power expenses, salaries and wages, designing charges, administrative expenses and water expenses that together constituted more than 6 per cent of the total variable cost followed by other items of expenditure. The total variable cost on an average was computed as Rs. 1711.56 lacs per annum. The scope lies for reduction of cost on the power component if power could be supplied to the plant continuously. Presently, the plant runs on diesel because of non-availability of continued power supply and can be minimized considerably if power supply provided continuously by the state electricity board.

Table 26.2: Processing Cost of Apple Juice Plant (Rs. In lakhs)

Sl.No.	Particulars	Year			Average
		2011-12	2012-13	2013-14	
I	**Variable cost**				
1	Raw material	1742.99	1427.14	1169.89	1446.67 (84.52)
2	Power expenses	30.89	25.59	21.23	25.90 (1.51)
3	Water expenses	0.62	0.52	0.43	0.52 (0.03)
4	Designing charges	19.29	15.98	13.26	16.18 (0.95)
5	Salaries and wages	23.43	21.3	19.36	21.36 (1.25)
6	Repair and maintenance	49.45	40.96	33.98	41.46 (2.42)
7	Administrative expenses	11.95	10.86	9.87	10.89 (0.64)
8	Other items of expenditure	167.98	147.64	130.08	148.57 (8.68)
a	Total variable cost	2046.6	1689.99	1398.10	1711.56 (100.00)
II	**Fixed costs**				
1	Interest on term loan and working capital	139.14	117.01	101.67	119.27 (52.63)
2	Depreciation	31.54	31.54	31.54	31.54 (13.92)
3	Tax provisions	96.17	74.69	56.58	75.81 (33.45)
b	Total fixed cost	266.85	223.24	189.79	226.63 (100.00)
	Total processing cost (a+b)	2313.45	1913.23	1587.89	1938.19

Source: Field survey, 2015; Figures in parentheses indicate per cent of total variable and fixed cost.

The fixed costs (Rs. 226.63 lacs) included the expenditure incurred on taxes in the form of toll tax, sales tax, licensing fee, and insurance premium, interest on working capital and depreciation on the plant. This cost constitutes about more than 11 per cent of the total processing cost. The major cost component of fixed cost included interest on working capital (52.63 per cent), tax provisions (33.45 per cent) and depreciation on plant (13.92 per cent). The average processing cost of Apple Juice Concentrate (CAJ) worked out to be Rs. 1938.19 lacs.

3. Economics of Apple Processing

Economics of apple processing is presented in Table 26.2. The average capacity utilization of plant worked out to 66.75 per cent and ranged in between 60 per cent and 70 per cent. Average revenue realized out of final product *viz.* Concentrated Apple Juice (@ Rs. 75/kg) worked out to Rs. 2151.21 lacs by processing raw apple culls. It can be observed from the table that the revenue realized had decreased continuously over the period of three years due to decline in capacity utilization.

On an average, plant realized a net return of Rs. 213.02 lacs per annum. The highest net return of Rs.270.21 lakh was in the year 2011-12, while it was lowest (Rs. 158.97 lakh) in the year 2013-14. The price of final product (CAJ) was Rs. 2151.21 lakh. Overall, the results indicate that the processing unit had not fully utilized the installed capacity. Utilization of the plant to its installed capacity will not only reduce the cost of CAJ production and increase the net returns, but also will benefit the producers (orchardists) by utilizing their apple culls.

Table 26.3: Economics of Apple Processing (Rs. in lacs)

Sl.No.	Particulars	2011-12	2012-13	2013-14	Average
1.	Capacity utilized (per cent)	73.21	66,55	60.50	66.75
2.	Apple juice recovery	2583.66	2123.11	1746.86	2151.21
3.	Total processing cost	2313.45	1913.23	1587.89	1938.19
4.	Net returns	270.21	209.88	158.97	213.02

Source: Field survey, 2015.

References

Chengappa, P.G. 2004.Emerging trends in agro processing in India. *Indian Journal of Agricultural Economics* 59(1): 55-74.

Deepa, D. 2008. Indian product, "Brief India fresh fruit sector", *Holy Higgins,* US Embassy.

Jairath, M.S. 1996. Agro-processing and infrastructure development in hilly area: A case study of fruit and vegetable processing. *Indian Journal of Agricultural Marketing* 10(2): 28-47.

NABARD Consultancy Services 2013.*A proposal for strengthening the apple value chain in Jammu and Kashmir.* NABCONS, Corporate Office, BandraKurla, Mumbai 29 pp.

Naqash, Farheen. 2015. A value chain analysis of apple in Jammu and Kashmir. *MSc. Thesis submitted to Sher-e-Kashmir University of Agricultural Sciences and Technology of Kashmir, Shalimar*, pp. 94.

Rosa, S.1998.*Outlook for concentrated apple juice production and trade in selected Countries*, World Horticulture Trade and US Export Opportunities. 4: 26-32.

Shah, G. H. 1999.Present status and future potential of fruit processing industry in J&K state, paper presented at seminar on *'Fruit Processing in J&K'* organised by FICI New Delhi and J&K SIDCO on 5-6-1999, held at International Convocation Centre, Srinagar (J&K).

Shaheen, F.A. and Gupta, S.P. 2004. Economics and potentials of apple processing industry in Kashmir Province of Jammu and Kashmir. *Agricultural Marketing* XLVII (3): 31-34.

Chapter 27

CA Storage of Apple

S. A. Wani, Farheen Naqash, F. A. Shaheen and Haris Manzoor

Network project on Market Intelligence (ICAR),
School of Agricultural Economics and Horti-Business Management,
Sher-e-Kashmir University of Agricultural Sciences and
Technology of Kashmir, Srinagar
E-mail: dr.shabirwani@rediffmail.com

1. Introduction

Since most processors cannot use the whole harvest they receive, some fruit is stored, for short term. Other fruit is stored refrigerated in a temperature range of 1 to 4°C, depending on the cultivar. The next level of storage is controlled atmosphere (CA). CA storage usually consists of a modified atmosphere, 2 to 3 percent oxygen and 1 to 4 percent carbon dioxide, at a reduced temperature. The exact specifications are adjusted to the cultivar being stored. Apples can maintain quality under these conditions for 4 to 6 months. Only the highest quality apples destined for the fresh market are placed in CA storage. However, many times the fresh market price will drop to the point that CA apples will be dumped to a processing market. Apples from CA storage should be allowed to "normalize" for a few days before processing. These apples from refrigerated storage are capable of producing good quality processed product. Processors take into consideration that different qualities of juice or apple sauce can be manufactured from the same cultivar, depending on the type of storage, time of storage and stage of maturity when processed.

Advances in controlled atmosphere technology have had a dramatic effect on apple storage logistics and opened up markets hitherto unavailable for fresh and processed apple products. This is an advantage not fully shared by other fruit crops whose shelf life extension by CA is much less.

(a) Storage Facilities

There are essentially three types of storage buildings for apples: air cooled storage, mechanically refrigerated storage and refrigerated and controlled-atmosphere (CA) storage.

(i) Air-cooled Storage

These storage houses cool by admitting cold night air (applicable climates) at inlets near the floor of an insulated building and forcing upper accumulated warm air out at outlets near the ceiling. Both openings are closed during the day. These storages are economical and effective in areas where the night air becomes cooler than the accumulated air in the storage house.

(ii) Mechanically Refrigerated Storage

For longer periods of storage than is afforded by the air-cooled storages, mechanical refrigeration is needed. This would become necessary to extend the season on fresh-market apples or to extend the cooling of processing apples because the volume is so large that they cannot be completed in the period of time afforded by the air-cooled storage situation. Specifications of room size, room construction, capacity, compressors, condensers, expansion coils, *etc.* for both mechanically refrigerated storage and CA storage can be found in Childers manual.

(iii) Controlled Atmosphere (CA) Storage

It is a widely used technique for long-term storage of freshly picked fruits and vegetables. Historically, CA storage has been the primary method for the long-term storage of apples. Through a biological process called respiration, apples take in oxygen and generate carbon dioxide, water, and heat. Controlled Atmosphere storage is an entirely natural process that reduces the effects of respiration to a minimum by controlling the environmental conditions surrounding the stored fruit. CA storage makes it possible for consumers to buy crisp, juicy apples year round. Many cultivars of apples can be preserved for a remarkable 9–12 months in CA storage, as opposed to only 2–3 months if using refrigerated storage (GOI, 2001; 2001b).

(b) Harvesting Apples at Optimum Maturity

For successful controlled atmosphere (CA) storage, harvest apples when they are physiologically mature but not ripe. Harvest each cultivar at the proper maturity to achieve maximum storage life and marketing season. Apples harvested too early are of poor colour and small size and have little flavour. They may fail to ripen or ripen abnormally, and the overall quality will be poor. Characteristics of immature apples that contribute to inadequate flavour development include high water loss, low sugar content, high acidity, low aroma volatile production and high starch content. Immature apples are also more likely to develop storage disorders such as superficial scald and bitter pit.

Harvesting apples too late can result in a short storage life. Such apples are too soft for long-term CA storage and are more susceptible to mechanical injury and disease infection. Over-mature apples may develop poor eating quality and off-flavours and are more susceptible to watercore and internal breakdown. For these reasons, determining optimum apple maturity for harvest is essential for maximizing storage life and quality, while minimizing post harvest losses. Numerous methods have been suggested for determining harvest date, but no single test is completely satisfactory, and some are too unpredictable, complicated or expensive.

Days after full bloom for a given cultivar provide an approximate date of harvest maturity. Confirm the date using tests such as internal ethylene concentration (IEC), starch-iodine staining, flesh firmness and soluble solids content (sugars). In general, an IEC of 1 ppm is considered to be the ultimate threshold above which fruit ripening and flesh softening are initiated and progress rapidly.

Complete harvest for long-term storage before 20 per cent of the apples has an IEC greater than 0.2 ppm. Using the starch-iodine test, apples destined for long-term storage should have 100 per cent of the core tissue starch degraded (no stain) with greater than 60 per cent of the flesh tissue still having starch present (stain). It is important to note that not all apples mature and ripen in the same manner each year. Often there will be a need to compromise between correct maturity and the required firmness and sugar levels for market (Khan *et al*, 2006).

(c) Guidelines for Placing Apples into CA Storage

Segregate apples into lots at harvest by their storage potential. The following types of apples are not suitable for long-term storage because of their potential for internal breakdown (or developing bitter pit):

☆ Large fruit from lightly cropped tree

☆ Fruit from excessively vigorous trees

☆ Fruit from young trees just coming into bearing

☆ Fruit from heavily shaded interior parts of trees

☆ Early-harvested fruit high in starch

☆ Fruit with a low number of seeds

After harvest, cool the apples as rapidly as possible. Fruit of the tree mature much faster; with warmer temperatures, fruit begin to ripen sooner. Try to get the harvest from each day into the cooler by nightfall without straining the capacity of your cooling system to the detriment of apples already pre-cooled and in storage.

When using CA storage, the quicker the apples are cooled and the desired atmosphere is achieved, the longer the apples will store and be of good quality upon removal. The longer it takes to adjust the oxygen (O_2) and carbon dioxide (CO_2) levels, the less effective the length of storage will be. The objective should be to cool the apples and achieve the desired atmosphere within 5 days of initial harvest. CA storage will not improve fruit quality - place only the best fruit in CA storage. If over or under mature or poor quality apples are put into CA, the result will be poor-quality apples upon removal. Successful CA storage begins with harvesting apples at the proper maturity, followed by rapid cooling and establishment of the CA, then proper maintenance of the desired temperature and atmosphere. In general, the standard CA recommendations range between 2.5 per cent -3 per cent O_2 and 2.5 per cent -4.5 per cent CO_2 at 0°C-3°C. Due to recent research using new storage technologies and strategies, cultivar-specific CA recommendations have been reviewed.

2. Situation Analysis of CA Storage

(a) Integrated Cold Chain in Jammu and Kashmir

The peak harvest season witnesses a glut in the market and depresses the price realisation. This is caused by the absence of viable infrastructure to pack, transport and store apple in a manner designed to preserve quality and release the same in the market when the prices are attractive. Most of the outflow of the apples takes place during September to December, being the period of peak harvesting arrivals. The farmers will benefit from arrangements that reduce the peak arrivals during September to December, and delay the marketing till March-June of the next year. This will be possible if adequate storage facility suitable for apples is created and more importantly the holding capacity of farmer is increased through credit for stored apple.

There are 18 operational Cold Storages (CS) with a total capacity of 49769 MT in the State of Jammu and Kashmir; all of which are located in Jammu. These cold storages are multipurpose with a part of their capacity (say 30 per cent, being about 15000 MT) used for storage of apples. There are 8 Controlled Atmosphere (CA) Storages in the Kashmir valley with a total capacity of 38000 MT (Table 27.1). The average capacity utilization is 60 per cent during peak season and the annual average capacity utilization is estimated to be around 40 per cent. The capacity is not fully utilized due to lack of awareness about grading, packing and storage as well as prevailing trading systems in the valley that focus on immediate sale after harvest. Due to seasonal nature of apple and no other commodity being stored in CA stores, capacity utilisation is low. Better scientific cultivation practices, training, pruning, pre-harvest foliar sprays, proper harvesting and handling practices are likely to improve the availability of produce suitable for CA storages. Although the importance and benefits of proper sorting, grading and packing is known to some of the growers, the same is done manually by the growers before bringing the same to the market yards. The different grades are mixed up in manual sorting and grading which adversely affects the price realization to the growers and also reduces the bargaining power of the growers due to mixing of different grades. None of the market yards in the valley are equipped with any sorting, grading and packing facilities. Electronic sorting, grading and packing facilities are available with the CA storages in the valley.

Table 27.1: Functional CA Stores in Kashmir Valley

Sl.No.	Name of the CA Store	Location	Capacity (MT)
1	Valley Fresh Cold Chain Pvt. Ltd.	Lassipora, Pulwama	5000
2	I-Fresh, Kehwa Square Pvt. Ltd.	Lassipora, Pulwama	5000
3	Golden Apple Agro Fresh Pvt. Ltd.	Lassipora, Pulwama	2000
4	Harshana Naturals Agri Serve Pvt. Ltd.	Lassipora, Pulwama	5000
5	Fruit Master Agro Fresh Pvt. Ltd.	Lassipora, Pulwama	5000
6	Shaheen Agro Fresh Pvt. Ltd.	Lassipora, Pulwama	5000
7	Kashmir Premium Apples Pvt. Ltd.	Lassipora, Pulwama	5000
8	FIL Industries Pvt. Ltd.	Rangreth, Budgam	5000

Source: Field Survey, 2015.

The requirements of storage in 2017-18 for even 25 per cent of incremental market arrivals is likely to be about 3 lakh tons. A study after examination of production and market clusters has proposed creation of cold store capacity of 1.6 lakh tons and CA store capacity of 3 lakh tons. This will form less than 10 per cent of the grade A and B apples produced in 2017-18. However, the proposed storage investments should be seen as proof of the concept which will induce more private investments riding on increased demand for storage. Refrigerated transport using reefer vans will be a necessity when the fruits are stored in CA stores. Under such situation we have to introduce 20 such reefer vans with a capacity of 10 tons each. When their functionality is established, more private sector investments will flow in refrigerated transport. Twenty seven automated sorting, grading and packing facilities each with capacity to handle 2 tons of apple per hour are also needed. When linked to CA stores or cold stores these lines would offer a complete packing cum storage solution to farmers. (NABCONS, 2013)

(b) Cold Stores and CAS around Delhi

There are approximately 100 cold stores and controlled atmosphere stores (CAS) which were set up recently in Industrial Growth Center at Kundli (60 units), Industrial Area Rai (30 units) and 10 units under construction, in Sonepat district of Haryana near Delhi–Haryana border, which is just 10 km from Tihri Khampur check post. Average capacity of each store is 1.50 lakh boxes (maximum up to 5 lakh boxes per store). Total storage capacity is 150 lakh boxes of apple or *3 lakh MT*. Currently 50 per cent or 1.50 lakh MT capacity is used for apples and 50 per cent (1.50 lakh MT) for other fruits and green vegetables, pulses, seeds and agri. commodities. Pattern of utilization among commodities may be changed within the maximum available capacity of 3 lakh MT depending upon demand and season. Of total capacity of 1.50 lakh MT currently earmarked/used for apple, 50 per cent (0.75 lakh MT) is used for Kashmir apple and 50 per cent (0.75 lakh MT) for apple arriving from Himachal Pradesh; this ratio may also be changed depending upon demand, requirement and profitability of storage. These cold stores/CAS units are owned by individual private entrepreneurs, companies, procurers, *etc.* and some of which are also set up by commission agents of Delhi market for storage of owned apple and renting out excess space to other wholesalers. During 2012, 10-12 new units are came up in Sonepat with average capacity of 2 lakh boxes each.

Besides, there are 7 cold stores/controlled atmosphere stores within principal market yard of Delhi Azadpur market. These cold stores and controlled atmosphere stores in Delhi, Kundli and Rai (Sonepat border) are instrumental in enhancing the buying power, control and hoarding (storage) of apple from Kashmir by commission agents and traders who store it for 2-3 months initially starting with July/August only to effect off-take of stored produce in October-November and December around Diwali festival in phases of staggered supplies to the local Delhi and outside markets at higher price. Some of them take out stored apple around May-June in off season to supply it to the metropolitan cities of Mumbai, Bangalore and other big cities like Ahmedabad and South India. In off season, stored apple sells at higher price up to Rs.2000/- per box whereas its seasonal price may be Rs.800/- per box at the most. Therefore, it is the commission agents, wholesale buyers, companies, organized retailers, procurers, traders, *etc.* who make profits out of stored apple in Kundli, Rai and Delhi. Grower rarely stores in these cold stores except under compulsion when so-called "excess" or unsold apple in auction is diverted to these cold stores at extra storage and transportation charges incurred by growers only to be sold at lower price later (or at the most same price) under distressed conditions.

Other concurrent major developments that took place during first decade of the new millennium were headways made by Indian corporate (industrial) sector by way of investment in apple business and marketing for organized retail trade. Major industrial companies entering in apple market around the year 2007 were Reliance, Bharti, Mahindra, Field Fresh, Adani, UniFruti, *etc.* They would buy directly from orchards as also in APMC markets (Parimpora in Srinagar, *etc.*). Bharti Group is now reportedly doing market research to re-enter Kashmir market for wholesale buying of apple. Presently, there is no domestic industrial/corporate sector company operating in Kashmir. Adani has shifted to Himachal Pradesh because Shimla/Kinnaur apple is regarded as cheaper and of better quality (crunchy, colour) which suits the requirements of organized retail business, more than what is possible with Kashmir apple (NABARD, 2013).

3. Costs and Returns from Storage

(a) Estimation Procedure

The gross return on storage may be defined as the increase in the price of the stored product at the time of storage till it is "de-stored" and either sold or consumed. The cost of storage should include the following:

1. The cost of the maintenance of the storage structure *i.e.*, depreciation, repairs, insurance and interest on sunk capital; or, alternatively, the rent paid for hiring the storage structure;

2. Interest on the value of the stored goods;

3. Value of the quantitative and qualitative loss during storage;

4. Risk premium for a possible price fall and damage during storage;

5. The cost of protective materials, for example, insecticides, pesticides. The costs and returns from the storage were determined to know whether it pays a farmer to store his farm produce and it was worked out with the help of the following formula, given by Acharya and Aggarwal, 2011:

$$NR = GR- C$$

where,

NR = Net returns to storage

GR = $P_1 - P_0$ (Gross Returns)

P_0 = Purchase price or market price at the time of storage

P_1 = Selling or market price at the time of de-storing

C = Cost involved in storage

NR = > 0, implies positive returns on storage

NR = <0, implies negative returns on storage.

(b) Economics of Apple CA Storage

CA storage was found to be an important link in the value chain which was considered and the economics of CA storage also worked out. The results depicted in Table 27.2 showed that more than 67 per cent of the total production cost was constituted by variable cost. The wages and salaries to staff and casual labourers turned to be the major cost component of the variable cost on which plant spends more than 98 lacs. The total variable cost on an average was computed as Rs. 626.70 lacs. The second important cost item includes utilities which account for 80.07 lacs. The costs on waxing of fruit (@ 50 paisa/kg) and advertising expense accounts for 28.34 lacs and 24.69 lacs, respectively. The fruits are purchased @Rs. 23/kg and costs about 388.13 lacs on an average. The other cost components are administrative expenses, rent and insurance, pre-operative expenses and repairs and maintenance which together accounted for about 7.01lakhs.The fixed costs included interest on working capital and term loan and depreciation on the plant.

Table 27.2: Cost of Apple Storage, 5000 MT Cold Storage (Rs. in lakhs).

Sl.No.	Particulars	YR 1	YR 2	YR 3	YR 4	YR 5	YR 6	Average
	Capacity Utilization	50 per cent	60 per cent	70 per cent	70 per cent	70 per cent	70 per cent	65 per cent
1	Revenue	981	1177.2	1373.4	1373.4	1373.4	1373.4	1275.3
2	**VARIABLE COSTS**							
A	Purchase of fruit	345	362.25	379.5	396.75	414	431.25	388.13
B	Waxing of fruit	25	26.25	27.56	28.94	30.39	31.91	28.34
C	Utilities	54.10	64.92	90.35	90.35	90.35	90.35	80.07
D	Salaries and wages	75.74	90.89	106.04	106.04	106.04	106.04	98.47
E	Repair and maintenance	1	0.7	0.8	0.8	0.8	0.8	0.82
F	Rent and insurance	1.5	1.2	1.25	1.25	1.25	1.25	1.28
G	Advertising expenses	14.72	23.54	27.47	27.47	27.47	27.47	24.69
H	Pre-Operative expenses	5	0	0	0	0	0	0.83
I	Administrative expenses	2.4	2.88	4.8	4.8	4.8	4.8	4.08
	Total Variable cost	524.46	572.64	637.77	656.40	675.10	693.86	626.70
3	Gross Profit	456.54	604.56	735.63	717.00	698.30	679.54	648.60
4	**FIXED COSTS**							
A	Interest on Term loan	135.32	135.32	108.26	81.19	54.13	27.06	90.21
B	Interest on working capital	16.17	16.36	16.58	16.58	16.58	16.58	16.47
C	Depreciation	218.74	218.74	218.74	218.74	218.74	218.74	218.74
5	Total fixed cost	370.23	370.42	343.57	316.50	289.44	262.38	325.42
6	Total storage cost	894.69	943.06	981.34	972.90	964.54	956.24	952.13
7	**NET PROFIT**	86.31	234.14	392.06	400.50	408.86	417.16	323.17
8	Return to variable costs	1.87	2.06	2.15	2.09	2.03	1.98	2.03
9	Return to total costs	0.10	0.25	0.40	0.41	0.42	0.44	0.34

Source: Field Survey, 2015.

This cost constitutes about 34.17 per cent of the total production cost. The major cost component of fixed cost is depreciation on plant (@11 per cent per annum) which amounts on an average about 218.74 lacs followed by interest on term loan (90.21 lacs) and interest on working capital (16.47 lacs) both at the interest rate of 15 per cent per annum. The capacity utilized by the plant had increased from 50 per cent to 70 per cent. The average capacity utilization of plant worked out 65 per cent. On an average, the revenue released was Rs. 1275.3 lacs. It can be observed that the revenue has increased continuously over the period of first three years and then remained constant. The gross profit on an average worked out to be 648.60 lacs with a net profit of about 323.17 lacs.

(c) Effect of cold storage capacity on prices

It was observed that both absolute and relative price variability decreased for apple crop after the promotion of cold storages in Pulwama market for the entire period. However, increased arrivals were witnessed particularly during the storage period. After setting up of cold storages, the tendencies of price fluctuations were minimal. This indicated that with the promotion of cold storages, fluctuations in prices were reduced and this has helped in achieving the price stabilization. These results were in close proximity with the findings of Jairath, 2000, 2002, 2004.

(d) Costs and Returns from Storage

The costs and returns from storage were also worked out to study whether it is profitable for farmer to store his farm produce or not. The results from Table 27.2 revealed that the net returns from storage turned greater than zero in all the three cases, hence, reflecting positive returns on storage. It was observed that the cold storages were utilizing 50 per cent of their capacity. Majority of the promoters complained that lack of continuous availability of power supply accompanied by high tariff rate was a major constraint for the poor utilization of the installed capacity. Lack of technical guidance from the National Horticulture Board (NHB) was another problem encountered by the promoters of CA stores. Besides this there was complete absence of publicity as well as awareness programme on CA store. In the absence of such programmes, it becomes all the more difficult for the entrepreneurs to take full benefit of the CA Stores. Time taken for the appraisal of projects, sanctioning and disbursement of term loans and release of subsidy by banks (both partial as well as full) were some other constraints faced by the promoters of CA stores. There is also a strong need to launch a campaign on CA stores with a focus to educate farmers which varieties of apple are to be grown and stored to meet the consumer demand.

4. Conclusion

There are problems of inadequate CA store facilities for fresh fruits. It has been observed from the field studies that no CA store was found in the Baramulla district-an important producing hub for apple, although a small percentage of the fruit was kept in the CA stores located at other places in the valley. The quantity that finds accommodation in the CA stores does not suffice the need of the customers. On account of non-availability of required number of CA stores, many growers after harvesting apple, store them in orchards with huge spoilage because of dampness, high temperature and humidity problems. Overall, 10 to 25 per cent of fruits get destroyed in this way. Besides, the quality of the produce is also affected adversely due to lack of cold storage facilities and the chances for damages to the fruits. The grower's capacity to hold stock for better prices during off-season also gets considerably reduced when storage facilities is minimal. These facts compel the growers to dispose off their produce in the shortest possible time. This situation naturally leads to a slump in the market prices and thereby the growers fail to harvest good prices. Extended storages facilities in the potential areas at the block/ tehsil level may prolong the marketing season and may prove beneficial to the growers. An important concern CA stores and the processing plants face that they incur high power expenses. The scope lies for cost reduction if power could be supplied to the plants continuously at a realistic rate. At present, the processing plant runs on diesel because of non-availability of continued power supply.

References

Acharya, S. S. And Aggarwal, N.L. 2011. *Agricultural Marketing in India*. Oxford and IBH Publishing Co. Pvt. Ltd. New Delhi, pp 555.

Govt. of India(GOI)2001."*Study on implementation of capital investment scheme on cold storages and storages for horticultural produce*", Directorate of Marketing and Inspection, Department of Agriculture and Cooperation, Ministry of Agriculture, Govt. of India, Faridabad, July.

Govt. of India (GOI) 2001b. "*Study on Implementation of capital investment scheme on cold storages and storages for horticultural produce*", Directorate of Marketing and Inspection (DMI), Department of Agriculture and Cooperation, Ministry of Agriculture, Govt. of India, Faridabad, July.

Jairath, M.S. 2000. Agricultural marketing infrastructure in arid India. *Agricultural Situation in India* 56(3): 125-131

Jairath, M. S. 2002. *Effect of agricultural marketing infrastructure on price. A case of cold storage*", National Institute of Agricultural Marketing, Jaipur.

Jairath, M. S. 2004. Effect on prices of horticultural commodities- A case study of cold storages. *Agricultural Marketing* XLVII (3): 11-15.

Khan, F. A., Rather, A. H., Qazi, N. A., Bhat, M. Y., Darzi, M. S., Beigh, M. A. and Ahmad, I. 2006. Effect of modified atmosphere packaging on maintenance of quality in apple. *Journal of Horticulture Sciences* 1(2) : 135-137.

NABARD, 2013.*Price Spread, marketing and financing system of apple in Kashmir (Draft report)* NABARD, J&K Regional Office, Jammu180pp.

NABARD Consultancy Services, 2013. *A proposal for strengthening the apple value chain in Jammu and Kashmir*. NABCONS, Corporate Office, Bandra Kurla, Mumbai 29 pp.

Chapter 28

Growth, Instability and Economic Optima in Apple Cultivation

Farheen Naqash, S. A. Wani and F. A. Shaheen

School of Agricultural Economics and Horti-Business Management,
Sher-e-Kashmir University of Agricultural Sciences and
Technology of Kashmir, Srinagar
E-mail: naqashfarheen@gmail.com

1. Introduction

Apple is an ideal value chain in the temperate horticulture sector and is a profitable product for all value chain participants (Weinberger and Thomas, 2007; Zbanca and Negritu, 2013). Apples in India are mainly grown in three mountainous states of North India *viz.* Jammu and Kashmir, Himachal Pradesh and Uttaranchal at an altitude of 4000 to 11000 feet. Jammu and Kashmir and Himachal Pradesh have roughly equal acreage under apple, but J&K has the highest average yield and accounts for more than 65 per cent of total apple production, hence important for economic growth (Masoodi, 2003). Jammu and Kashmir state being endowed with natural advantage of topography, climate and enormous diversity of agro-climatic niches has immense scope for horticultural development (Swarup and Sikka, 1987; Deepa, 2008). The apple cultivation in Jammu and Kashmir is an old age activity and around 200 varieties of apple were used to be cultivated in the state. Kashmir apple has lived upto its reputation for being one of the choicest fruits in India (Wani *et al.*, 2015). Amongst all other fruit and field crops apple has found a better reception with the growers due to higher returns and ability to stand transportation stress (Mir, 2014).

The apple crop involving around half a million households dominates the horticultural industry and has an important role in economic scenario of the state (Malik and Choure, 2014). The trends in the apple production showed that the acreage has been increasing at a faster rate in the last five years and the farmers see more potential for the fruit. But the continued cultivation of apple in districts that are not best endowed naturally such as Budgam, Anantnag, Kulgam, *etc.*, results in sub-optimal application of scarce resources (Ahmed, 2013). Increasing acreage seems to be a spontaneous response to the need for higher incomes, rather than improved productivity and efficiency.

Although apple production in the state is increasing with positive growth momentum but there is not a significant growth in exports. Weak production and supply chain along with poor marketing strategies, low transparency in the marketing system have together completely eroded incentive for producers to improve quality and productivity of apple. The low quality of apple is linked to large acreage under senile plantations; planting of varieties requiring cross pollination; shortage of quality planting material;low planting density; low quality of farm inputs; lack of irrigation; inappropriate pruning practices; poor orchard management – including disease and pest control and inadequate extension services (Lone, 2014).

2. Conceptual and Analytical Framework

The methodology adopted for the selection of study area, sampling design, data collection, analytical framework and concepts and estimation procedures used in the cost analysis of the study have been discussed under the following heads.

(a) Sampling Design

Apple forms the most important fruit crop in J&K state. A combination of secondary and primary data was collected followed by quantitative and qualitative assessment for comprehensive analysis to achieve the desired results and objectives of the study. District Baramulla of the Kashmir valley was delineated because of having maximum area (24952 ha.) under apple cultivation with the production of 423637 M.T. during 2014-15. Moreover, district experiences tremendous inclination of the farming community towards diversification of agriculture through apple cultivation (Hakeem *et al.,* 2006). One block *viz.* Pattan with the largest area/ production from the selected district was selected to ensure wider coverage of the sample. Multistage Random Sampling was used to select the 75 farmers from 5 villages with 15 randomly selected farmers from each village. Primary data collection was followed by the personal interview method using pre-structured schedules.

(b) Cost-Benefit Analysis

The cost of cultivation has been worked out by using the variable and fixed cost components and standard cost concepts. The variable cost includes value of hired and family labour, owned and hired machinery labour, seed, manure, fertilizer, pesticides, irrigation, interest on working capital and other miscellaneous expenses. The fixed cost includes rental value of owned land, rent paid for the leased land, depreciation on implements and farm buildings and interest on the fixed capital.

The gross return has been calculated as:

Gross return = Total production x average price

The net returns over different cost concepts have been calculated as the difference between the gross return and particular cost. The benefit cost ratios for full bearing apple orchard was calculated by dividing the gross return by total cost. The costs and gross returns for apple were estimated by taking sample of apple growers representing full bearing orchards. These economic parameters were valued at current prices and therefore represented present values of respective parameters. Benefit-cost ratio was then worked out by dividing average annual gross returns with average annual cost, using the following formula:

$$BCR = \frac{\text{Average of annual return}}{\text{Average of annual costs}}$$

The net return in fruit cultivation for selected fruit growers was computed by summing up the annual differences of gross returns and gross costs which estimated at current prices during the full bearing stage of apple orchard, using the following formula:

Net Returns = Gross income - Gross cost

(c) Compound Annual Growth Rate (CAGR)

Growth rate was used to measure the past performance of economic variables. CAGR is a business and investing specific term for the smoothed annualized gain of an investment over a given time period. CAGR is not an accounting term, but remains widely used, particularly in growth industries or to compare the growth rates of two periods because CAGR dampens the effect of volatility of periodic returns that can render arithmetic means irrelevant. CAGR is often used to describe the growth over a period of time of some element of the business.

$$CAGR\,(t_0,\,t_n) = \left(\frac{V(t_n)}{V(t_0)}\right)^{\frac{1}{t_n - t_0}} - 1$$

$V(t_0)$: start value, $V(t_n)$: finish value, $t_n - t_0$: number of years.

(d) Econometric Model (Multiple Linear Regression Model)

Multiple regression analysis was carried out to know the factors influencing the apple production. Production function was estimated for per hectare to measure returns to various factors of production. Some of the non-strategic collinear variables were dropped from the analysis to improve the precision of regression parameter. Based on the goodness of fit (R^2), the linear regression model of the following form was used:

$$Y = \beta_0 + \sum_{i=1}^{n} \beta_i X_i + U_i$$

where,

Y = Gross revenue from apple cultivation (Rs ha^{-1})

β_0 = Intercept

β_i = Regression coefficient of ith independent variable (i = 1....n)

X_1 = Expenditure incurred on fertilizers (Rs ha^{-1}.)

X_2 = Expenditure incurred on plant protection (Rs ha^{-1})

X_3 = Expenditure incurred on manures (Rs ha^{-1})

X_4 = Expenditure on irrigation (Rs ha^{-1})

X_5 = Expenditure incurred on total labour (Rs ha^{-1})

U_i = Random term (i = 1.....n)

The significance of regression coefficient was tested by employing student't' test as follows:

$$t\,cal. = \frac{\beta_i}{SE(\beta_i)}$$

where,

SE (β_i) = Standard error of regression coefficient

(e) Concepts and Estimation Procedures Used in Cost Analysis

In the following section, different terms were used in the study and various estimating procedures followed are outlined and discussed.

(i) Operational Land Holding

The operational land holding of a farm family was obtained as a sum of total owned and leased in operated area, but excluded the owned land leased out during survey period.

(ii) Fixed Costs

Fixed costs may be defined as those costs which do not affect the volume of output, even if the latter is zero. The fixed costs remain unchanged, irrespective of production. In case of fruit crop farming it includes depreciation on building and farm equipment's, interest on fixed capital, land rent and Govt. taxes if any.

(iii) Depreciation

Depreciation is the loss of value of an asset due to its use, wear and tear and time. It represents the amount by which a farm asset decreases in value. Annual depreciation on building and farm equipment's used was calculated by straight-line method.

(iv) Interest on Fixed Capital

Interest on fixed capital assets was worked out at the rate of 8 per cent per annum. This rate of interest on the fixed investment was charged on the assumption that if farmers had invested their funds in terms of deposits for a period of three years, they could have earned 8 per cent interest from the bank (Lead Bank *i.e.* Jammu and Kashmir Bank Ltd.).

(v) Rental Value of Land

The rental value of land was taken. @ Rs. 22000 ha^{-1} $annum^{-1}$ or the actual rent, prevalent in the area.

(vi) Variable Costs

The variable costs are those for which the variable factors are responsible and are thus dependent in total magnitude upon the volume of output. Variable costs vary with the output. In case of fruit crop farming variable cost included the cost of planting material, manure and fertilizers, irrigation, human labour, plant protection chemicals and miscellaneous expenditure.

(vii) Interest on Working Capital

Interest on working capital was charged as per the current bank rate of 4 per cent on savings charged half yearly.

(viii) Miscellaneous Expenditure

The expenditure incurred on the minor repairs of shed, implements, hand tools, irrigation structures, and other minor costs were included in miscellaneous expenses. The joint costs were apportioned and were allocated on the basis of crop-wise.

(ix) Human labour

The amount of labour used on various operations was estimated by recording the amount of time spent on different operations. The cost of hired labour was based on the actual wage rate prevailing in the study area. The cost of family labour was calculated on the basis of average wage paid to a permanent labour in the area. All the types of labour *viz.*, male, female and child used in the different operations were converted in to man-equivalent days. A man-day of eight hours was taken equivalent to 1.5 women workday and 2 workdays of child.

(x) Cost Concepts

Cost A: It includes expenses on planting material, manures and fertilizers, plant protection chemicals; irrigation charges; miscellaneous expenses and interest on working capital

Cost B: Cost A plus interest on fixed capital

Cost C: Cost B plus imputed value of family labour

(xi) Return Concepts

Farm business income = Gross income – Cost A

Family labour income = Gross income – Cost B

Net income over Cost C = Gross income - Cost C

3. Growth and Variability of Apple at Disaggregate Level

High growth and low instability in production is pre-requisite for sustainable agricultural performance. There is a growing concern that with technological change in production, variability has increased. Since the magnitude of growth and instability in production has serious implications for policy makers, the period-wise growth and level of instability in area, production and yield of apple in the various districts of Kashmir region were estimated by using time series data from 2001-02 to 2014-15.

Table 28.1: Growth and Variability in Area, Production and Yield of Apple at Disaggregate Level

Sl. No.	Districts	Area			Production			Yield		
		CGR	R2	CV	CGR	R2	CV	CGR	R2	CV
1	Srinagar *	6.66	0.96	28.43	6.41	0.84	29.51	0.23	0.001	13.62
2	Budgam	4.71	0.80	19.04	1.86	0.71	27.06	-2.73	0.19	21.17
3	Baramulla **	1.86	0.89	7.82	0.23	0.0001	19.34	-1.60	0.16	17.08
4	Kupwara	3.04	0.95	12.62	6.91	0.48	34.88	3.51	0.24	27.22
5	Anantnag***	5.93	0.97	23.87	4.23	0.69	21.22	-1.83	0.28	13.64
6	Pulwama #	7.15	0.95	28.26	6.91	0.68	31.75	-0.23	0.003	15.81
	Kashmir	4.71	0.98	≤0.000	3.28	0.43	21.20	-1.37	0.16	14.31

Source: Directorate of Horticulture (P and M), Kashmir; Significant at 5 per cent level; Includes districts of * Ganderbal, ** Bandipora *** Kulgam and #Shopian

Based on the annual compound growth rates, the fruit crops can be classified into four categories: Category A (high growth rate) – growth rate of 5 per cent or above, category B: (moderate growth) - growth rate of >1 and 5 per cent, category C (slow growth) - growth rate upto 1 per cent and category D (negative growth rate) - growth rate of < 0 per cent. Similar classification has been followed by Cuddy and Della (1978), Deb et al. (1999) and Shaheen and Shiyani (2004).

In order to confirm this hypothesis, the variability was also computed and classified into <6 per cent - low variability, 6 per cent -10 per cent - moderate variability, >10 per cent - high variability. For period 2001-14 the apple has registered moderate growth with high instability both in area and production front while as negative growth with high instability on the productivity front in the valley which may be attributed to the many factors (Table 28.2).

Table 28.2: District-wise Growth and Variability Scenarios

Sl.No.	Districts	Area	Production	Yield
1	Srinagar	High growth and high variability*	High growth and high variability*	Low growth and high variability
2	Budgam	Moderate growth and high variability*	Moderate growth and high variability	Negative growth and high variability
3	Baramulla	Moderate growth and moderate variability*	Low growth and high variability	Negative growth and high variability
4	Kupwara	Moderate growth and high variability*	High growth and high variability*	Moderate growth and high variability
5	Anantnag	High growth and high variability*	Moderate growth and high variability*	Negative growth and high variability
6	Pulwama	High growth and high variability*	High growth and high variability*	Negative growth and high variability
Total Kashmir		Moderate growth and high variability*	Moderate growth and high variability*	Negative growth and high variability

The Table 28.1 represents the growth in area and production under apple for the last decade (2001-02 to 2014-15) in Kashmir valley. Apple has shown a remarkable growth in area with a percentage increase of 4.71 per cent compound annual growth rate (CAGR) in the valley. However, the maximum growth rate in terms of area under apple cultivation has been seen in the Pulwama district (7.15 per cent) followed by district Srinagar (6.66 per cent) with a significant trend. The performance of apple crop over the period in the valley is more pronounced on production front with CAGR of 3.28 per cent. The districts Pulwama (6.91 per cent) and Kupwara (6.91 per cent) have shown a significant trend in growth rate in terms of production of apple over the years. In terms of yield of apple, Kupwara and Srinagar districts have shown a positive trend as compared to other districts of the valley, which depict a negative trend in growth rate over the years. Overall, yield depicted a negative trend because most of the area brought under cultivation is in transition phase. It is general hypothesis that production instability has increased due to technology transfer.

4. Costs and Return Structure in Apple Cultivation

The cost of cultivation is of wide interest to the users of cost data and assumes importance in the area of planning. The utility of data on the cost of cultivation of horticultural commodities for planning assumes importance as it guides the planners about the area where it is economical to produce and the regions which would accordingly be most suitable for the development of industries based on the horticultural raw material. At the micro level, it enables the farm management experts to study the efficiency of the various cultivation practices and alter the crop planning by providing information regarding their profitability. This helps the experts to make practical recommendations for farm planning aimed at better allocation of existing resources and introduction of improved agronomic practices which would increase the efficiency of apple production (Malik and Heijdra, 2011; Wani *et al.,* 1994).

Cost structure, output and return from apple crop grown in the study area has been discussed under cost of plantation, cost of maintaining the orchard in non-bearing stage and expenditure incurred during the bearing stage. For estimating the costs and returns for apple crop, it has been assumed that (i) first bearing starts from 8th year (ii) the major operations and input requirement vary in 8-15, 16-25, 26-40 and above 40 years. The above mentioned groups are based on the physiological growth and productivity pattern of the plant. The analysis in this section has been divided into three parts (Wani *et al.,* 1993), *viz.* (a)establishment cost (b) cost during non-bearing stage (other than plantation cost) (c)bearing stage costs.

In apple, initial investment is quite heavy for reasons of the cost involved in land development, digging of pits, application of manure and fertilizers and material cost. Growers have to incur cost on maintenance for about 8 years without any economical returns. Farmers can take intercrop from the orchard upto 5th year after which this practice becomes uneconomical due to competition for nutrition and shade effects. In this study, it is assumed that income and expenditure incurred in the cultivation of intercrops are equal.

(a) Establishment Cost

The item wise cost incurred on the apple plantation in the first year is presented in Table 28.3. It is imperative to examine the resource position of the growers before deciding to establish an orchard. This included costs on plant material, pit digging, tree planting, manures and fertilizers, fencing and other aspects of initial expenditure. The establishment costs were found to be Rs. 114240 per hectare of which fencing alone accounted for 27.48 per cent (Rs. 31400 per hectare) of the total establishment cost, followed by plant material cost (26.27 per cent). Manures and fertilizers constituted 14.70 and 4.07 per cent respectively of the total establishment cost.

(b) Maintenance Cost of Non-bearing Orchards

The expenditure on labour, manure, fertilizers, pesticides, interest on working capital, fixed capital and accumulated establishment cost, rental value of land and depreciation includes maintenance costs of non-bearing orchards. The results presented in Table 28.4 showed that under the variable costs, costs on plant protection chemicals, labour utilization, fertilizer cost and manure cost were higher for the non-bearing orchards which contributed to about 14.99, 12.24, 10.64 and 9.78 per cent respectively of the total variable cost as compared to other costs.

Table 28.3: Establishment Cost of Apple Plantation in the Sampled Farms (Rs ha⁻¹)

Sl.No.	Cost Component	Cost	Sl.No.	Cost Component	Cost
1	Plant material	30000 (26.27)	5	Fertilizers	4640 (4.07)
2	Pit digging	12000 (10.50)	6	Fencing	31400 (27.48)
3	Tree planting	12000 (10.50)	7	Transportation	2400 (2.10)
4	Manure	16800 (14.70)	8	Miscellaneous costs	5000 (4.38)
		Total cost 114240 (100)			

Figures in parentheses indicate per cent of total costs.

Among the fixed cost items, rental value of land and depreciation on fixed capital were found to be the main components responsible for large proportion of fixed cost in the total cost during the non-bearing age with 16.83 and 12.76 per cent respectively.

Table 28.4: Maintenance Cost of Non-bearing/Bearing Age Orchards (Rs.ha⁻¹)

Items	Cost of Non-bearing Orchards	Cost of Bearing Orchards
A. Variable cost		
Labour and energy component for sprays	16000 (12.24)	16000 (7.68)
Manure	12800 (9.78)	6000 (2.87)
Fertilizers	13900 (10.64)	16500 (7.91)
Hoeing and fertilizer application	5000 (3.82)	8000 (3.83)
Plant protection chemicals	19600 (14.99)	62775 (30.13)
Irrigation	3000 (2.29)	4000 (1.92)
Miscellaneous cost	3000 (2.29)	23000 (11.05)
Interest on working capital	5864 (4.49)	10902 (5.23)
Total variable cost	79164 (60.54)	147177 (70.62)
B. Fixed cost		
Depreciation on fixed capital	16688 (12.76)	25000 (12.00)
Rental value of owned land	22000 (16.83)	22000 (10.56)
Interest on fixed capital	3760 (2.87)	3760 (1.80)
Total fixed cost	42448 (32.46)	50760 (24.36)
Interest on accumulated establishment cost	9139 (7.00)	10460 (5.02)
Total maintenance cost	130751 (100)	208397 (100)

Figures in parentheses indicate per cent of total costs.

(c) Maintenance Cost of Bearing Orchards

Maintenance cost of bearing orchards in the study area are presented in Table 28.4. This included input on labour, fertilizers and pesticides, interest on working and fixed capital, and depreciation on fixed capital. The maintenance cost of bearing orchards was found to be Rs. 208397 per hectare. Another feature of the cost structure was that the maximum expenditure was incurred on pesticides (30.13 per cent), which included the HMO/Dormant spray (7.68 per cent). Labour was also the most expensive component (7.68 per cent) along with fertilizers (7.91 per cent) and other miscellaneous costs (pruning and training, watch and ward *etc.*) (11.05 per cent). No expenditure was incurred on the land revenue and taxes, as there is no taxation in case of un-irrigated lands now. The inputs bore a direct relationship with the age of orchard. It might be because of the fact that with the advancement of age, the tree canopy also increases, increasing the expenses on account of pesticides, fertilizers, weeding and hoeing. The lesser the amount of inputs is directly related to the returns from the orchards. With the advancement in age of the orchard the quality and quantity of fruits produced is considerably reduced, lowering the overall returns from the orchards, thus, discouraging the use of inputs like fertilizers, pesticides, *etc.*

Among the fixed costs, depreciation (12 per cent) and rental value of land (10. 56 per cent) accounted for maximum share of the fixed costs.

(d) Returns from Bearing Age Orchards

The returns from the bearing orchards were calculated on per hectare basis so as to present the actual picture of the economics of orchard raising in apple. The results revealed that the orchards exhibited the gross returns of Rs. 703125 and the net returns of Rs 494063 (Table 28.5). The average production per hectare was found to 1875 boxes of apple, where one box contains 18 kg of fruit, and the cost per box was found at an average of Rs. 375. The net returns from apple can be increased if the extension services strengthened to educate the people about the proper input use which was found below merit during the course of investigation. The average production cost per kg was found to be Rs. 6.19 with the benefit cost ratio of Rs. 3.39.

Table 28.5: Returns from Apple Orchards and Regression Coefficient Estimates

Returns from Apple Orchards during Bearing Stage		Regression Coefficients Explaining Determinants on Revenue		
Particulars	Cost (Rs ha^{-1})	Independent Variables	Estimated Coefficients	P-value
Total production cost (Rs.)	208397	Fertilizers	0.36	0.015*
Average production (kg)	33750	Pesticides	9.03	0.045*
Gross returns (Rs.)	703125	Manures	1.30	0.000*
Net returns (Rs.)	494728	Irrigation	0.05	0.983
Average production cost per kg	6.17	Labour	0.24	0.008*
Output/input ratio	**3.37**	Adjusted R^2 (per cent)	92.10	

*Significant at 5 per cent level of significance.

(e) Estimates of Regression Analysis

To quantify the factors that determine the role of various variables in apple cultivation, regression analysis was used. The estimates of regression function depicted in Table 28.5, revealed that plant protection chemicals and manures were the most significant and positive determinants of revenue from apple cultivation. Irrigation at the farm level also had a positive contribution to the improvement of revenue of apple, however its coefficient turned statistically insignificant. The pesticide level was found an important determinant of apple revenue due to the fact that more application of pesticides reduces the chance of losses in apple on scientific lines. The analysis further revealed that the irrigation and the labour component were also used efficiently on the farms. The positive and significant coefficients indicated that revenue from the apple cultivation can be generated more by using efficiently these factors. The value of coefficient of adjusted R^2 shows that the exogenous variables specified in the model explained large variation (92.10 per cent) in total revenue.

5. Conclusion

Apple production and marketing is an important economic pursuit and source of livelihood to 35 lakh people of the state of J&K. The state in recent years has given lot of attention to the development process of apple industry. However, there exists wide and marked gap in productivity of apple as compared to major apple producing countries of the world. This study being a humble stride to study apple production system of the state, emphasizes that the cost of cultivation is of wide interest to the users of cost data and assumes importance in the area of planning. The utility of data on the cost of cultivation of horticultural commodities for planning assumes importance as it guides the planners about the area where it is economical to produce and the regions which would accordingly be most suitable for the development of industries based on the agricultural raw material. High growth and low instability in production is pre-requisite for sustainable agricultural performance. There is a growing concern that with technological change in production, variability has increased. Since the magnitude of growth and instability in production has serious implications for policy makers, the period-wise growth and level of instability in area, production and yield of apple in the various districts of Kashmir region were estimated by using time series data from 2001-02 to 2014-15.The relative peace in the state has made it possible for farmers to focus on improving their livelihoods. The apple sector has the potential to influence several households and improve their economic prospects. New market players have to be invited in, resources found for investments, change in policy and support systems from the government and building capacities in individuals and institutions for effective and remunerative participation in the value chain. Some of the measures required to improve prospects of apple farmer can be taken at farm level and enterprise level. But a number of measures that are critical for ensuring an equitable return to the farmers have to be taken at a sector level in close coordination with the government. Hence, the revamp of apple sector has to be planned with a mix of investments, capacity building, innovations and committed institutional leadership.

References

Ahmad, N. 2013. *Problems and prospects of temperate fruits and nut production scenario in India vis-à-vis international scenario.* Central Institute of Temperate Horticulture, Srinagar.

Cuddy and Della, V. 1978. Measuring the Instability in time series data. *Oxford Bulletin of Economics and Statistics* February, 1978.

Deb, U. K., Joshi, P. K. and Bantilan, M. C. S.1999. Impact of modern cultivars on growth and relative variability in sorghum yields in India. *Agricultural Economics Research Review* 12(2): 84-106.

Deepa, D. 2008. Indian product, "Brief India fresh fruit sector", *Holy Higgins* US Embassy.

Hakeem, A.H., Peer, F.A., Tantray, A.M. and Ghani, I. 2006. Adoption level of production recommendations in apple cultivation by fruit growers of Baramulla district. *SKUAST Journal of Research* 8: 168-174.

Lone, R. A. 2014. Horticulture sector in Jammu and Kashmir economy. *European Academic Research* 2(2): 2405-2432.

Malik, V. K. and Heijdra, H. 2011. *A value chain analysis of apple from Jumla*, Ministry of Agriculture and Cooperatives, Department of Agriculture.

Malik, Z.A. and Choure, T. 2014. Horticulture growth trajectory evidences in Jammu and Kashmir (A lesson for apple industry in India). *Journal of Business Management and Social Sciences Research* 3: 7-10.

Masoodi, M.A. 2003. *Agriculture in Jammu and Kashmir - a Perspective.* Mohisarw Book Series, Srinagar p.195.

Mir, S.M. 2014. Problems of apple industry in J&K with special reference to Sopore town. *International Journal in Management and Social Science* 2(3): 33-46.

Shaheen, F. A. and Shiyani, R. L. 2004. Growth and Instability in area, production and yield of fruit crops in Jammu and Kashmir- A Disaggregate Analysis. *Agricultural Situation in India* 59: 657-663.

Swarup, R.K. and Sikka, B.K. 1987. *Production and Marketing of Apples.* Mittal Publications Delhi, pp. 72.

Wani, M.H., Wani, S.A. and Mir, N.A. 1993. Economic analysis of different age orchards in apple. *Agricultural Situation in India* 48: 657-660.

Wani, M.H., Bhat, A.R. and Mir, N.A. 1994. Economic viability of apple orchards in Kashmir. *Agricultural Situation in India* 49: 659-662.

Wani, S.A., Wani, M.H., Bazaz, N.H. and Mir, M.M. 2015. *Commodity profile of apple*, Network Project on Market Intelligence-SKUAST-K, Shalimar, pp. 40

Weinberger, K. and Thomas A.L. 2007. Diversification into horticulture and poverty reduction: A research agenda. *World Development* 35(8): 1464-1480.

Zbanca, A. and Negritu, G. 2013.Feasibility of investments for planting and maintenance of apple orchards by applying various technologies. *Scientific Papers Series Management, Economic Engineering in Agriculture and Rural Development* 13(1): 465-467.

Chapter 29

Pesticide Application on Apple: Issues of Marketing and Ecological Implications

S. H. Baba, M. H. Wani and S. A. Wani

School of Agricultural Economics and Horti-Business Management,
Sher-e-Kashmir University of Agricultural Sciences and
Technology of Kashmir, Srinagar
E-mail: drshbaba@gmail.com

1. Introduction

The potential yield loss due to weeds, diseases and pre- and post harvest pests is significant both in quantitative and monetary terms (Puri *et al.*, 1999; Singh, 1999). Pesticide application has been an essential ally in the farmers struggle to protect crops. Despite their higher use, losses throughout the production system remain high owing to various negative externalities like rejection of agricultural export due to the presence of high pesticide residues, pesticide-related health hazards (WHO, 1990, WRI, 1998) and the extent, severity and frequency of associated environmental problems. In 2012, the total consumption of technical grade pesticides by weight was estimated at 1789 metric tonnes (Anonymous (a), 2015) in which shares of companies by status vary significantly. Pesticide market of the state in terms of value was estimated around ' 400 million and the calculated shares of the MNCs and NCs were 33.6 per cent and 52.3 per cent, respectively. Fungicides alone accounted for 71.1 per cent of total pesticide sale in the state followed by insecticides (15.4 per cent) and acaricides (7.7 per cent). The pesticides applied on apple together constituted about 83 per cent of total value of agro-chemicals utilized in the state (Baba *et al.,* 2012a). Pesticides coupled with other input technologies have enabled the farmers to enhance the apple productivity in J&K during the last three decades. However, excessive/indiscriminate uses of pesticides not only increase the cost of apple cultivation but, also resulted in many human health problems and environmental contaminations. These problems got more accentuated with the availability of spurious/sub-standard chemicals and existence of chain of functionaries including unlicensed players between firms and farmers in the markets. In this backdrop, this chapter discusses the entire issue of pesticide application in apple in relation with existing delivery system and associated environmental hazards with feasible policy options.

2. Existing Delivery System of Pesticides and Role of Traders

There are many activities in apple cultivation that require huge investment; the variable cost inputs for managing one kanal of average age orchard was estimated approximately at Rs 5000.00; in which pesticides alone comprises about 54 per cent. Farmers used pesticides frequently and spray their crop more than eight times. Generally only one spray of insecticides was done in the study area while the applications of fungicides constitute major proportion of total chemical application on apple.

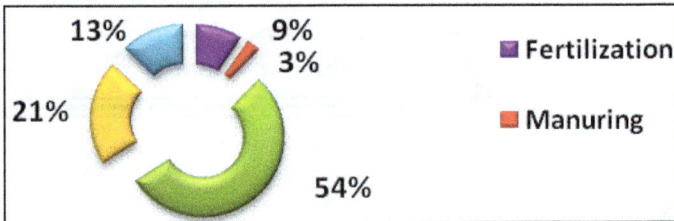

Figure 29.1: Composition of Cost of Cultivation of Apple.

Fungicides alone constituted about 47 per cent of total cost incurred on pesticides followed by dormant sprays and acaricides. The application of pesticides as per scientific recommendation could reduce cost on this input significantly.

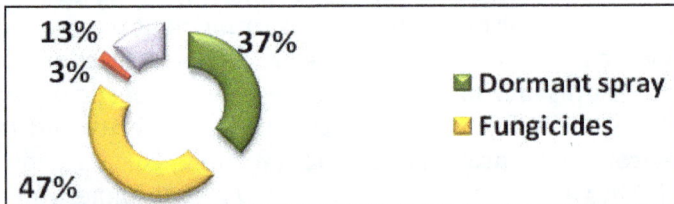

Figure 29.2: Composition of Cost of Pesticides.

a) Existing Delivery System of Pesticides and Role of Traders

There is an extensive network of pesticide companies that popularise and promote agro-chemicals in rural areas. At present there are 7 multinational (MNCs) and 15 national/generic companies (NCs) in pesticide trade. The dealers prefer to promote products of those companies that give maximum incentives. Besides, unlicensed dealers and retailers who are not completely aware of the toxicity of pesticides also sell them. Pesticide companies sell major portion of their supplies (68 per cent) through distributors to farmers. About 11 per cent chemicals are sold to institutions for direct consumption while another 16 per cent are sold to co-companies existing in the markets which in turn manipulate products to qualify their own standards and deal with distributors and institutions.

After distributors' level, agro-chemicals undergo a number of ownerships before reaching the ultimate consumers; leaving a wide room for malpractices and adulteration. To sum up, as the number of intermediaries increased, the farmer had to pay more prices for pesticides and the company's share in it decreased

b) Exploitative Role of Traders/Dealers

i. Mis-guidance of Traders/Dealers and Poor Adoption of Scientific Pesticide Packages

The farmers in the study area were guided by trader-cum-contractors or unlicensed dealers and their choice/brand preference of chemicals were steered by these players. Accordingly, the scientific pesticide spray schedule (Annex.I) released by SKUAST-K in collaboration with line departments has depicted poor adoption at the field. The farmers use agro-chemicals indiscriminately without consideration of age of orchards, number of sprays, and compatibility of chemicals. It was observed that all the dormant spray oils were being sprayed at more than recommended levels and showed a technological gap of about 70 per cent. At farmers' level, fungicides and insecticides/acaricides were applied 20 per cent and 48 per cent more than recommendations, respectively (Table 29.1).

Table 29.1: Technological Gaps in Pesticide Application

Chemical	Gap (per cent)
Dormant oil	70.0
Fungicides	20.0
Insecticides/acaricides	48.0

Famers were asked about their perception of using current level of pesticides and as high as 84 per cent of the farmers showed concern about availability of spurious pesticides. They used chemicals at more than recommended level because they perceived that the pesticides available in the markets have lower efficacy (70.50 per cent) than it actually should have by standards (Table 29.2). The non-availability of recommended chemicals was also brought out by over 60 per cent of farmers. Indebtedness to traders could be a more significant reason that encourage supply of spurious/sub-standard chemicals. It was observed that fruit growing temperate regions have received only 14 per cent total credit requirement from institutional sources (Baba *et al.*, 2014b). There were other reasons farmers reported for using pesticides at current level.

Table 29.2: Reasons for using Current Level of Pesticides

Reason	Per cent
Less efficacious chemicals	70.5
Spurious/sub-standard pesticides	83.5
Indebtedness and traders' guidance	68.5
Scarcity of skilled labour	19.5
Ignorance about chemicals	25.5
Higher incidence of diseases/pests	41.5
Non-availability of recommended chemical	62.2

ii. Supply of Spurious/Sub-standard Pesticides

Traders and unlicensed players advance loan to resource poor farmers, either in form of cash or kind (fertilizers and pesticides) against standing crops. In this process farmers are being exploited by these functionaries in two ways: i) by advancing kind loan as pesticides, often suspected for their quality, at relatively higher prices, and ii) through distress sales of farmers' produce. It was observed that contractors (unlicensed functionaries) quoted 15 per cent more price to pesticides compared to prices offered by registered retailers. Not only this on entering into contract with farmers, they quoted lower than actual yield per unit area of standing crop. It was observed that 53.5 per cent of the pesticide dealers offered pesticides on credit, advanced either as cash or kind (fertilizers and pesticides).

A good proportion of pesticides available to farmers were spurious or sub-standard. On an average over 34 per cent of agro-chemicals available in the market are spurious/sub-standard, as worked out on the basis of responses obtained from stakeholders including famers. The proportion of sub-standard/spurious chemicals varies from 12.5 to 62.2 per cent of total quantity of Triadimefon and Hexaconazole pesticides purchased in the season, respectively. About 29 per cent of dealers/retailers were found not possessing registered license in the surveyed area and scenario further aggravated this problem. It was reported that the sale of unlabelled and spurious pesticides is highly prevalent in apple growing regions especially in district Baramulla. As high as 29 per cent of farmers have applied unidentified/unlabelled pesticides as one of the sprays on their orchards on the recommendations of traders. Moreover, the use of banned chemicals and pesticides not present in spray schedule were also seen prevalent in the study area. Another problem is the use of incompatible combination of fungicides and insecticides (like Captan, Ziram with Chloropyriphos and Fenzaquin). Application of spurious/sub-standard pesticides, their wrong formulation and combination of incompatible chemicals would further prompt ignorant farmers to go for yet higher doses of pesticide which may have a long term negative implication on crop yield and environment.

c) Socio-ecological Implications of Pesticide Application

It was observed that pesticides were applied without adequate understanding of pest ecology, economies, insect/pest specific pesticides, their formulation/ methods of application, and other precautionary measures. The intensive use of pesticides had significantly prevented apple yield losses to the tune of 41 per cent, however, the benefits of pesticides were offset to some degree by costs imposed by them in mitigating environmental hazards including health problems. Pesticides are necessarily poisons and hence they have adverse effects on any organism having physiological functions similar to the target organism. Over use of pesticides has brought about a decline in the bio-diversity of non-target organisms in these regions. About 23 per cent of the cultivators in these districts reported a significant decline in populations of beneficial organisms. According to them, populations of beneficial organisms/natural enemies of pests like honey bees (pollinators), ladybird beetles, spiders, other parasitoids and in particular populations of birds and earthworms. There are several cases of animal/poultry poisoning reported by respondents though they prevent animals, particularly cattle from entering pesticide sprayed orchards. Pesticide runoffs that reach nearest water bodies have detrimental effect on fish, and aquatic plants, which are part of the food web and play an important role in maintaining the eco-balance. Many minor poisoning cases are not reported to doctors or there is no systematic monitoring of poisoning cases in these regions but manifestations of hazards such as vomiting, nausea, excessive salivation, and blurred vision, headache, and disturbances in consciousness were reported from study area. The indiscriminate application of pesticides has resulted in deaths in the J&K state in 2012-13 (Anonymous (b), 2015). Studies have shown an incidence of deadly ailments like brain cancer among population exposed to lethal pesticides (Bhat *et al.,* 2010). These ailments in human health accrue either due to direct exposure to pesticides or owing to the use of improper and inadequate safety measures while spraying. Responses of stakeholders revealed direct contamination of environment as a result of heavy pesticide applications. About 11 per cent of cultivators have shown concern about pesticide residues in their produce at harvesting stage owing to untimely application of chemical on their orchards. The indiscriminate application of pesticides strains ecological set up in other ways including development of resistance and resurgence of insect pests.

3. Policies Implications

To combat the problems of pests, farmers apply larger quantities of pesticides and try irrational combinations, which has resulted in pesticide related environmental hazards more than its beneficial outcomes. Moreover, sale of spurious/sub-standard chemicals and existence of unlicensed dealers has been a major concern for stabilization of growth of agriculture and apple sector in particular. Based upon discussion, the study arrives at few pragmatic policy suggestions:

☆ There is a need for effective regulation of pesticide trade. Since availability of spurious/sub-standard pesticides in the market is a major concern, therefore, there is a need of a strict legislation in order to prevent marketing of such chemicals. All pesticides need to be labeled in local/common language and should contain information regarding proper handling of these toxic chemicals.

☆ To prevent the distortion of trade practices, it is essential to eradicate black sheep from the pesticide trading system. Government even needs to prescribe a minimum educational qualification and also specialized competence in the field of plant protection for obtaining the license to trade.

☆ There is a need to establish input check posts around each production centre equipped with chemical testing facilities and it should be mandatory that each imported container of pesticides should undergo registration at this check post with sample based testing.

☆ Development in pesticide application technology has to be kept in pace with pesticide development. The personal safety devices or equipment are not available in the market; it is essential to devise safety equipment suitable for local climatic conditions.

☆ Integrated Pest Management that encompasses the use of resistant varieties, biological control methods and modifying agronomic practices for keeping pest population below the economic injury level and also minimising environmental contamination could be a viable option. These practices would help to switch to production of organic apple in future. Accordingly the farming community needs to be educated and provided necessary assistance so that they could readily appreciate the need to change from chemical intensive farming to eco-friendly farming techniques, which in turn will ensure long-term food security and environmental safety.

☆ Companies/R and D institutions need to promote/identify those chemicals, which are efficient and environmentally safe. Companies should also rule out the play of misguiding representatives/dealers.

☆ Since an effective response is required to various WTO negotiations to become globally competitive, therefore, the emphasis would be on better quality and higher productivity. Strengthening of extension services to disseminate safe methods of pesticide application/formulation and organization of farmer's workshops would help to produce apple safe from residues and to ease the strain on ecology.

References

Anonymous (a) (2015). *Ministry of Statistics and Programme Implementation*, Government of India, (http//:www.indiastat.com).

Anonymous (b) (2015). Lok Sabha Unstarred *Question No. 2014*, Dated 09-03-2010 and Lok Sabha Unstarred *Question No. 2848*, Dated 13-03-2015, (http//:www. indiastat.com).

Baba, S. H., Wani, M. H. and Malik, Hilal A. (2012b) *Fruit economy linkages and role in employment generation and rural upliftment in Jammu and Kashmir*. Final Report of Horticulture Mini-Mission Sponsored Research Project (Project Report # 04). Division of Agricultural Economics and Marketing, SKUAST-K, Shalimar 191 121 (J&K)

Baba, S. H., Wani, M. H., Wani, S. A., Zargar, Bilal A. and Kubrevi, S. S. (2012a) Pesticide delivery system in apple growing belt of Kashmir Valley. *Agricultural Economics Research Review* 25(Conf. No.): 435-444.

Baba, S.H., Wani, M.H., Wani, S.A., Zargar, B.A. and Malik, H. A. (2014a) Imperatives for sustenance of agricultural economy in the mountains: A prototype from Jammu and Kashmir. *Agricultural Economics Research Review* 27 (2): 243-257.

Baba, S.H., Wani, M.H., Wani, S.A., Zargar, B.A. and Qammer, N.A. (2014b) Institutional credit to mountain agriculture: Issues of structural changes and impact in Jammu and Kashmir. *Agricultural Economics Research Review* 27 (Sept. Oct): 111-122.

Bhat, A.R., Wani, M.A. and Kirmani, A. R.(2010) Brain cancer and pesticide relationship in orchard farmers of Kashmir. *Indian Journal of Occup Environ Med.* Sep-Dec; 14(3): 78–86.

Puri, S.N., Murthy, K.S. and Sharma, O.P. (1999) Pest problems in India – Current status. *Indian Journal Plant Protection* 27(1 and 2): 20-31.

Singh, S. P. (1999) Biological control in India. *Indian Journal of Plant Protection* 27(1 and 2): 126-138.

Wani, M.H., Baba, S.H. and Yousuf, Shoaib, (2009), Market economy of apple in Jammu and Kashmir. *Ind. Jour. Agril. Mktg* 22(2): 42-58.

World Health Organization (WHO) (1990), *Public Health Impact of Pesticides Used in Agriculture*, Geneva.

Annexure 1: SKUAST-K Pesticide Spray Schedule for Apple - 2017

Sl.No.	Fungicides	Formulations	Average Market Price (Rs)	Name of the Company
		FUNGICIDES		
1.	Azoxystrobin11 per cent + Tebuconazole 18.3 per cent	Custodia (29.3 SC) (P1)	3400	M/S ADAMA India Pvt Ltd.
2.	Bitertanol	Baycor (25 WP) (P2)	2700	M/S Bayer Crop Science Ltd.
3.	Kresozim methyl	Ergon (44.3 SC) (P3)	4750	Rallis India Ltd
4.	Metiram + Pyraclostrobin	Cabrio Top (60 WG) (P4)	1800	M/S BASF India Ltd.
		Chemtop (60 WG)*	3850	M/S Cheminova India Ltd
5.	Captan	Captaf (50 WP) (P5)	460	M/S Ralis India Ltd.
		Deltan (50 WP)*	485	M/S Coromandel AgricoPvt. Ltd.
		Kohicap (50 WP) (P6)	460	M/S Fungicide India Ltd.
		Jaicap (50 WP)*	-	M/S Jai Chemicals Ltd.
		Captan (50 WP) (P7)	500	M/S Jai Chemicals Ltd.
		Captax (50 WP) (P8)	450	M/S India Pesticides Ltd
		PANTHER (50 WP) (P9)	470	M/S HPM Chemicals and Fertilizers Ltd
6.	Captan + Hexaconazole	WAVE (75 WP) (P10)	1150	M/S Fungicide India Ltd.
		Boxer (75WP)*	-	M/S Anu Products Ltd
		Rely (75 WP)*	1140	M/S Cheminova India Ltd
		Horse Power (P11)	-	M/S Bomageri Crop Science Ltd
7.	Carbendazim +Mencozeb	SAATHI (75 WP) (P12)	550	M/S FIL Industries Ltd
8.	Chlorothalonil	Tata Ishaan (75 WP) (P13)	1050	M/S Rallis India Ltd
9.	Difenaconazole	Score (25 EC) (P14)	3700	M/S Syngenta India Ltd.
		Scale (25 EC) (P15)	3500	M/S Nagarjuna Ltd.
		Rubigan –D (25 EC)(P16)	3600	M/S FIL Industries Ltd.
		KARARA (25 EC) (P17)	-	M/S Agro Life Science Ltd
10.	Dithionon	Tata Shan (75 WP)*	-	M/S Rallis India Ltd.
11.	Dodine	Super Star (65 WP) (P18)	1700	M/S FIL Industries Ltd
		Super Star (40 WP)*	-	M/S FIL Industries Ltd
		Noor (65 WP) (P19)	1600	M/S Indofil Industries Ltd.
12.	Flusilazole	Governor (40 EC) (P20)	6500	M/S FIL Industries Ltd
		Cursor (40EC)*	6800	M/S DhanukaAgritech Ltd

Sl.No.	Fungicides	Formulations	Average Market Price (Rs)	Name of the Company
13.	Hexaconazole	Contaf (5 EC) (P21)	525	M/S Tata Rallis India Ltd.
		Anvil (5 EC) (P22)	460	M/S Syngenta India Ltd.
		Control (5 EC)*	-	M/S Maghmani Industries, Chandigarh
		Titan (5 EC) (P23)	475	M/S Sudarshan Chemical Ltd.
		Envil (5 EC) (P24)	530	M/S FILL India Ltd.
		Mainex (5EC) (P25)	425	M/S ADAMA India Pvt Ltd
		Krizole (5EC) (P26)	475	M/S KrishiRasayan Export Pvt Ltd
14.	Mancozeb	Indofil (M-45) 75 WP (P27)	330	Indofil Chem. India Ltd
		Kohinoor M-45 (75 WP) (P28)	330	M/S Fungicide India Ltd.
		Jai M-45 (75 WP)	-	M/S Jai Chemicals Ltd.
		Mancozebflowable 35 SL(P29)	-	M/S Indofil Chemicals India Ltd.
		Dithane M-45 (75 WP) (P30)	338	M/S Dow Agro Science Mumbai
		Manfil (75 WG) (P31)	275	M/S Indofil Chemicals India Ltd.
		Hindustan M-45 (75 WP)	-	M/S Hindustan Pulverizing Mills Ltd.
		Macoban (75 WP) (P32)	300	M/S ADAMA India Pvt Ltd
		MOUNT (75WP)	280	M/S Bharat Insecticide Ltd
15.	Myclobutanil	Grapple (10 WP) (P33)	1575	M/S Fungicide India Ltd.
		Boon (10 WP) (P34)	1500	M/S Indofil Chemicals India Ltd.
		Index TM (10 WP) *	1600	M/S NagarjunaAgri-Chemicals Ltd.
		Systhane (10 WP) (P35)	1650	M/S Dow Agro Science Mumbai Ltd.
		Myclomain (10WP) (P36)	-	M/S ADAMA India Pvt Ltd
		Bonas (10 WP) *	-	M/S Godrej Agrovet Ltd
		Inyst (10WP) (P37)	-	M/S Biostat India Ltd
		Limpid (10WP) *	1500	M/S Cheminova India Ltd
		Revolve (10WP) *	1500	M/S Bharat Insecticide Ltd
16.	Propineb	Antracol (75 WP) (P38)	550	M/S Bayer Crop Science Ltd.
		Proximain (70WP) (P39)	480	M/S ADAMA India Pvt Ltd
		Scale (70WP) (P40)	550	M/S KrishiRasayan Export Pvt Ltd
		Filprostar (70WP) *	470	M/S Fungicide India Ltd.
		Sway (70WP) *	540	M/S Cheminova India Ltd

Sl.No.	Fungicides	Formulations	Average Market Price (Rs)	Name of the Company
17.	Tebuconazole	Folicur(P41)	-	M/S Bayer Crop Science Ltd.
18.	Ziram	Ziride (80 WP) *	-	-
		Cuman-L (27 W/V) (P42)	380	M/S Syngenta India Ltd.
		Ziron (27 W/V) *	250	-
		Zed-78 (80 WP) (P43)	470	M/S Fungicide India Ltd.
		Zirex-L (27SC) (P44)	250	M/S Fungicide India Ltd.
19.	Trifloxystrobin + Tebuconcazole	Nativo (75 WG) (P45)	6200	M/S Bayer Crop Science Ltd.
20.	Zineb	Indofil Z-78 (75 WP) (P46)	465	M/S Indofil Chemicals Ltd.

INSECTICIDES/ACARACIDES				
1.	**Oil**	Formulations	Average Market Price (Rs)	Name of the Company
	`Horticulture mineral oils for dormant spray	i. ATSO*	150	M/S KrishinaAntioxidents Pvt ltd
		ii.Duatek*	145	M/S Raj Petro specialties Pvt. ltd
		iii.ORCHOL-13*	145	M/S R G Industries
		iv.ORCHOL-TSO*	-	M/S RG Industries
		v. Arbofine*	158	M/S Total oil India Ltd
		vi. HP Tree spray oil*	120	M/S Hindustan Petroleum Corp Ltd
		vii.MAK all season HMO(P47)	160	M/S Bharat Petroleum Corp India Ltd
		viii.Petrostar HMO*	-	M/S Petro star Pvt. Ltd
		ix. Servo(P48)	140	M/S India Oil Corporation Ltd
	Horticulture Mineral oils for summer spray	i.ATSO*	150	M/S Krishina Antioxidants Pvt. ltd
		ii.Arobfine extra*	158	M/S Total oil India Ltd
		iii.HP HMO*	150	M/S Hindustan Petroleum Corp Ltd
		iv.ORCHOL-13*	145	M/S R G Industries
		v.Duatek*	145	M/S Raj Petro specialities pvt ltd
		vi.Power plant DF*	-	M/S Green Planet Bioproducts
		vii.MAK all season HMO*	160	M/S Bharat Petroleum Corp India Ltd
		viii.Petrostar HMO*	-	M/S Petro star Pvt Ltd
		ix. Servo*	140	M/S India Oil Corporation Ltd

2.	Insecticides	Formulations	Average Market Price (Rs)	Name of the Company
	Chlorpyriphos	Coroban (20 EC)(P49)	350	M/S Coromandel AgricoPvt. Ltd.
		Kohiban (20 EC)(P50)	330	M/S Fungicide India Ltd.
		Dursban (20 EC)(P51)	320	M/S Denocil Crop Protection Ltd. M/S Dow Agro Science Mumbai Ltd
	Dimethoate	Rogor (30 EC)(P52)	385	M/S Cheminova India Ltd
	Quinalphos	Ekalux (25 EC)(P53)	650	MS Syngenta India Ltd.
	Thiocloprid	Alanto (240 SC)(P54)	2350	M/S Bayer Crop Science Ltd.
3.	Acaricides	Formulations	Average Market Price (Rs)	Name of the Company
	Abamectin	Vertimec (1.8 EC)(P55)	-	M/S Syngenta India Ltd.
		Abacin (1.9 EC)*	-	M/S Crystal India Ltd
	Dicofol	Colonal S (18.5 EC)(P56)	-	M/S Indofil Chemicals Mumbai
	Ethion	Tope (50 EC)*	600	M/S Fungicide India Ltd.
		Lazor (50EC)*	-	M/S Cheminova India Ltd
		Tafethion (50EC)(P57)	500	M/S Rallis India Ltd
	Fenazaquin	Magister (10 EC)(P58)	2400	M/S Denocil Crop Protection Ltd.
		Majestic (10 EC)(P59)	2400	M/S Fungicide India Ltd.
	Fenpyroximate	Sedna (5 SC)(P60)	-	M/S Rallis India Ltd.
	Hexythiazox	Maiden (5 EC)(P61)	2100	M/S Biostadt India Ltd.
		Karadite (5SC)*	-	M/S Godrej Agrovet India Ltd
	Milbemectin 1 EC	Milbeknock (1 EC)(P62)	2400	M/S NagarjunaAgri- Chemical Limited
	Propargite	Omite (57 EC)(P63)	1300	M/S Crompton Specialties Asia Pacific Pvt. Ltd.
	Spiromesifen	Oberon (240 SC)(P64)	4000	M/S Bayer Crop Science India Ltd.
4	Insecticide/ Acaricide	Formulations	Average Market Price (Rs)	Name of the Company
	Clothianidin	Danitop (50 WDG)*	-	M/S NagarjunaAgri- Chemical Limited

Chapter 30

Price Forecasting and Co-Integration of Apple Markets in India

S. A. Wani, S. H. Baba, M. M. Mir and Haris Manzoor

Network Project on Market Intelligence (ICAR),
School of Agricultural Economics and Horti-Business Management,
Sher-e-Kashmir University of Agricultural Sciences and
Technology of Kashmir, Srinagar
E-mail: dr.shabirwani@rediffmail.com

1. Introduction

During the first few five year plans, priority in India was assigned to achieve self-sufficiency in food-grains production. However, over the years in the state of Jammu and Kashmir, horticulture emerged as an important and growing sub sector of agriculture, offering a wide range of choices to the farmers for crop diversification. It also provides ample opportunities for sustaining large number of agro industries which generate substantial employment opportunities. With agriculture and allied sectors finding alternate ways of increasing productivity of crops, horticulture as a sub-sector, is a revelation, showing remarkable signs of progress in the state. Horticulture sector has emerged as an important sector for diversification of agriculture and has established its credibility in improving farm income through increased productivity, generating employment and in enhancing exports besides providing household nutritional security. The focussed attention on investment in horticulture during the last two decades has been rewarding in terms of increased production and productivity of horticultural crops with manifold export potential.

Apple (*Malus pumila*) is commercially the most important temperate fruit and is fourth among the widely produced fruits in the world after banana, orange and grapes. There are more than 7,500 known cultivars of apples, resulting in a range of desired characteristics. Different cultivars are bred for various tastes and uses, including cooking, fresh eating and cider production. Apples in India are mainly grown in three mountainous states of North India. The apple cultivation in Jammu and Kashmir is an old age activity and around 200 varieties of apple were used to be cultivated in the state. Jammu and Kashmir contributes around 65 per cent of total apple production in the country and the productivity reached 13.07 metric tonnes per hectare which is highest in India and is comparable to China. The horticulture sector in the state contributed Rs. 6000 crores towards state gross domestic product during 2013-14 of which apple alone accounted for about 4500 crores (Economic Survey, 2014).

The increased production of temperate fruits in general and apple in particular is of no avail unless there is a satisfactory marketing system. Marketing of apple involves considerable risk due to its perishability, affordability, availability, convertibility, alternate uses and high price sensitivity. Temperate apples are novelty of the region, invites attention of the huge Indian market, but at the same time its trade is highly exploitative. The increase in per-capita income has made people to shift to the cities leading to expanded urbanization and far spreading demand for fruits to make rich their food basket. This in turn increased the market demand for Kashmir apple. Thus growing demand for temperate fruits from Kashmir in domestic markets provided basis to extend incentives to growers to adopt this trade as the core sector of the region's economy. The producers (apple growers) could live up to an incentive, if their share in the consumer's rupee is improved. Unfortunately the presence of large number of intermediaries in the channels of distribution of temperate fruits not only discourage growers from Kashmir but also has reduced the market efficiency. For the development of fruit industry it is thus imperative to develop market infrastructure which could help to explore the possibilities of reducing the marketing costs/margins involved in marketing and diverting it towards the producers.

Inspite of tremendous progress in production and productivity of apple in the state, there are various issues pertaining to marketing which needed to take care of. The state does not have specialized markets, and these fruits are traded in distant markets such as the Delhi, Ahmadabad, Bengaluru, Mumbai and Kolkata besides other major terminal markets of India. This leads to a rise in post harvest costs, including marketing cost, forcing the growers to opt for pre-harvest disposal to meet their financial obligations and avoid climatic risks. Apples are characterized by strong seasonality and perishability. This induces competition and affects their prices and quantities supplied. In this background it is imperative to analyse existing marketing system and price behaviour of apple in different major markets in India.

2. Data and Methodology

Both the primary and secondary data has been used in the study. Primary data was collected from sample respondents in major fruit belts of the valley which included Sopore, Shopian, Pulwama and Baramulla. Secondary data with respect to area, production and daily data on wholesale prices from 2005 to 2015 of different markets was collected from Directorate of Horticulture Planning and Marketing and Directorate of Economics and Statistics, Government of J&K.

Apple comes from every corner of the state and is disposed to different primary/secondary wholesale markets. Apple is marketed almost in every major primary wholesale markets of the country. However, based on the highest volume of the apple receipts (arrivals), eight terminal markets (Delhi, Mumbai, Kolkata, Bangalore, Ahmadabad, Amritsar, Sopore and Parimpora) were selected. Furthermore, three prominent apple species were considered for the study *i.e.* Red Delicious, American and *Maharaji*. Daily data on wholesale prices from 2005 to 2015 of different markets was later averaged to obtain weekly wholesale prices set from September to December every year for each identified market.

(a) Stationarity in Data Set

The time series data has to be checked for stationarity before conducting any statistical analysis. A time series is said to be stationarity, if its properties are unaffected by a change of time origin, that is if the joint probability distribution associated with k observation Z_{t1}, Z_{t2}........ Z_{tk}, made at any set of time t_1, t_2 ....t_k respectively. Z_{t1+k}, Z_{t2+k},........ Z_{tk+m} made at any set of time t_{1+m}, t_{2+m}.....t_{k+m}. Thus for a discrete time series to be strictly the joint distribution of any set of observations must be unaffected by shifting all the time of observation forward or backward by any integer amount m. Stationarity implies that the series remains at a fairly constant level overtime. If a trend exists in the time series then series is not stationary. The time series should show constant variation in its fluctuations over a period of time.A time series must satisfy the stationarity conditions in order to be eligible for the application of this methodology.

(b) Augmented Dickey–Fuller (ADF) Test

The stationarity of a series can be tested using a unit root test, the most widely used being Augmented Dickey-Fuller unit root test. It would test the null hypothesis that the series has a unit root, *i.e.* non stationary. The test is applied by running the regression of the form given in Equation (2):

$$Yt = \beta1 + \delta Yt\text{-}1 + \alpha i \Sigma mi=1 \, \Delta Yt\text{-}i + \varepsilon t \qquad \qquad ...(2)$$

Where, ȧt is a pure white noise error-term and

$$\Delta Yt\text{-}i = (Yt\text{-}1 - Yt\text{-}2).$$

(c) Error Correction Model

The co- integration analysis reflects the long-run movement of price indices, although in the short run they may drift apart. Johansen's (1988) multivariate co-integration approach was used to examine co-integration between two price indices. Before conducting co- integration test, it is mandatory to perform stationarity test. Augmented Dickey-Fuller (ADF) unit root test (Dickey and Fuller, 1979) was performed in this study to check stationarity for both the series.

A co integrated system can be written as:

$$\Delta y_t = \sum_{i=1}^{k} \Gamma_i \Delta y_{t-i} + \alpha \beta' y_{t-k} + \varepsilon_t$$

...(1)

where y_t is the price series, Δy_t is the first difference *i.e.* ($\Delta y_t = y_t - \Delta y_{t-1}$), and the matrix $\alpha \beta'$ is n x n with rank ($0 \le r < n$), which is the rank of linear independent co integration relations in the vector space of matrix. The Johansen's method of co integrated system is a restricted maximum likelihood method with rank restriction on matrix $\Pi = \alpha \beta'$. The rank of Π can be obtained by using λ_{trace} statistic. The test statistic can be given as:

$$\lambda_{trace} = -T \sum_{i=r+1}^{n} \ln(1 - \overline{\lambda_i}) \forall r = 0,1,\ldots n-1$$

...(2)

Where, T is the total number of observations, $\overline{\lambda_i}$'s are estimates of the Eigen values representing the strength of the correlation between the first difference part and the error-correction part. Now the null and alternative hypothesis for testing co-integration rank, are, H_0: rank of $\Pi = r$ and H_1: rank of $\Pi > r$ respectively. Where, r is the number of co-integrating vectors. This test is carried under the condition that the co integrating equation has only intercept (no trend) and the original price series follows a trend since the mean and variance are non-constant over a period of time (non-stationary).

(d) Auto-Regressive Integrated Moving Average

As noted, the BJ methodology is based on the assumption that the underlying time series is stationary or can be made stationary by differencing it one or more times. This is known as the ARIMA (p, d,q) model, where d denotes the number of times a time series has to be differenced to make it stationary. In most applications d = 1 – that is, we take only the first differences of the time series. Of course, if a time series is already stationary, then an ARIMA (p, d,q) becomes an ARMA (p,q) model. The practical question is to determine the appropriate model in a given situation. From the fitted model is white noise; if they are, we can accept the chosen model, but if they are not, we will have to start afresh. That is why the BJ methodology is an iterative process. The ultimate test of a successful ARIMA model lies in its forecasting performance, within the sample period as well as outside the sample period.

(e) Vector Auto Regressive Model

A VAR model describes the evolution of a set of k variables (called *endogenous variables*) over the same sample period ($t = 1,....,T$) as a linear function of only their past values. The variables are collected in a $k \times 1$ vectory_t, which has as the ith element, $y_{i,t}$, the observation at time "t" of the ith variable. For example, if the ith variable is GDP, then $y_{i,t}$ is the value of GDP at time t.

A *p-th order VAR*, denoted VAR (p), is

$$y_t = c + A_1 y_{t-1} + A_2 y_{t-2} + ... + A_p y_{t-p} + e_t$$

Where the l-periods back observation yt−l is called the l-thlag of y, c is a $k \times 1$ vector of constants (intercepts), Ai is a time-invariant $k \times k$matrix and et is a $k \times 1$ vector of error terms satisfying

$E(e_t) = 0$— every error term has mean zero;

$E(e_t e'_t) = \Omega$— the contemporaneous covariance matrix of error terms is dinu123 (a $k \times k$ positive-semi definite matrix);

$E(e_t e'_{t-k}) = 0$ for any non-zero k — there is no correlation across time; in particular, no serial correlation in individual error terms.[1]

A pth-order VAR is also called a VAR with p lags. The process of choosing the maximum lag p in the VAR model requires special attention because inference is dependent on correctness of the selected lag order.In the present investigation, VECM model was also used for forecasting the apple price in different markets.

(f) Autocorrelation

The autocorrelation of a random process describes the correlation between values of the process at different times, as a function of the two times or of the time lag. Let X be some repeatable process, and i be some point in time after the start of that process. (i may be an integer for a discrete-time process or a real number for a continuous-time process.) Then X_i is the value (or realization) produced by a given run of the process at time i. Suppose that the process is further known to have defined values for meanμ_i and variance σ_i^2 for all times i. Then the definition of the autocorrelation between times s and t is

$$R(s,t) = \frac{E[(X_t - \mu_t)(X_s - \mu_s)]}{\sigma_t \sigma_s}$$

Where "E" is the expected value operator. Note that this expression is not well-defined for all-time series or processes, because the variance may be zero (for a constant process) or infinite (for processes with distribution lacking well-behaved moments, such as certain types of power law). If the function R is well-defined, its value must lie in the range [−1, 1], with 1 indicating perfect correlation and −1 indicating perfect anti-correlation.

3. Key Results

The sections that follow presents importance of the crop in the region in terms of output, major markets in terms of arrivals, price pattern in the selected markets, data used for co-integration, testing stationarity, Causality among prices besides the key results and findings

(a) Marketing System

Apple is produced by large number of smallholders scattered around the valley whereas, the consumers are located throughout the country. Small produce, lack of knowledge of marketing system and less liquidity potential *etc.*, prevent them to undertake direct marketing of apple. Accordingly majority of farmers are compelled to dispose of their produce through different functionaries (Figure 30.1). Among functionaries pre-harvest contractors are the major players and who purchased standing crop at blooming or fruiting stage at relatively lower prices. Only 14 per cent of the farmers were found selling produce directly to wholesalers. The marketing system for apple is highly complex and is composed of different marketing channels for distribution of apple in different markets. In each channel varying number of functionaries are involved and numerous specialized business activities called marketing functions are to be performed by them.

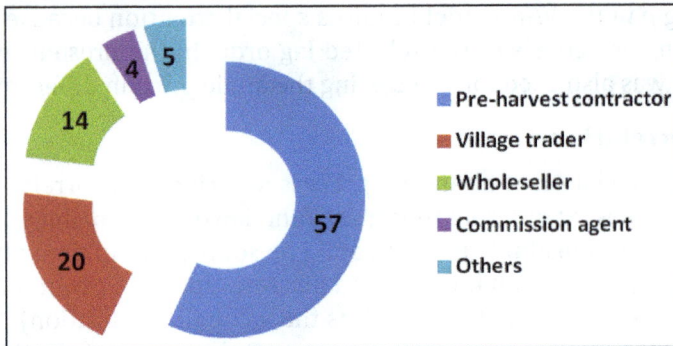

Figure 30.1: Sale of Apple through different Functionaries (per cent).

(b) Market Practice

The price discovery process in the local markets within the state are not transparent. The markets do not have a price dissemination mechanism and it is difficult to know the prevailing prices on any given day. While there is some understanding of the fees payable to the market intermediaries, there are no norms and enforcement of such norms. High commissions and fees payable to intermediaries tend to get blurred with other fees and charges; often adjusted in the price thereby making the realisation uncertain. While APMC law is passed in the State, it has not been implemented. The law needs amendments in line with the changes suggested by the Central Government and adopted by several states. This has led to the proliferation of unlicensed traders, agents acting in the market with non-transparent auction procedures. The APMCs do not earn revenues through collection of *Mandi* tax and are unable to improve the conditions in the *Mandi*. The major destination of Kashmir apples is the Azadpur *Mandi*, which is a buyers' market and designed to be so. Manipulation of prices by traders in the *Mandi* is resorted through a) stopping apple trucks at the border of entry in to Delhi, use of cold stores to alter supply of apples in the *Mandi*, keeping away small buyers with artificially high price quotes and later reducing prices to low levels to benefit preferred buyers and use of proxies in auctions. The markets within the state are comparatively better in price determination and transparency. Growers with pre-harvest contracts (PHC) access the markets easier, but lose out on full benefit of market prices on account of their taking money in advance. Free growers find it difficult to enter markets even when the demand is brisk and the commission agents prefer their 'captive growers' with PHC. Setting up satellite markets has helped growers (especially the free ones) in marketing. Farmers who market apples through cooperatives realise higher prices. Trade margins range from 42 to 73 per cent in the different channels of marketing. Price discovery by grower would be more realistic and effective if he is able to hold back and store his produce for some time. The farmer needs to have conditions (local storage and financial capacity to hold) under which distress selling can be checked (NABCONS, 2013)

(c) Major Markets in Terms of Arrivals

Apple is marketed in all major primary wholesale markets of the country however, based on the highest volume of the apple receipts; eight markets (Delhi, Mumbai, Kolkata, Bangalore, Ahmadabad, Amritsar Sopore and Parimpora) were selected. Out of the total exported apple around 48 percent goes to Delhi market followed by Ahmadabad 7 percent, Mumbai 6 percent and Kolkata 6 percent respectively.

(d) Arrival and Price Pattern in Lead and Lag Markets

Apple is available in the market from September to December every year and January and February in the following year which means that the fruit is available for almost six months in a year in the market. However during the months of September to December, there is maximum arrival in the market which subsequently becomes lesser and lesser in the coming months. During the study two market were identified as lead and lag market. The major chunk of the production goes to the Delhi, identified as lead market. Similarly Parimpora market was identified as lag market. In order to get an exact behaviour of price and arrival pattern in lead and lad market, daily data of price and arrivals were plotted. It was found that both lead and lag market followed the same pattern (Figures 30.2 and 30.3).

(e) Price Behaviour and Market Integration

The study analyses the interdependence of price changes of apple in eight different major terminal markets in India. The criteria for selecting the markets was

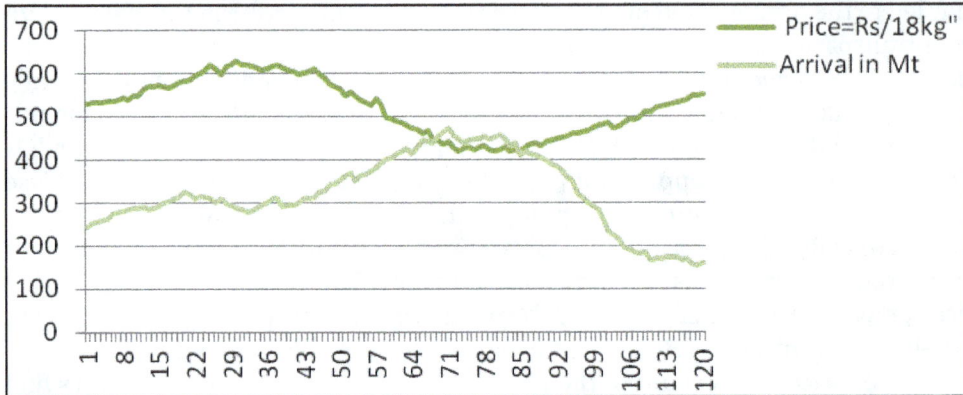

Figure 30.2: Arrival and Price Pattern in Lead Market Delhi.

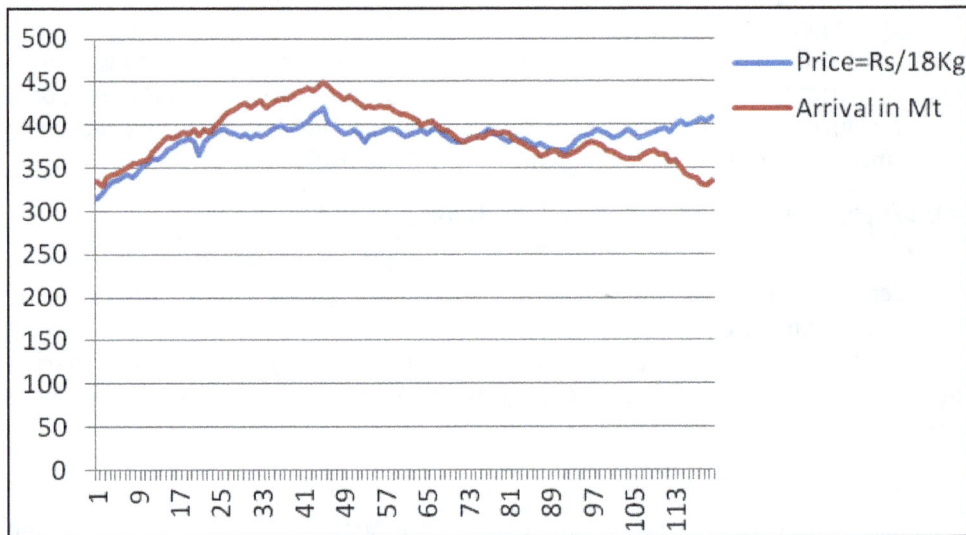

Figure 30.3: Arrival and Price Pattern in Lag Market Parimpora.

the geographical distribution like, local, northern, western, southern and eastern markets alongwith the volume of transactions. In addition, availability of data is also an important factor for selecting the markets. Price variability is the major component of market risk for both producers and consumers (Schumpeter, 1999). Government at the national level plays an important role in administering agriculture prices in India through various market intervention mechanisms. However, these strategies are mostly in vogue for agricultural commodities especially for some food grains most of which are not perishable like fruits and vegetables, which too have reduced overtime (Jha and Srinivasan, 1999; Ramaswamy, 2002; Chand, 2003; Golettie *et al.,* 2005). Usually under this spectrum commodity group from hill and mountainous states are mostly the sufferers, despite having potential of pushing growth in agriculture beyond predicted values. It is important to note that more than 90 per cent of market surplus in fruits and more than 60 per cent of vegetables are sold in open market arrangements in these states. Under such a condition the discovery of price behaviour under various market situations becomes important for risk management. The paper looks into the issues for the presence of co integration among the markets. It also, examines the possibility of causal linkages among the different markets.

(f) Market Efficiency

The market efficiency evaluation under co-integration analysis recognizes that the time series prices for various markets are usually non-stationery variables. (Shen and Wang, 1990; Fortenbery and Zapata, 1993; Wang and Ke, 2005) and if these series are found to be non-stationery then it becomes necessary to test for co-integration, as a pre-condition for market efficiency and un-biasedness (Kallar *et al.*, 1999) and also finding of no-integration of markets is normally interpreted to imply market in-efficiency. ADF test was applied at level and first difference to check the stationarity of this series. Table 30.1, presents the results of unit root test for three commercial varieties (Red Delicious, American and Maharaji) with two grades in selected markets. The results revealed that the null hypothesis of the unit root cannot be rejected for all the price series. Therefore, we could conclude that all the prices in selected markets are non-stationery. Apple prices in different major Indian markets become stationery at first difference.

Table 30.1: Augmented and Dickey-Fuller Unit Root Test on Market Prices

Market		Test Statistics	Mackinnon p-value	Critical Value	
				1 per cent	5 per cent
Parimpora	Level	-2.838	0.1835	-4.178	-3.512
	Ist difference	-5.853	0.000	-4.187	-3.516
Sopore	Level	-2.784	0.2029	-4.178	-3.512
	Ist difference	-6.607	0.000	-4.187	-3.516
Delhi	Level	-2.624	0.2688	-4.178	-3.512
	Ist difference	-6.352	0.000	-4.187	-3.516
Mumbai	Level	-3.273	0.0709	-4.178	-3.512
	Ist difference	-7.802	0.0000	-4.187	-3.516
Bangalore	Level	-3.766	0.0184	-4.178	-3.512
	Ist difference	-6.728	0.0000	-4.187	-3.516
Amritsar	Level	-3.139	0.0972	-4.178	-3.512
	Ist difference	-7.700	0.0000	-4.187	-3.516
Ahmadabad	Level	-2.419	0.3698	-4.178	-3.512
	Ist difference	-6.458	0.0000	-4.187	-3.516
Kolkata	Level	-3.436	0.0468	-4.178	-3.512
	Ist difference	-6.820	0.000	-4.187	-3.516

(g) Co-integration among Markets

The trace test results presented in Table 30.2 determine the number of co-integrating vectors. The figures documented in the table revealed that there are four co-integrated vectors at 5 per cent level of significance. The discussion is suggestive of the fact that even if there is geographical dispersion of markets, which are spatially segmented, the prices are linked together indicating that all the market locations are in the same economic market system. The flow of apple produce from one market to another market may affect the price levels in destiny market and accordingly the supply of the produce has to be taken care of so, that each market receives optimum volume of the produce as per the quantum of demand.

Table 30.2: Johansen's Co-integration Test for Selected Apple Varieties/Grades/Markets

Maximum Rank	Eigen Value	Trace Statistics	Critical Value (5 per cent)	Max. Statistics	Critical Value (5 per cent)
0	-	430.41	156.0	163.51	51.42
1	0.97567	266.90	124.24	88.76	45.28
2	0.86698	178.14	94.15	78.58	39.37
3	0.83236	99.56	68.52	57.68	33.46
4	0.73041	41.88	47.21	22.84	27.07
5	0.40495	19.04	29.68	13.89	20.97
6	0.27067	5.15	15.41	4.97	14.07

Trace test indicates 4 co-integrating equations at the 0.05 level.

(g) Causality in Various Markets/Varieties and Grades

The co-integration tests performed indicated the existence of long run relationship among the prices across the 7 markets only. Since the direction of the relationship among price series and market is equally important, therefore, Granger Causality Tests were also performed. The results presented in Table 30.3 revealed that the uni-directional causality in the Delhi market which affect to the prices of apple in Kolkata, Sopore, Parimpora, Amritsar and Ahmadabad. These uni-directional relationships where prices of one market affects the prices of other market without having a reciprocal impact on the prices would imply that the market for such varieties/grades is not very efficient in terms of influencing the prices of the other markets and also would increase prices in such markets. Similarly, the bi-directional causation was observed in Parimpora and Kolkata, Ahmadabad and Sopore, Kolkata and Sopore and between Ahmadabad and Kolkata. It should be noted here that the Granger causality results may vary for different number of lags or time horizon included in the models. Granger causality always has to be tested in the context of some model. In the specific case of the test function, the model has p past values of each of the two variables in the bi-variate test. A conventional way to choose p for this model would be to try this regression with various values of p and use keep track of the AIC or BIC for each lag length. Then

Table 30.3: Granger Causality Test Statistics for Selected Apple Markets

Causality	χ^2	Prob.	Relationship
SOPORE does not Cause PARIMPORA	6.949	0.1390	P → S
PARIMPORA Cause SOPORE	45.702	0.0000	
DELHI Cause PARIMPORA	25.506	0.0000	D → P
PARIMPORA does not Cause DELHI	3.3248	0.505	
AHMADABAD Cause PARIMPORA	3.010	0.0000	A ↔ P
PARIMPORA Cause AHMADABAD	58..601	0.0000	
KOLKATA Cause PARIMPORA	17.877	0.001	P ↔ K
PARIMPORA Cause KOLKATA	22.894	0.000	
MUMBAI Cause PARIMPORA	15.102	0.004	M → P
PARIMPORA does not Cause MUMBAI	1.194	0.879	
AMRITSAR Cause PARIMPORA	4.1478	0.386	P → A
PARIMPORA does not Cause AMRITSAR	64.224	0.000	
DELHI cause SOPORE	19.768	0.001	D → S
SOPORE does not Cause Delhi	2.5196	0.641	
AHMADABAD cause SOPORE	30.478	0.000	Ah ↔ S
SOPORE cause AHMADABAD	26.037	0.000	
KOLKATTA cause SOPORE	44.189	0.000	K ↔ S
SOPORE cause KOLKATA	12.97.1	0.011	
MUMBAI cause SOPORE	28.411	0.000	M → S
SOPORE does not cause MUMBAI	2.415	0.660	
AMRITSAR cause SOPORE	14.736	0.005	A ↔ S
SOPORE cause AMIRTSAR	51.335	0.000	
AHMADABAD does not cause DELHI	3.084	0.544	D → AH
DELHI cause AHMADABAD	15.927	0.003	
KOLKATTA cause DELHI	6.3923	0.172	D → K
DELHI does not cause KOLKATA	23.834	0.000	
MUMBAI does not cause DELHI	8.7023	0.069	M → D
DELHI cause MUMBAI	7.5057	0.111	
AMRITSAR cause DELHI	5.605	0.231	D → A
DELHI does not cause AMRITSAR	37.578	0.000	
KOLKATTA cause AHMADABAD	21.393	0.000	AH ↔ K
AHMADABAD cause KOLKATA	69.235	0.000	
MUMBAI cause AHMADABAD	30.381	0.000	M → AH
AHMADABAD does not cause MUMBAI	1.8136	0.770	
AHMADABAD cause AMRITSAR	55.358	0.000	AH ↔ A
AMRITSAR cause AHMADABAD	12.093	0.017	
MUMBAI cause KOLKATA	23.315	0.000	M → K
KOLKATA does not cause MUMBAI	4.9629	0.291	
AMRITSAR cause KOLKATA	2.1441	0.709	K → A
KOLKATTA cause AMRITSAR	24.276	0.000	
AMRITSAR does not cause MUMBAI	7.1078	0.130	M → A
MUMBAI cause AMRITSAR	35.37	0.000	

run the test again using the value of *p* which had the lowest Information criteria in the regressions. In the present investigation, it was found that the AIC values are minimum at 3 to 6 lags. Accordingly, specific lag length was used for testing the causality with the help of VAR model. This implies that the markets by and large have enough ability to predict subsequent prices among them. The results of the study are, therefore, quite useful to various stakeholders like producers, traders, commission agents and policy makers. The results will be helpful in framing well oriented policy on market intervention schemes for an open commodity market in apple.

(h) Price Forecast

Daily price data of apple for seven important markets of Delhi, Mumbai, Amritsar, Kolkata, Ahmadabad, Parimpora, Bangalore and Sopore were tabulated and finally data was arranged fortnightly in order to carry out forecasting for the benefit of producers so that they can choose time period for disposal and *mandi* of choice where their produce will fetch good prices. Further it will help them to diversify their market option and to some extent their harvesting time schedule for better returns. In order to carry out the price forecast daily price data of apple of different markets from the year 2005 to 2014 was taken into account and subsequently forecasting was done using different forecasting models. However, VAR and VECM model yielded better results and are presented in Table 30.4.The results of forecasted price of apple were validated then with actual price of apple during the season. Overall, accuracy percentage turned very high and varied from 86 per cent to 91 per cent for different markets.The results of price validation of a few markets are presented in Table 30.5.

Table 30.4: Price Forecast of Apple for the Year 2015-16

Date	Delicious (Rs./FC)		American (Rs./FC)		Maharaji (Rs./FC)	
	Grade I (Min-Max)	Grade II (Min-Max)	Grade I (Min-Max)	Grade II (Min-Max)	Grade I (Min-Max)	Grade II (Min-Max)
Parimpora (Mandi)						
16th to 30th sep.	701-803	424-535	595-693	316-387	309-388	169-255
Ist to 15th Oct.	680-841	406-560	584-707	311-392	308-395	154-274
16th to 31th Oct.	664-869	393-580	579-719	306-397	306-400	142-288
Ist to 15th Nov.	652-895	383-598	575-730	302-402	304-404	131-300
16th to 30th Nov.	642-919	374-614	572-741	298-406	302-407	122-310
Ist to 15th Dec.	666-941	367-629	570-752	294-411	300-411	113-319
16th to 31th Dec.	658-960	360-643	568-762	291-414	298-415	111-312
1st to 15th Jan	652-980	354-657	567-771	288-418	297-418	103-319
Sopore (Mandi)						
16th to 30th sep.	688-786	521-637	650-812	351-471	-	-
Ist to 15th Oct.	665-820	506-661	623-858	328-500	-	-
16th to 31th Oct.	649-847	495-682	602-895	311-523	-	-

Date	Delicious (Rs./FC)		American (Rs./FC)		Maharaji (Rs./FC)	
	Grade I (Min-Max)	Grade II (Min-Max)	Grade I (Min-Max)	Grade II (Min-Max)	Grade I (Min-Max)	Grade II (Min-Max)
Ist to 15ᵗʰ Nov.	636-871	486-699	585-927	297-542	-	-
16ᵗʰ to 30ᵗʰNov.	625-892	479-716	570-955	285-560	-	-
Ist to 15ᵗʰ Dec.	647-912	472-731	558-981	275-576	-	-
16ᵗʰ to 31ᵗʰ Dec.	638-931	467-745	548-1006	265-591	-	-
1ˢᵗ to 15ᵗʰ Jan	631-949	462-759	538-1029	257-605	-	-
Delhi (Mandi)						
16ᵗʰ to 30ᵗʰsep.	834-990	479-634	733-857	597-721	442-526	317-389
Ist to 15ᵗʰ Oct.	802-1043	450-669	719-888	579-751	428-547	311-407
16ᵗʰ to 31ᵗʰ Oct.	777-1082	429-697	709-913	567-776	417-564	306-419
Ist to 15ᵗʰ Nov.	757-1116	412-721	700-935	557-798	409-578	301-430
16ᵗʰ to 30ᵗʰNov.	741-1148	397-743	693-955	550-818	402-591	297-440
Ist to 15ᵗʰ Dec.	763-1177	384-763	688-974	544-837	395-603	293-448
16ᵗʰ to 31ᵗʰ Dec.	751-1205	372-781	683-992	538-855	390-614	290-457
1ˢᵗ to 15ᵗʰ Jan	739-1231	361-799	679-1009	534-873	385-625	287-465
Bangalore (Mandi)						
16ᵗʰ to 30ᵗʰsep.	1074-1198	796-952	764-931	520-692	614-801	441-587
Ist to 15ᵗʰ Oct.	1047-1244	773-992	741-975	490-729	582-846	420-594
16ᵗʰ to 31ᵗʰ Oct.	1029-1283	757-1024	720-1010	468-758	558-882	409-604
Ist to 15ᵗʰ Nov.	1016-1317	745-1053	703-1040	449-783	539-913	402-616
16ᵗʰ to 30ᵗʰNov.	1005-1348	735-1079	689-1067	433-806	523-941	397-627
Ist to 15ᵗʰ Dec.	1047-1378	726-1103	676-1091	419-827	509-967	392-639
16ᵗʰ to 31ᵗʰ Dec.	1040-1405	719-1126	665-1115	406-846	497-992	388-650
1ˢᵗ to 15ᵗʰ Jan	1034-1433	714-1148	655-1137	394-865	486-1015	384-660
Kolkata (Mandi)						
16ᵗʰ to 30ᵗʰsep.	1046-1165	488-598	666-797	531-643	-	-
Ist to 15ᵗʰ Oct.	1021-1216	472-623	643-826	512-670	-	-
16ᵗʰ to 31ᵗʰ Oct.	1004-1257	459-643	628-851	498-692	-	-
Ist to 15ᵗʰ Nov.	992-1293	449-661	615-874	487-711	-	-
16ᵗʰ to 30ᵗʰNov.	984-1301	441-676	605-894	478-729	-	-
Ist to 15ᵗʰ Dec.	1026-1358	434-691	597-913	471-745	-	-
16ᵗʰ to 31ᵗʰ Dec.	1021-1388	428-705	589-931	464-760	-	-
1ˢᵗ to 15ᵗʰ Jan	1016-1416	422-718	583-947	458-774	-	-
Mumbai (Mandi)						
16ᵗʰ to 30ᵗʰsep.	809-939	625-825	-	-	-	-
Ist to 15ᵗʰ Oct.	801-991	592-869	-	-	-	-
16ᵗʰ to 31ᵗʰ Oct.	784-1018	569-905	-	-	-	-
Ist to 15ᵗʰ Nov.	774-1046	551-937	-	-	-	-
16ᵗʰ to 30ᵗʰNov.	764-1071	535-966	-	-	-	-
Ist to 15ᵗʰ Dec.	795-1094	521-993	-	-	-	-

Date	Delicious (Rs./FC)		American (Rs./FC)		Maharaji (Rs./FC)	
	Grade I (Min-Max)	Grade II (Min-Max)	Grade I (Min-Max)	Grade II (Min-Max)	Grade I (Min-Max)	Grade II (Min-Max)
16th to 31th Dec.	789-1116	509-1018	-	-	-	-
1st to 15th Jan	783-1138	499-1042	-	-	-	-
Ahmadabad (Mandi)						
16th to 30thsep.	933-1049	622-742	-	-	-	-
Ist to 15th Oct.	913-1099	605-775	-	-	-	-
16th to 31th Oct.	897-1137	594-802	-	-	-	-
Ist to 15th Nov.	886-1172	585-826	-	-	-	-
16th to 30thNov.	878-1204	579-848	-	-	-	-
Ist to 15th Dec.	916-1233	573-868	-	-	-	-
16th to 31th Dec.	911-1262	569-888	-	-	-	-
1st to 15th Jan	908-1289	566-906	-	-	-	-
Amritsar (Mandi)						
16th to 30thsep.	523-626	341-444	-	-	-	-
Ist to 15th Oct.	496-652	321-470	-	-	-	-
16th to 31th Oct.	480-676	306-489	-	-	-	-
Ist to 15th Nov.	467-696	293-506	-	-	-	-
16th to 30thNov.	455-715	282-521	-	-	-	-
Ist to 15th Dec.	467-732	273-534	-	-	-	-
16th to 31th Dec.	458-749	264-547	-	-	-	-
1st to 15thJan	449-764	256-559	-	-	-	-

Table 30.5: Validation of Forecasted Price in Apple

Delicious Grade I				
Market	Actual Price (Rs/18Kg)	Forecasted Price (Rs/18Kg)	Validation (Percentage Change)	Accuracy (Per cent)
Parimpora	730	783	7	93
Delhi	860	947	11	89
Mumbai	879	920	8	93
Amritsar	614	589	7	93
Ahmadabad	914	1043	15	85
Overall average				**91**
Delicious Grade II				
Parimpora	457	491	8	92
Delhi	674	569	15	85
Mumbai	665	748	13	87
Amritsar	492	404	18	82
Ahmadabad	652	710	9	91
Overall average				87

4. Conclusion and Policy Implications

The apple cultivation in Jammu and Kashmir is an old age activity and around 200 varieties of apple were used to be cultivated in the state. Jammu and Kashmir contributes around 65 per cent of total apple production in the country with the productivity of more than10 tonnes per hectare which is highest in India and comparable to China. The horticulture sector in the state contributed Rs 6000 crores towards state GDP of which apple alone accounted for about 4500 crores.

An attempt was made in this study to examine the existing marketing spectrum and market integration of apple across different major markets in India employing both primary and secondary data. The results revealed that the state has made a tremendous progress on the front of area expansion and increasing production. The marketing system of apple was unorganized characterized by numerous marketing channels and existence of number of intermediaries which deprive farmers of real benefits. Furthermore, the major markets of apple have either one way or two way causations. Forecast of prices made on available data revealed its steady increase towards upcoming season. The findings of the study put forth few policy implications as:

☆ The study emphasized upon scientific management to improve productivity of apple orchards across different regions of the state.

☆ Farming community of the state has a predominance of smallholders and they lack resources and initiatives to be spent on marketing and production. Their resource position has to be supplemented as credit from external sources especially through government institutions wherein group lending should be given priority.

☆ Cooperative movement from production to marketing phase is need of the hour and accordingly cooperative marketing is to be encouraged/ popularised in the state. There is a need to strengthen existing cooperatives through effective leadership.

☆ The situation also demands place, time and form utilities for this crop should be addressed which otherwise results in the poor returns to the growers.

☆ The forecasting of price during different periods of its availability in the market could help farmers in disposing of their fruit to the markets where the prices would be better and remunerative.

There is need to strengthen market intelligence to help farmers and trading community to find remunerative markets. Although this initiative has already been taken by the Division of Agricultural Economics and Marketing, SKUAST-K under, "Network Project on Market Intelligence" with financial assistance from Indian Council of Agricultural Research (ICAR) but it has to be launched as a regular feature. The price forecasts should be disseminated through variety of media and at the same time further R and D efforts are required to assess optimum supply that should enter each wholesale market to contain price level forecasted for each market.

References

Chand, R. 2003. *Domestic reforms for trade liberalization: Analysis and approaches*, Paper presented at the workshop on "Analysis of Trade Liberalization for Poverty Alleviation", April 21-25, Colombo, Sri Lanka.

Economic Survey, 2014. *Directorate of Economics and Statistics*, Jammu and Kashmir Government, Srinagar 876 pp.

Fortenbery, T. R. and H. O. Zapata, 1993. An examination of co- integration relations between futures and local grain markets.*The Journal of Futures Markets* 13 (1993): 921-932.

Golettie, F., A. Raisuddin, and N. Farid, 1995.Structural determinants of market integration: The case of rice market in Bangladesh. *Developing Economics* Vol. 33 (2): 185-202

Jha, Shikha and P.V. Srinivasan,1999.Grain price stabilization in India: Evaluation of policy alternatives. *Agricultural Economics* (21): 93-108.

Johansen, S., 1988. A statistical analysis of co -integration vectors. *Journal of Economic Dynamics and Control* (12): 231-54.

Kellard, N., New bold, P., Rayner, T., Ennew, C. 1999.The relative efficiency of commodity futuresmarkets. *The Journal of Futures Markets* (4): 413-432.

NABARD Consultancy Services 2013. *A proposal for strengthening the apple value chain in Jammu and Kashmir.* NABCONS, Corporate Office, Bandra Kurla, Mumbai 29 pp.

Ramaswami, Bharat. 2002. Efficiency and equity of food market interventions. *Economic and Political Weekly* (Mumbai), March (23): 1129-1135.

Schumpeter, J.A. 1999.The creative response in economic history.*Journal of Economic History* (2): 149-59.

Shen, C. and L. Wang 1990.Examining the validity of a test of futures market efficiency: A comment.*The Journal of Futures Markets* (10): 195-196.

Wang and B. Ke 2005. Efficiency tests of agricultural commodity futures markets in China.*The Australian Journal of Agricultural and Resource Economics* (49): 125-141.

Index

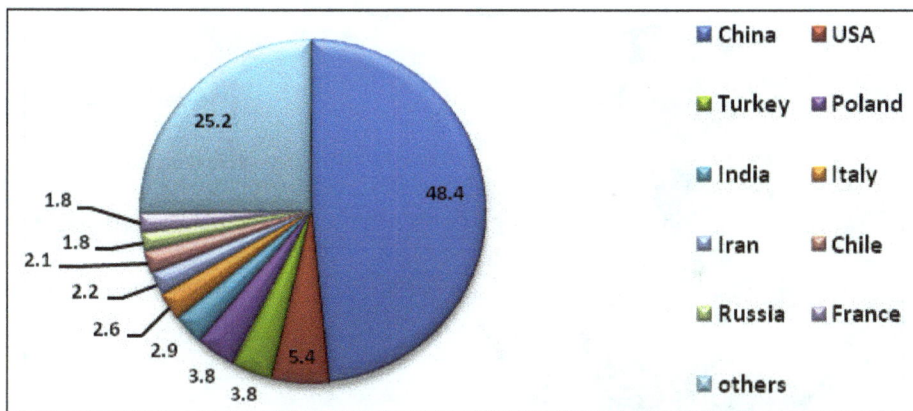

Figure 1.1: Top Ten Apple Producing Countries. (p. 2)

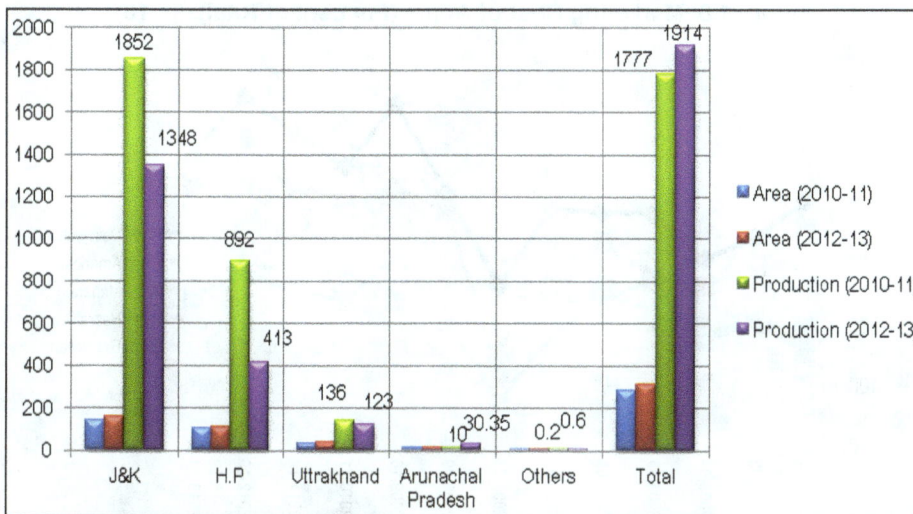

Figure 1.2: State-wise Area and Production of Apple during 2011-12 and 2012-13. (p. 3)

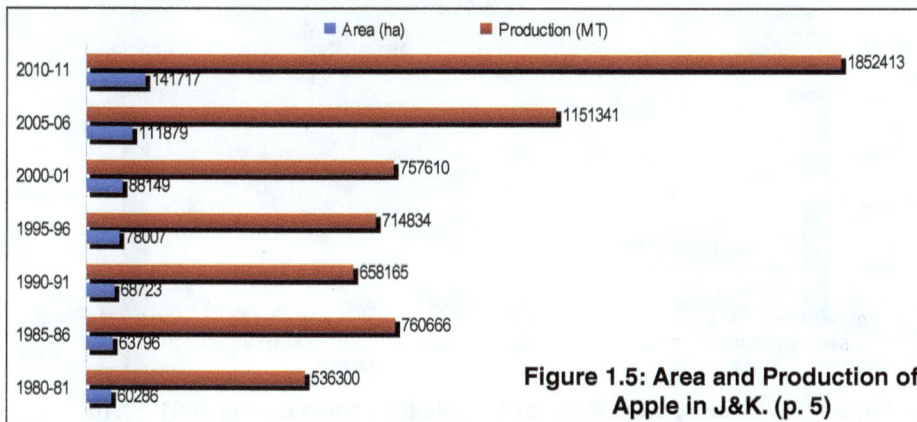

Figure 1.5: Area and Production of Apple in J&K. (p. 5)

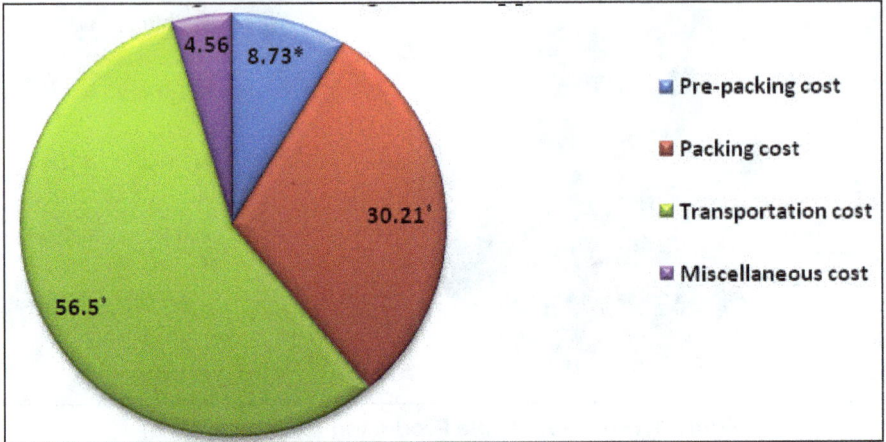

Figure 1.6: Marketing Cost of Apple (Per cent of Total). (p. 10)

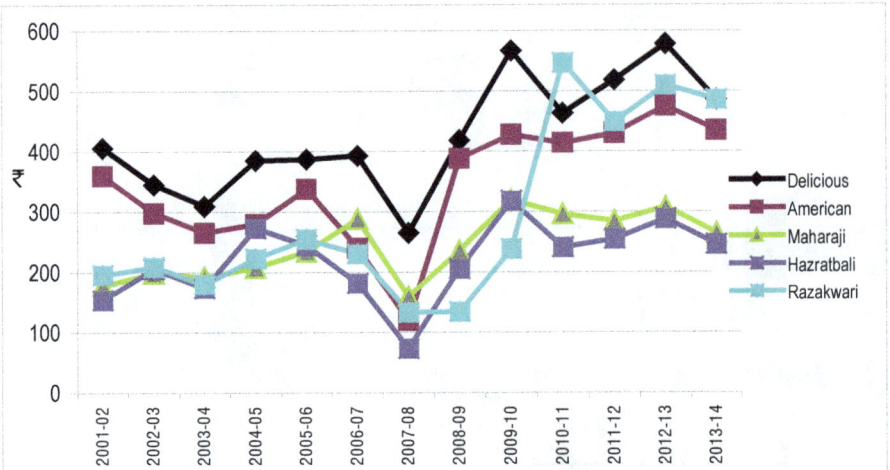

Figure 1.7: Variety-wise Wholesale Rates of Apple (Rs per box). (p. 11)

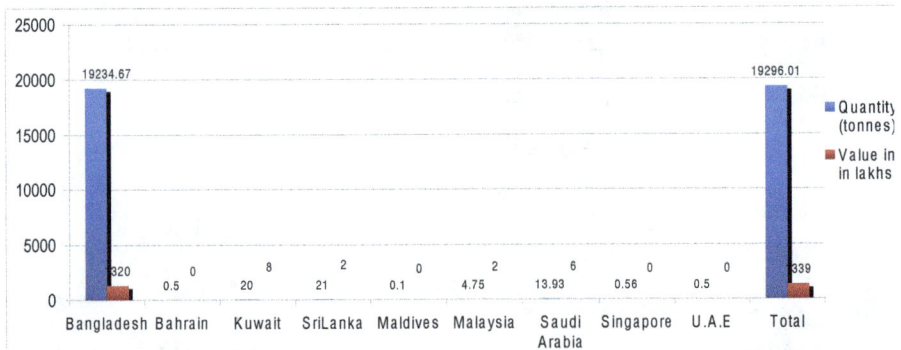

Figure 1.8: Country-wise Export of Apple from India during 2001-02. (p. 12)

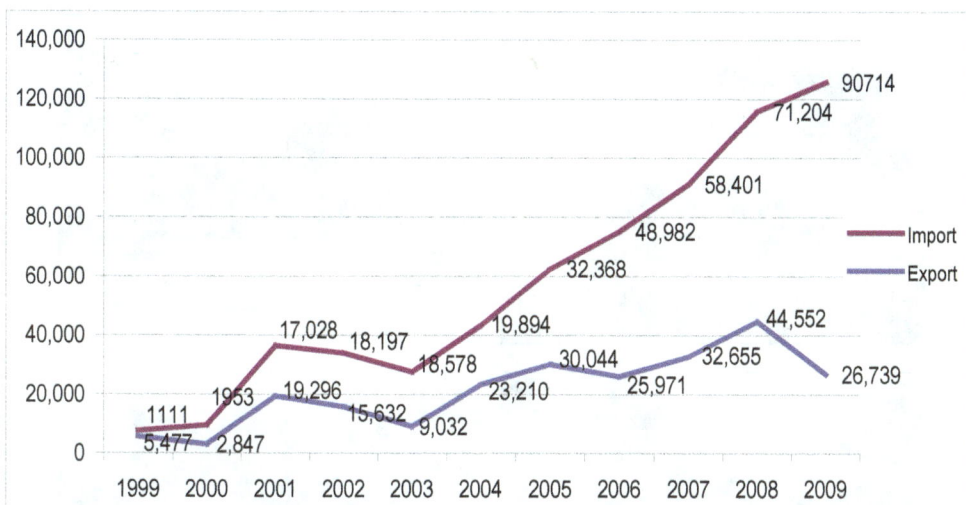

Figure 1.9b: India's Share in Global Trade of Apple (metric tonnes). (p. 13)

Plate 6.1: Established Apple Nursery. (p. 75)

Plate 6.2: Introduced Rootstocks of Apple in Nursery at CITH-Srinagar. (p. 78)

Plate 6.3: (a) Apple Seedling Block for Budding at Zainapora Nursery; (b) Apple Rootstock Block for *in situ* Grafting at CITH-Srinagar. (p. 80)

Plate 8.2a-k: Various Training Systems in Apple. (p. 126)

a. Central Leader System.

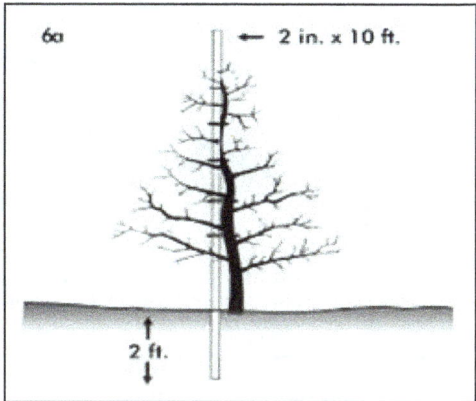

b. Slender Pyramid.

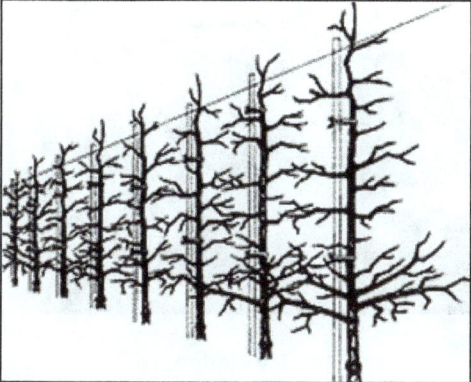

c. Vertical Axis.

d. Tall spindle.

e. Spindle Bush.

f. Double Leader.

(p. 127)

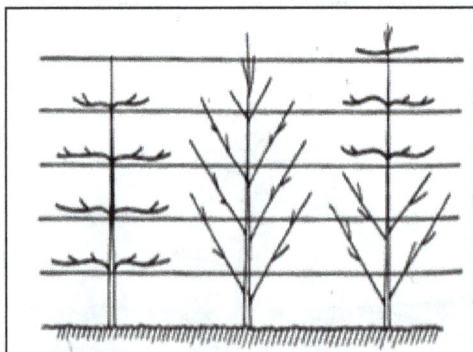

g. Palmette Systems.

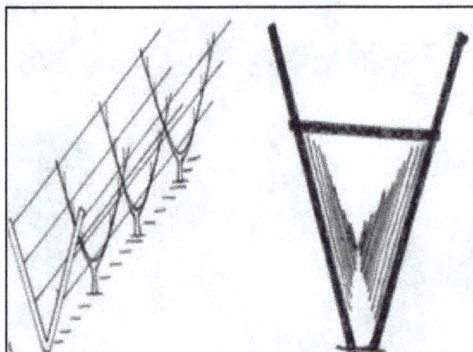

h. V/Y System.

j. Tatura Trellis.

i. V Hedge.

k. Super Spindle.

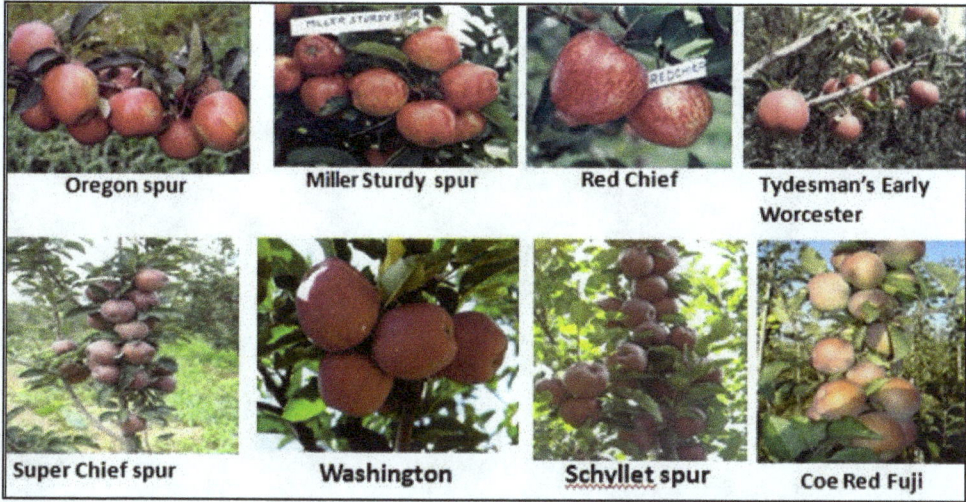

Oregon spur

Miller Sturdy spur

Red Chief

Tydesman's Early Worcester

Super Chief spur

Washington

Schyllet spur

Coe Red Fuji

(p. 137)

(p. 140)

High Density Apple (3m x 3m). (p. 141)

(A) *In vitro* **Shoot Establishment, (B) Shoot Elongation, (C) Shoot Multiplication, (D) Rooting, (E) Hardening and (F) Transfer into Soil. (p. 142)**

| Old and senile apple tree | Rejuvenated apple tree |

(p. 143)

Half-moon Water Harvesting
System. (p. 146)

Trench Water Harvesting System.
(p. 146)

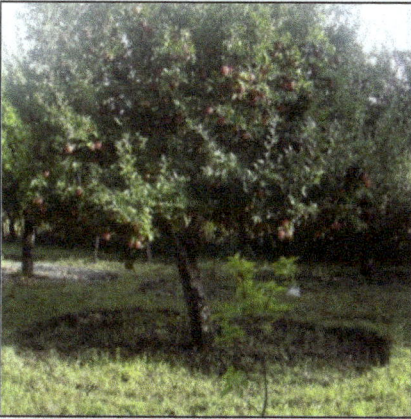

Full Moon Water Harvesting System.
(p. 147)

Cup and Saucer Water Harvesting
System. (p. 147)

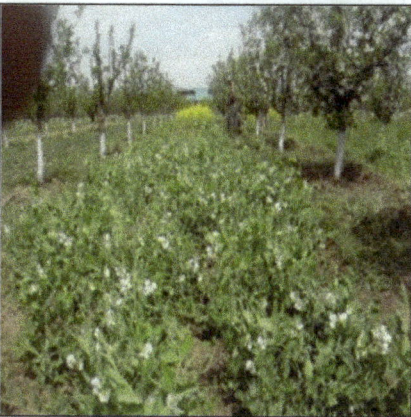

Apple + Peas (p. 148)

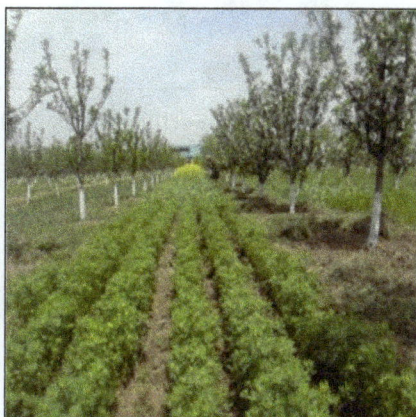

Apple + Methi. (p. 148)

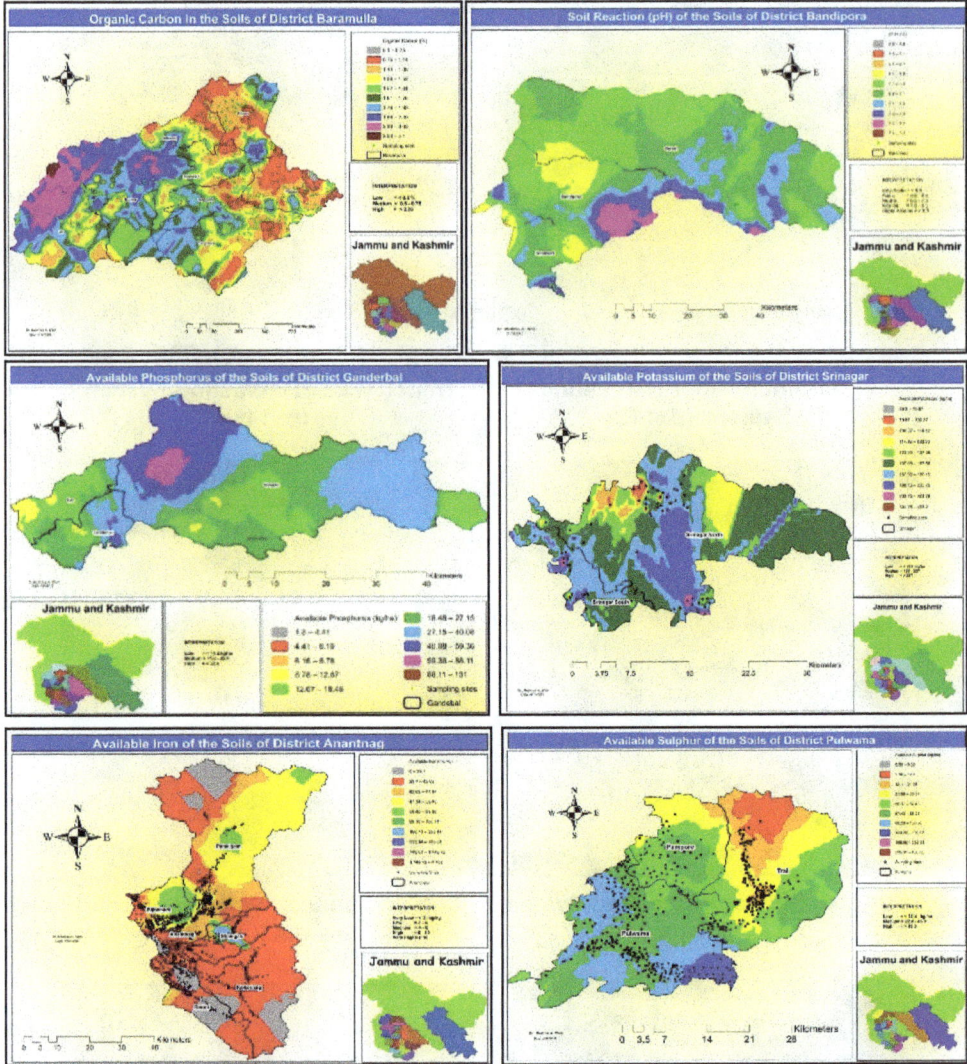

Figure 10.1: Soil Fertility Maps of different Districts of Kashmir. (p. 154)

(p. 156)

Figure 10.3: Saffron Cultivation in Pampore Kerewa. (p. 158)

Figure 10.4: Apple Trees Dying of Root Rot. (p. 160)

Figure 10.5: White Root Rot of Apple. (p. 161)

Figure 10.6: Collar Rot or Crown Rot. (p. 162)

Figure 10.7: Seedling Blight. (p. 163)

Figure 10.8: Crown Gall of Apple Trees. (p. 164)

Figure 10.9: Hairy Root of Apple Tree. (p. 165)

Plate 16.1: Initial Velvety Brown Scab Lesion on under Surface of Leaf.

Plate 16.2: Scab Lesions on Leaf Petiole.

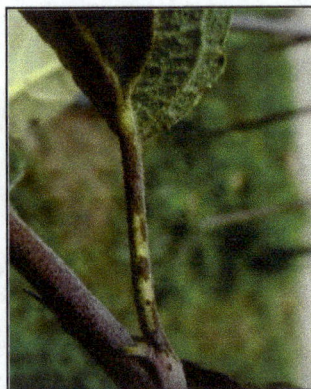

Plate 16.3: Sheet Scab Covering the Entire Leaf Surface.

(p. 251)

Plate 16.4: Light Olive Green Scab Lesions on Young Fruit.

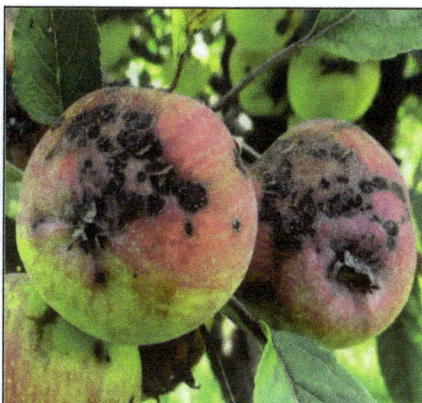

Plate 16.5: Black Corky Lesions on Mature Fruit.

(p. 251)

Plate 16.6: Circular to Irregular Light Brown Non-sporulating Lesions.

Plate 16.7: *A. mali* Outbreak Leading to Severe Pre-mature Leaf Fall.

(p. 252)

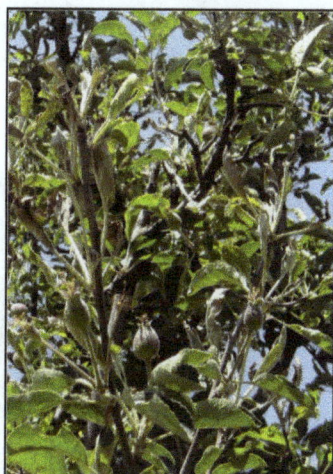

Plate 16.8: Powdery Mildew Symptoms on Terminal Growth of Apple.

Plate 16.9: Marssonina Blotch on Apple Leaves.

(p. 254)

Plate 16.10: Brown Spots with Greyish Centre of Frog Eye Spot on Apple Leaf.

(p. 255)

Plate 16.11: Sooty Blotch as Dark Filmy Smudge/Shades, and Fly Speck as Small Circular Black Glistering Dots.

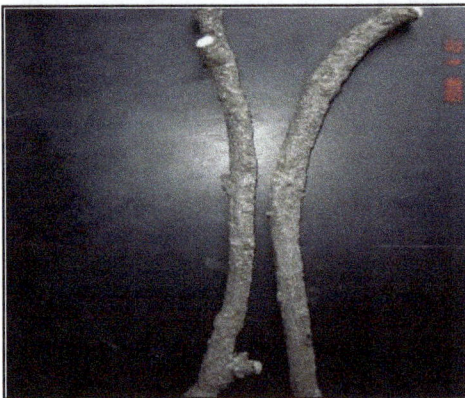

San Jose Scale Infestation on Apple Twigs and on Apple Fruit. (p. 263)

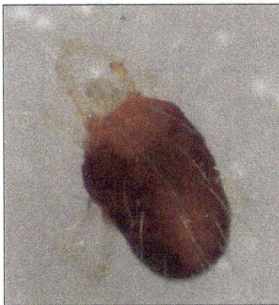

Adult Mite, *Panonychus ulmi.* (p. 264)

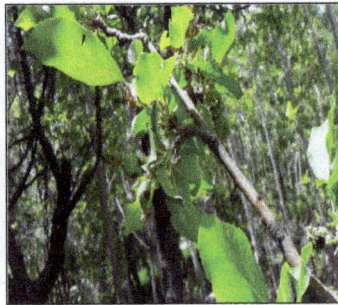

Fruit Injury by Codling Moth. (p. 266)

Boring on Fruit by the Larva. (p. 266)

A: Male

B: Female.
(p. 267)

Males trapped in delta traps.

Wooly Apple Aphid Infestation on Branches. (p. 269)

Adult Tent Caterpillar. (p. 270)

Tents made by the Tent Caterpillar. (p. 270)

Zig-zag Galleries made by Stem Borer. (p. 271)

Emergence of Bark Beetle and Life Stages of *Scolytus nitidus*.
A: Egg stage, B: Larval stage (I and V instars), C: Pupal stage, D: Adult stage. (p. 273)

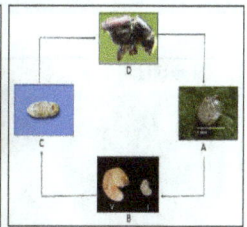

Leaf Miner Infestation (p. 274)

Figure 19.1 (p. 290)

Figure 19.2 (p. 292)

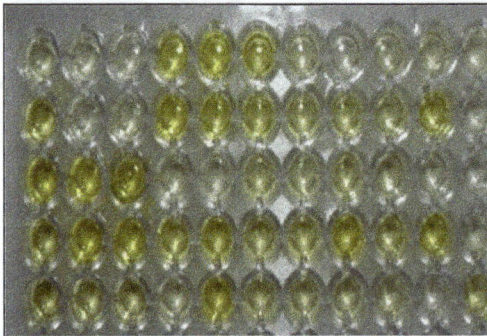

Figure 19.3 (p. 292)

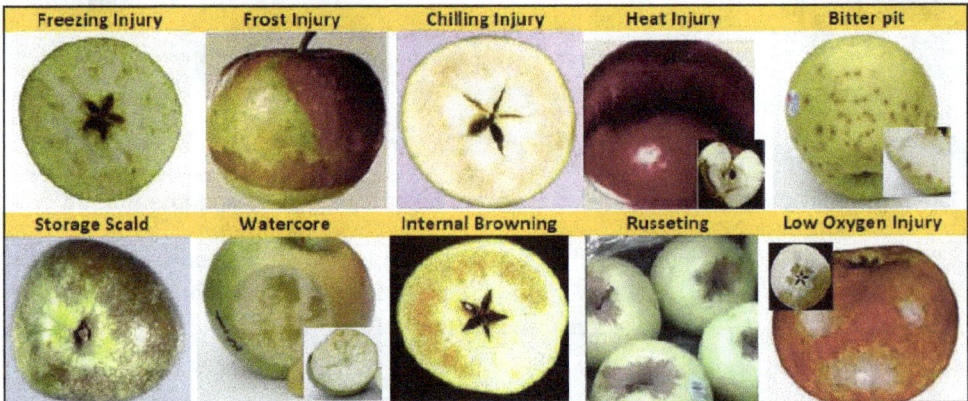

Figure 21.3: Major Physiological Disorders of Apple. (p. 318)

Figure 22.1: Apple Harvester. (p. 332)

Figure 22.2: Power Tiller Operated Leaf Collector. (p. 332)

Figure 22.3: 'Pluk-O-Trak' Harvesting Golden Delicious Apple in France. (p. 334)

Figure 22.4: Instrumented Glove. (p. 335)

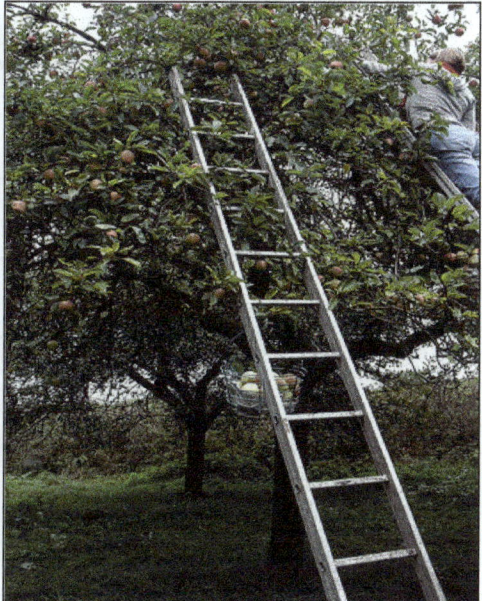

Figure 22.5: Ladder for Fruit Plucking. (p. 335)

(p. 338)

Figure 23.7: SKUAST-K Research Team. (p. 348)

Figure 23.8: Commercial Trial with 1-MCP on Kashmiri Apples. (p. 348)

Figure 23.9: Commercial Apple Storage CA Rooms in Kashmir. (p. 348)

Figure 24.1: Traditional Practice of Heaping. (p. 354)

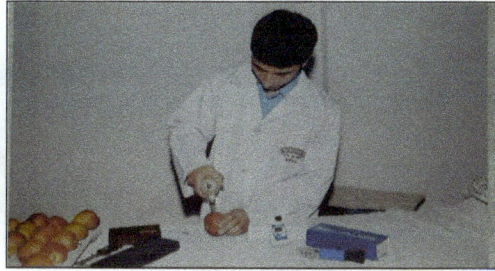

Fig 24.2. Determination of Fruit Firmness using Penetrometer. (p. 358)

(i)

(ii)

Figure 24.4: Packaging of Apple (i) Corrugated Fiber Board; (ii) Net Packing. (p. 364)

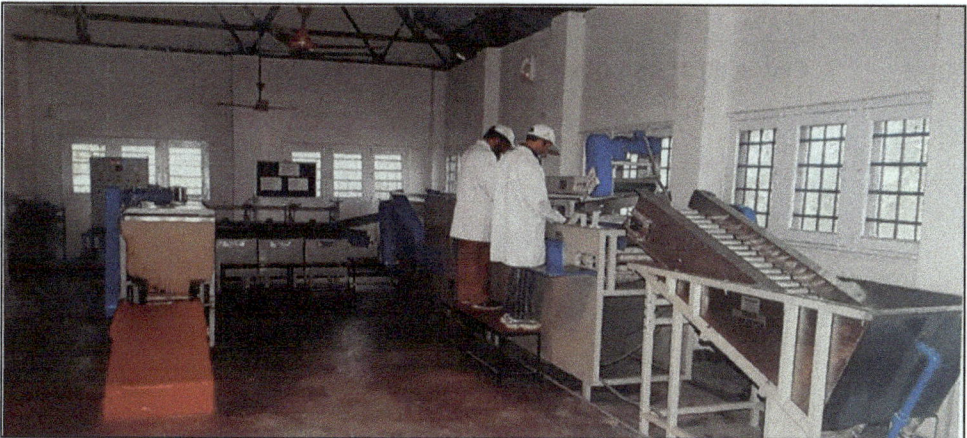

Figure 24.6: Internal View of Pack House for washing, Grading, Waxing and Packing of Fruit and Vegetables (*Division of Post harvest Technology, SKUAST-K Shalimar*). (p. 374)

Figure 25.3: Grading of Kashmiri Apple. (p. 397)

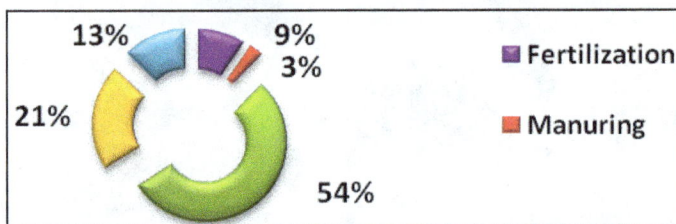

Figure 29.1: Composition of Cost of Cultivation of Apple. (p. 436)

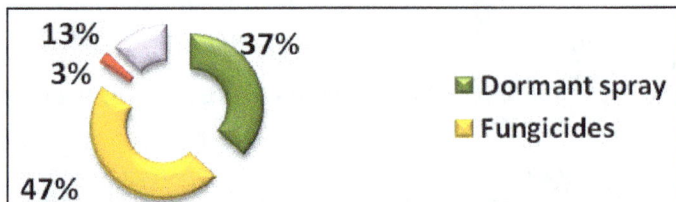

Figure 29.2: Composition of Cost of Pesticides. (p. 436)

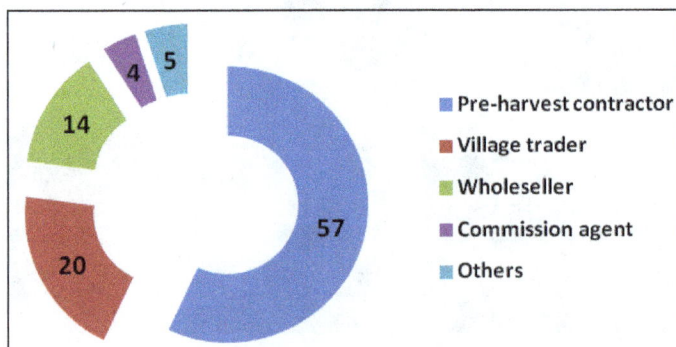

Figure 30.1: Sale of Apple through different Functionaries (per cent). (p. 452)